EMERGING AND REEMERGING VIRAL PATHOGENS

EMERGING AND REEMERGING VIRAL PATHOGENS

Volume 2: Applied Virology Approaches Related to Human, Animal and Environmental Pathogens

Edited by

Moulay Mustapha Ennaji

Laboratory of Virology, Microbiology, Quality, Biotechnologies/Eco-Toxicology and Biodiversity, Faculty of Sciences and Techniques, Mohammedia, University Hassan II of Casablanca, Casablanca, Morocco

ELSEVIER

ACADEMIC PRESS

An imprint of Elsevier

Academic Press is an imprint of Elsevier
125 London Wall, London EC2Y 5AS, United Kingdom
525 B Street, Suite 1650, San Diego, CA 92101, United States
50 Hampshire Street, 5th Floor, Cambridge, MA 02139, United States
The Boulevard, Langford Lane, Kidlington, Oxford OX5 1GB, United Kingdom

British Library Cataloguing-in-Publication Data
A catalogue record for this book is available from the British Library

Library of Congress Cataloging-in-Publication Data
A catalog record for this book is available from the Library of Congress

ISBN: 978-0-12-814966-9

For Information on all Academic Press publications
visit our website at https://www.elsevier.com/books-and-journals

Publisher: Andre Gerhard Wolff
Acquisition Editor: Linda Versteeg-buschman
Editorial Project Manager: Timothy Bennett
Production Project Manager: Sreejith Viswanathan
Cover Designer: Christian Bilbow

Typeset by MPS Limited, Chennai, India

Contents

4. Molecular Modeling of Major Structural Protein Genes of Avian Coronavirus: Infectious Bronchitis Virus Mass H120 and Italy02 Strains

KHADIJA KHATABY, YASSINE KASMI, AMAL SOUIRI, CHAFIQA LOUTFI AND MOULAY MUSTAPHA ENNAJI

5. Biological Databases in Virology

PRAMODKUMAR PYARELAL GUPTA, VIRUPAKSHA AJIT BASTIKAR, SANTOSH SUBHASH CHHAJED AND SHANKER LAL KOTHARI

6. Involvement and Roles of Long Noncoding RNAs in the Molecular Mechanisms of Emerging and Reemerging Viral Infections

MARYAME LAMSISI AND MOULAY MUSTAPHA ENNAJI

List of Contributors

Karam Allali Laboratory of Mathematics and Applications, University Hassan II of Casablanca, FST-Mohammadia, Casablanca, Morocco

Yassine Amraouza Laboratory of Foods, Environment and Health, Faculty of Sciences and Techniques-Gueliz, Cadi Ayyad University, Marrakech, Morocco

Saaid Amzazi Laboratory of Biochemistry and Immunology, Faculty of Sciences, Agdal, University of Mohammed V, Rabat, Morocco

Paula N. Arrías Laboratory of Molecular Virology, Institute of Biotechnology and Molecular Biology (IBBM), University of La Plata-CONICET, La Plata, Argentina

Virupaksha Ajit Bastikar Amity Institute of Biotechnology, Amity University, Mumbai, India

Mustapha Benhassou Laboratory of Virology, Microbiology, Quality, Biotechnologies/Eco-Toxicology and Biodiversity, Faculty of Sciences and Techniques, Mohammedia, University Hassan II of Casablanca, Casablanca, Morocco; Mohammed VI University of Health Sciences of Casablanca, Casablanca, Morocco; School of Medicine and Pharmacy of Casablanca, University Hassan II of Casablanca, Casablanca, Morocco

Hlima Bessi Laboratory of Virology, Microbiology, Quality, Biotechnologies/Eco-Toxicology and Biodiversity, Faculty of Sciences and Techniques, Mohammedia, University Hassan II of Casablanca, Casablanca, Morocco

Gabriela Calamante National Institute of Agricultural Technology (INTA), CONICET, Institute of Agrobiotechnology and Molecular Biology (IABiMo), Buenos Aires, Argentina

Santosh Subhash Chhajed Department of Pharmaceutical Chemistry, METS Institute of Pharmacy, Bhujbal Knowledge City, Nashik, India

Federico Dapiaggi Department of Chemistry, University of Milano, Milano, Italy

María Paula Del Médico Zajac National Institute of Agricultural Technology (INTA), CONICET, Institute of Agrobiotechnology and Molecular Biology (IABiMo), Buenos Aires, Argentina

Yassine El Mallali Unit of Biology and Medical Research, National Center for Energy Sciences and Nuclear Techniques/Rabat (CNESTEN), Rabat, Morocco

Mohammed El Mzibri Unit of Biology and Medical Research, National Center for Energy Sciences and Nuclear Techniques/Rabat (CNESTEN), Rabat, Morocco

Aissam El-Aliani Laboratory of Virology, Microbiology, Quality, Biotechnologies/Eco-Toxicology and Biodiversity, Faculty of Sciences and Techniques, Mohammedia, University Hassan II of Casablanca, Casablanca, Morocco; Unit of Biology and Medical Research, National Center for Energy Sciences and Nuclear Techniques/Rabat (CNESTEN), Rabat, Morocco

Moulay Mustapha Ennaji Laboratory of Virology, Microbiology, Quality, Biotechnologies/Eco-Toxicology and Biodiversity, Faculty of Sciences and Techniques, Mohammedia, University Hassan II of Casablanca, Casablanca, Morocco

Youssef Ennaji Laboratory of Virology, Microbiology, Quality, Biotechnologies/Eco-Toxicology and Biodiversity, Faculty of Sciences and Techniques, Mohammedia, University Hassan II of Casablanca, Casablanca, Morocco

M. Laura Fabre Laboratory of Molecular Virology, Institute of Biotechnology and Molecular Biology (IBBM), University of La Plata-CONICET, La Plata, Argentina

Débora Garanzini National Institute of Agricultural Technology (INTA), CONICET, Institute of Agrobiotechnology and Molecular Biology (IABiMo), Buenos Aires, Argentina; National Administration of Laboratories and Health Institutes Dr C.G. Malbrán, National Institute of Biological Production, Antirabic Vaccine Service, Buenos Aires, Argentina

Pramodkumar Pyarelal Gupta School of Biotechnology and Bioinformatics, DY Patil Deemed to be University, Navi Mumbai, India

Jamal Hafid Laboratory of Foods, Environment and Health, Faculty of Sciences and Techniques-Gueliz, Cadi Ayyad University, Marrakech, Morocco

Rahma Ait Hammou Laboratory of Virology, Microbiology, Quality, Biotechnologies/Eco-Toxicology and Biodiversity, Faculty of Sciences and Techniques, Mohammedia, University Hassan II of Casablanca, Casablanca, Morocco

Khalid Hattaf Centre Régional des Métiers de l'Education et de la Formation (CRMEF), Casablanca, Morocco; Laboratory of Analysis, Modeling and Simulation (LAMS), Faculty of Sciences Ben M'sik, University Hassan II of Casablanca, Casablanca, Morocco

Yassine Kasmi Laboratory of Virology, Microbiology, Quality, Biotechnologies/Eco-Toxicology and Biodiversity, Faculty of Sciences and Techniques, Mohammedia, University Hassan II of Casablanca, Casablanca, Morocco; Faculty of Science and Techniques Mohammedia, University of Hassan II Casablanca, Casablanca, Morocco; Moroccan Foundation for Advanced Science, Innovation and Research, Rabat, Morocco

Khadija Khataby Laboratory of Virology, Microbiology, Quality, Biotechnologies/Eco-Toxicology and Biodiversity, Faculty of Sciences and Techniques, Mohammedia, University Hassan II of Casablanca, Casablanca, Morocco; Society Biopharma, Rabat, Morocco

Shanker Lal Kothari Amity Institute of Biotechnology, Amity University, Jaipur, India

Maryame Lamsisi Laboratory of Virology, Microbiology, Quality, Biotechnologies/Eco-Toxicology and Biodiversity, Faculty of Sciences and Techniques, Mohammedia, University Hassan II of Casablanca, Casablanca, Morocco

Chafiqa Loutfi Society Biopharma, Rabat, Morocco

Tomás Masson Laboratory of Molecular Virology, Institute of Biotechnology and Molecular Biology (IBBM), University of La Plata-CONICET, La Plata, Argentina

Mahi M. Mohiuddin Department of Biology, McMaster University, Hamilton, ON, Canada

Oscar Ramón Pérez National Administration of Laboratories and Health Institutes Dr C.G. Malbrán, National Institute of Biological Production, Antirabic Vaccine Service, Buenos Aires, Argentina

Matías L. Pidre Laboratory of Molecular Virology, Institute of Biotechnology and Molecular Biology (IBBM), University of La Plata-CONICET, La Plata, Argentina

Stefano Pieraccini Department of Chemistry, University of Milano, Milano, Italy; Institute of Molecular Science and Technologies, CNR, Milano, Italy

Črtomir Podlipnik Faculty of Chemistry and Chemical Technology, University of Ljubljana, Ljubljana, Slovenia

Donatella Potenza Department of Chemistry, University of Milano, Milano, Italy

Víctor Romanowski Laboratory of Molecular Virology, Institute of Biotechnology and Molecular Biology (IBBM), University of La Plata-CONICET, La Plata, Argentina

Darkaoui Sami Moroccan Food Safety Office (ONSSA), Rabat, Morocco

Herb E. Schellhorn Department of Biology, McMaster University, Hamilton, ON, Canada

Amal Souiri Laboratory of Virology, Microbiology, Quality, Biotechnologies/ Eco-Toxicology and Biodiversity, Faculty of Sciences and Techniques, Mohammedia, University Hassan II of Casablanca, Casablanca, Morocco

Francesca Vasile Department of Chemistry, University of Milano, Milano, Italy

Noura Yousfi Laboratory of Analysis, Modeling and Simulation (LAMS), Faculty of Sciences Ben M'sik, University Hassan II of Casablanca, Casablanca, Morocco

Mustapha Zemzami Laboratory of Sanitary Control, Control Unit of Plants, Domaines Agricoles Maâmora, Salé, Morocco

Editor Biography

Prof. Dr. Moulay Mustapha Ennaji is a Moroccan and Canadian citizen and native of Marrakesh (Morocco). He is a scientist specialized in the fields of virology, hygiene, and microbiology. He got a master of science in 1986 and a PhD in virology in 1993 from Armand Frappier Institute, University of Quebec (Canada). Between 1991 and 1993, he completed a postdoctorate at the Canadian Red Cross. From 1993 to 1995, he was a research associate (RA) and from 1995 to 1996 a research officer (RO) at the National Council of Research of Canada (CNRC). He was also a visiting researcher at the University of California, Irvine, United States, and abroad lecturer at the Histochemistry Institutes of Paris, France. He was a guest researcher of the Franklin Foundation in USA NIH Bethesda.

He was recruited in 1996 to the Faculty of Sciences and Techniques Mohammedia (FSTM), falling under University Hassan II of Casablanca (UH2C), as a lecturer and enabled professor where he was the head of the biology department from 1997 to 2000. He is currently professor of higher education (PES C) in the same faculty. Being a scientist who is concerned on the research development, he had been giving numerous conferences and lectures on virology, cancerology, hygiene, and microbiology since 1986 at many Moroccan, Canadian, and American universities.

Between 2005 and 2010 he was appointed director of virology, hygiene, and microbiology and coordinator of the consortium of Biomedical and Environmental Sciences laboratories at UH2C-FSTM. From 2010 to present, he is the director of the Laboratory of Virology, Microbiology, Quality and Biotechnology/Eco-toxicology and Biodiversity; leader of Virology Oncology and Medical Biotechnology Team; and deputy director of the Research Centre of Natural Resources and Food (rensa) of UH2C. He was also responsible for the master programs in biotechnology and biomedical technologies (2000–03), DESA of microbiology and bioengineering (2005–10), and master of science and technology (MST Microbiology, Applied Virology, and Bio-industry Engineering and MST of Livings) (Immuno-Virology and Applied

Microbiology) from 2010 to 2015. He is a member of the Council of the Center of Doctoral Studies (CEDoc) at FSTM-UH2C since 2008. Between 2005 and 2010, he was deputy head of the UFR DESA Biomedical Sciences and from 2000 to 2005 deputy leaders of the UFR PhD in health and environment. From 2005 to present, he is deputy head of Life and Environment Sciences Doctoral UFR. From 2010 to 2015, he was national expert at CNRST and member of the National Commission for scholarships. Previously, from 2012 to 2014, he was also a UNESCO expert on governance reform of university systems.

Throughout his career, he was rewarded with 24 awards. He has organized numerous national and international meetings in the fields of virology, microbiology, and hygiene.

At present, he is the vice president of the Moroccan Association of Biosafety, Cancer and Microbiology.

Preface

The whole book, regarding emerging and reemerging viral pathogens, originally consists of two volumes: Volume 1 concerns the fundamental and basic virology aspects of human, animal, and plant pathogens. As completion of the abovementioned fundamentals of viral pathogens, Volume 2 comes to share applied topics of viral pathogens and its complications. The main theme of Volume 2 is applied virology approaches related to human, animal, and environmental pathogens. The topics cover most emerging viral disease in human, animal, and plant; provide the most recent advanced tools and techniques in molecular virology and in computational and modeling viruses; create awareness that the manifestation of new transmissible pathogens is a global risk; and highlight the need to adopt shared policies for the prevention and control of infectious diseases. And it also concerns researchers in virology, micromolecular biology, and pharmaceutical sciences. Volume 2 entitled "Applied virology approaches related human, animal, and environmental pathogens" is organized into 18 chapters; each chapter includes definitions and key concepts as well as wide description of the subject. An abstract is also provided for each chapter. The contents of each chapter are supported by schematic figures and tables that are needed. The editor starts with general introduction explaining the phenomena of emergences and reemergence of the diseases: evolution of the concepts, risk factors, and state of the art. Subjects of the contents reported are viral dynamics in human immunodeficiency virus as emergent viral infection, CRISPR/Cas9 genome-editing technology against emerging and reemerging virus, infectious bronchitis virus in poultry, biological databases in virology, mathematical modeling in virology, scientific advances in the diagnosis of emerging and reemerging viral human pathogens, computational and bioinformatics tools in virology, roles of long noncoding RNAs in the molecular mechanisms of emerging and reemerging viral infections, designing antiviral substances targeting the Ebola virus vp24 protein, application of nanodiagnostics in viral infectious diseases, baculovirus-derived vectors for immunization and therapeutic applications, biotechnology of recombinant veterinary vaccines against rabies, epidemiology and ecology of emerging viruses in two freshwater lakes of the Northern Hemisphere, epidemiology and genetic variability of rabies and norovirus, and finally risk assessment and biosecurity considerations in control of emergent plant viruses. Hope we covered main applied virology domains that will impact the quality of life of coming generations.

Professor Dr. Moulay Mustapha Ennaji B.Sc., M.Sc., Ph.D., RSM (CCM)
Book Editor

Editorial Management Committees

EDITORIAL MANAGEMENT COMMITTEE

(*Continued*)

(Continued)

ASSISTANCE SUPPORT MANAGEMENT

Rihab Bouseettine	Laboratory of Virology, Microbiology, Quality, Biotechnologies/Eco-Toxicology and Biodiversity, Faculty of Sciences and Techniques, Mohammedia, University Hassan II of Casablanca, Casablanca, Morocco
Imane Saif	Laboratory of Virology, Microbiology, Quality, Biotechnologies/Eco-Toxicology and Biodiversity, Faculty of Sciences and Techniques, Mohammedia, University Hassan II of Casablanca, Casablanca, Morocco
Meryam Lamsisi	Virology Laboratory, FSTM, University Hassan II of Casablanca, Casablanca, Morocco
Fatima Ezzahra Rihane	Laboratory of Virology, Microbiology, Quality, Biotechnologies/Eco-Toxicology and Biodiversity, Faculty of Sciences and Techniques, Mohammedia, University Hassan II of Casablanca, Casablanca, Morocco
Rahma Ait Hammou	Laboratory of Virology, Microbiology, Quality, Biotechnologies/Eco-Toxicology and Biodiversity, Faculty of Sciences and Techniques, Mohammedia, University Hassan II of Casablanca, Casablanca, Morocco

Editorial

The unpredictable phenomenon of emergence and reemergence of some diseases is untimely related to various pathogens such as parasites, bacterial, and viral agents. Among these viral diseases the zoonoses are currently a hot topic in the scientific, socioeconomic, and political areas.

Globally, viral emergence and reemergence occur as a result of the accumulation of spontaneous or continuous genetic variations within viral genomes, making them more virulent and infecting a wide spectrum of hosts including humans. This visible part of the iceberg is linked to a multitude of complex and interrelated factors, including societal and environmental ones.

In fact, the unprecedented rapprochement between human and animal species was usually incriminated as a major cause in the evolution of this phenomenon.

The World Health Organization (WHO) recognizes that 60% of the human pathogens come from the animal and that 75% of the pathogens responsible for emerging and reemerging animal diseases show a potential transgression to cross the interspecies barrier establishing favorable conditions for genetic exchange. Such exchange leads to the emergence of new highly pathogenic variants and strains.

Therefore any public health prophylactic strategy requires a holistic approach taking into account the interaction between human, animal, and environment. The international community has made this globalized approach through the newly established concept of "One Health, One World" and makes the old bipolar concept of "separated human and animal health issues" obsolete.

The viral zoonotic diseases have constituted variable threats. Some of them are responsible for benign clinical forms, and others constitute a major source of fear and panic worldwide impacting global public health. Moreover, the viral acute respiratory diseases [highly pathogenic influenza avian (H5N1, H7N9, and H7N7), coronavirus, and Middle East Respiratory Syndrome] have periodically caused international alerts and forced WHO to take sanitary measures to limit the extension of outbreaks.

Furthermore, the arthropod-borne virus (arbovirus) diseases (West Nile, Chikungunya, Zika, Dengue, Rift Valley fever) generate about one million deaths each year, and their distribution depends on environmental and social factors. The phenomenon of global warming, the diversity of mosquito vector species, the precariousness of lifestyle, and the intensification of international traffic are certainly determinant factors of diseases maintenance and spread to unscathed geographical areas.

The viral zoonosis transmitted by bat species adds a new challenge to the global health community. The role of these animals in the emergence of many serious and highly mediated viral infectious diseases has been demonstrated: Hendra virus since 1994, Nipah virus since 1997, and Ebola and Marburg virus in 2014.

Moreover, the zoonotic diseases caused by domestic animals and poultry (dogs, cats, birds, etc.) also occupy a preponderant place in terms of (1) viral health risks such as rabies, influenza, retroviral diseases, and coronavirus disease and (2) bacterial and parasitic diseases such as leptospirosis, leishmaniasis, piroplasmosis, and echinococcosis (hydatidosis). A public awareness and prophylactic measures can limit the spread of contagion.

Viral food and waterborne zoonotic diseases including those related to rotavirus, poliovirus, calicivirus, hepatite A and E viruses, norovirus, coronavirus, enterovirus, Norwalk virus, poliovirus, coxsackie viruses A and B as well as to the pathogenic bacteria of water origin (endogenous or exogenous) (e.g., Salmonella, Listeria, and Brucella) were closely correlated to failures of hygiene, sanitary, biosafety, and biosecurity measurements application. The adoption of sanitary practices in all the events and stages of agro-food production (from barn to the table) is the only way to provide a healthy and harmless food and protect against food and waterborne illnesses.

In hospitals the emergence of nosocomial viral diseases, such as human immunodeficiency virus, and hepatitis A, B, C, has also been reported. The antiviral resistance is actually a major health concern that requires a concerted response from health professionals.

The extreme diversity of emerging and reemerging viral pathogens, the change of human lifestyle, the globalization of travel, the business exchanges, and tourism potentiate the risk of emergence of highly pathogenic zoonotic diseases.

Promoting intersectorial collaboration allowed the unification of the health and safety policy. The crosscutting ecological and health data at the national and global level are effective means for sustaining good health in human, animal, and ecosystem.

Although the phenomenon of emergence and reemergence concerns a range of pathogens other than viruses such as bacteria and parasites, it is regularly reported in the literature. This work is an update on emerging and reemerging viral diseases. It is presented in two volumes in which the basic knowledge in virology relating to the emergence and reemergence viral pathologies (Volume I: Fundamental and basic virology aspects of human, animal, and plant pathogens) and the virological applications related to the health threats (Volume II: Applied virology approaches related to human, animal, and environmental pathogens) are listed.

Moulay Mustapha Ennaji B.Sc., M.Sc., Ph.D., RSM (CCM)
Book Editor

Graphical Abstract

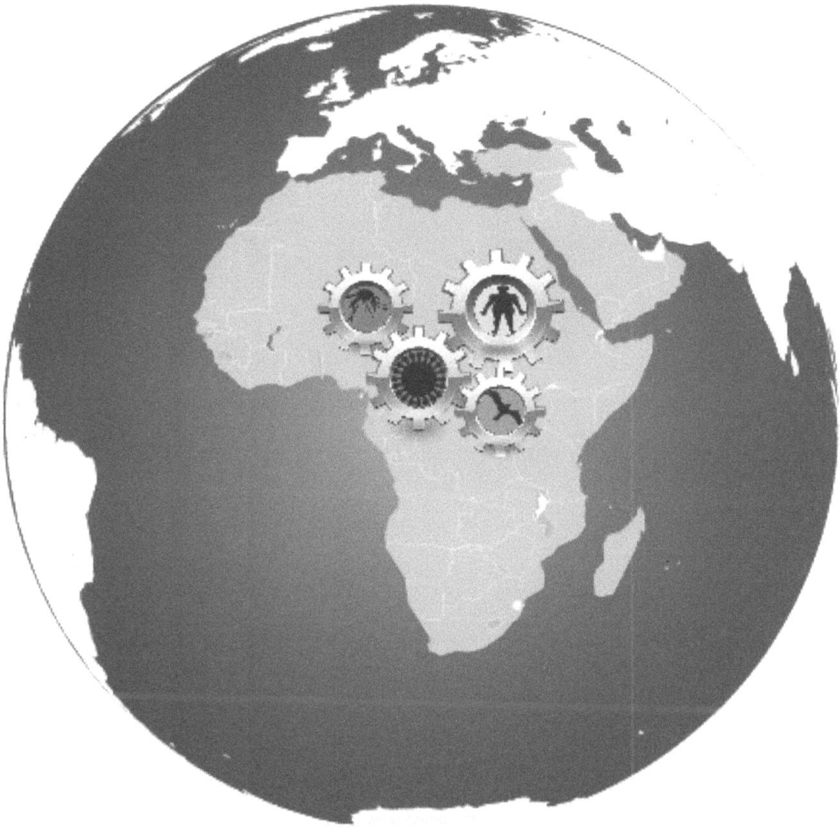

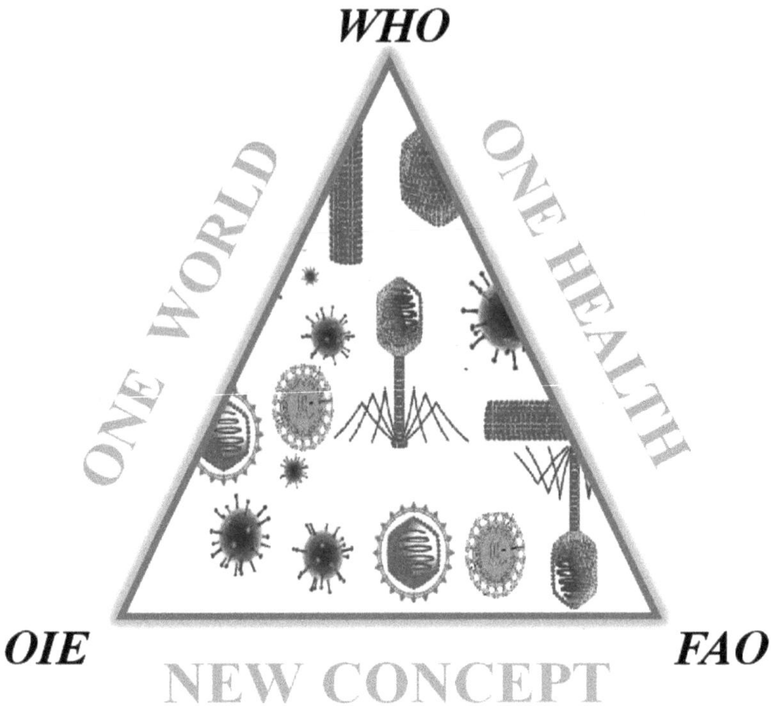
WHO
ONE WORLD
ONE HEALTH
OIE
FAO
NEW CONCEPT

CHAPTER

1

Phenomena of Emergences and Reemergences of the Diseases: Evolution of the Concepts, Risk Factors, and State of the Art

Moulay Mustapha Ennaji

Laboratory of Virology, Microbiology, Quality, Biotechnologies/ Eco-Toxicology and Biodiversity, Faculty of Sciences and Techniques, Mohammedia, University Hassan II of Casablanca, Casablanca, Morocco

According to scientific data, 60% of the pathogens affecting humans come from the animal kingdom. Also, among 75% of the agents are recognized as responsible of emerging animal diseases, can cross the interspecies interface and infect humans. This percentage reflects the health risk incurred when humans are in contact with domestic animals such as poultry, companion animals (dogs, cats, etc.), or wild animals (bats, wild boars, game, etc.) (World Health Organization (WHO), 1959; International Organization of Epizootics (OIE), 2007).

The characterization of zoonoses has evolved over time. They were initially defined by the WHO (1959) experts as "diseases or infections that are naturally transmitted from vertebrate animals to humans and vice versa". This definition was the subject of controversy between several authors (Ashford, 2003; Hubalek, 2003). Hubalek agreed to adopt a

Emerging and Reemerging Viral Pathogens
DOI: https://doi.org/10.1016/B978-0-12-814966-9.00001-9

broader definition emphasizing the nature of transmissible pathogens from animals to humans and reciprocally. The current definition has now become broader: "Zoonoses are diseases, infections or parasitic infestations caused by transmissible agents (bacteria, viruses, parasites, fungi or prions) that develop in at least two vertebrate species, including human" (Savey and Dufour, 2004).

The list of these zoonotic diseases is continually updated due to the emergence and reemergence of new pathogens, which presents a permanent challenge for both public and animal health.

The word "emergence," often used in the literature and the media, is defined as the sudden appearance of an unknown disease, while the "reemergence" reports a disease already known by the scientific community and reappearing again (Cohen, 1998). For example, HIV emerged in humans in 1982, according to specialist virologists; this virus circulated previously in nonhuman primates (chimpanzees and gorillas) and was transmitted to humans by pathogens of animal origin (Drevet, 2012). Also, the reemergence was seen in enteroviruses, especially poliomyelitis (Peigue-Lafeuille et al., 2014).

As new diseases appear for humanity, the word emergence becomes more and more used. It is agreed, according to several authors, that at the beginning, the idea of the emergence of diseases was closely linked to the birth of the modern epidemiology of the 19th century. The idea was based on the work of two researchers "Louis Pasteur" and "Robert Koch" who put the basis of modern science through the microbial theory that attributed the pathogenic effect to a specific microorganism. The word "emergence" was particularly used during the outbreak of the cholera pandemic in 1832 that ravaged Paris and London and resulted in several thousand deaths (Belongia, 2002; Drotman, 1998; Fagherazzi-Pagel, 2008).

Then the epidemic of typhus (salomiales) in 1848 in Germany was described by the German researcher Rudolf Virchow who has highlighted the role that environment can play on the disease spread and contamination (Cicolella, 2010).

The idea of emergence is already reported since 1930, in a book published by researcher Charles Nicolle, who received Nobel Prize in medicine in 1939. In the 1960s, the scientific community focused on emerging diseases in the animal world and their transmission to humans, researchers were talking about the emerging pathogenic concept to describe newly isolated or observed pathogens, as is the case for poliovirus causing *poliomyelitis* manifested by nerve signs leaving paralytic limb squeal (Bodian, 1955).

At the beginning of 1990, the American researcher "Stephen S. Morse" was one of the first scientists to defend the emergence of new highly pathogenic viruses (Chastel, 2000; Morse, 1995a,b).

Lederberg et al. (1992) have brought new perspectives by focusing on aspects that promote and trigger emergence, such as the environment, the political, and social issues.

In addition, several emergence and reemergence events have been reported: the major influenza pandemics were the Spanish in 1918 that caused 40 million deaths, Asian in 1957 that caused 4 million deaths, Hong Kong in 1968 with 2 million deaths, and most recently in 2009 the Mexican pandemic flu. Then, the recurrent emergence and reemergence of H5N1, the highly pathogenic avian influenza in 1997, H7N9 in 2013 and acute respiratory disease (SARS and MERS) in 2002 and 2012, respectively (Kilbourne, 2006; Choffnes and Mack, 2015).

The instability of some regions and the nonrigorous application of vaccine programs have prevented the eradication of many diseases such as polio and influenza (Gonzalo, 2017).

RNA viruses such as poliovirus, vesicular stomatitis virus, human immunodeficiency virus type 1, foot-and-mouth disease virus, highly pathogenic avian influenza, and vector-borne viral hemorrhagic fever diseases [that include serious viral diseases associated with bleeding as dengue, Ebola hemorrhagic fever, Marburg hemorrhagic fever, Crimean–Congo hemorrhagic fever, Rift Valley fever (RVF), and yellow fever] are the main issues of emergence and reemergence, due to their intrinsic capacity for genetic change. In fact, mutations, recombination, and rearrangement of the genomic segments, as well as the combination of these events, produce genetically and antigenetically diverse viral strains that result in considerable potential for viral emergence and reemergence (Domingo et al., 2002).

Moreover, this phenomenon of emergence and reemergence also affects microorganisms like bacteria. Tuberculosis is a real example of this bacterial reemergence, and it often reappears following the anarchic use of antibiotics making it responsible bacterium refractory to the usual treatments. The precariousness of the sanitary system, the poverty associated with the increase of slums, and the environment pollution are all contributing factors. Also, cholera that is best known by its reemergence often correlated to the decrease of access to drinking water following the social precariousness or in case of natural disaster as is the case during an earthquake in Haiti (Stamm and Mudrak, 2013). In addition, the reemergence of Malaria in several illustrious continents is an eminent example of parasite reemergence (Sharma, 1996).

Emerging diseases, depending on their nature, can be divided into two broad categories: new diseases and known diseases.

Thus new diseases are unknown diseases caused by previously unknown pathogens in humans. This is the case of SARS, hantavirus, AIDS, and bovine spongiform encephalopathy, which was unknown before 1986.

The known diseases correspond to diseases that circulate in animals but new to humans in this case, the pathogens have mutated and acquired the ability to cross the interspecies barrier, thus inducing changes in virulence and adaptation to new hosts such as human. In this case, three possibilities are evoked as follows:

- Pathology without nosological or clinical identity such as Legionnaire's disease.
- Diseases that have occurred with a quantitative or qualitative increase related to the increase in virulence.
- Change of the geographical territory: certain known disease in a region when it gains of the new territory becomes emerging for the new geographical areas.

The conditions favoring the emergence of diseases, often called emergence factors, have become hot topics. They are multiple and intermingled factors that act together in synergy. We can distinguish them as follows:

- Sociological and ecological factors produce diverse and changing environments in which certain pathogens, especially viruses, have wide possibilities of being selected. In this way, new human, animal, and plant viruses have appeared periodically and, obviously, will continue to emerge. Urbanization is now becoming a determining factor in the resurgence and emergence of viral diseases. Thus overcrowding promotes the spread of disease and creates unsanitary conditions (Smolinski et al., 2003; Howard and Fletcher, 2012).
- Climatic factor: the creation of dams and the increase of the global temperature promote the multiplication of the vectors notably mosquitoes which gain new territories thus conveying diseases. As is the case for RVF disease, which is a viral zoonosis (mainly affecting ruminant domestic animals that can be transmitted to humans), which is manifested by a fever and sometimes (1%–3% of cases) hemorrhage (Morse, 1995a,b).
- Intensive crops and industrial farms: those activities have often been the source of many zoonotic diseases; for example, maize fields in Argentina have favored the proliferation of rats responsible for hemorrhagic fever in humans (Maclachlan et al., 2016).
- Drug-related factors: Anarchic and excessive use of antibiotics and antivirals make virulent agents resistant to treatments adopted in hospitals.
- The reduction of wild areas and human–animal interface are factors often cited by experts as favoring the emergence of diseases. In fact, deforestation and fragmentation of forests have favored human contact. New pathogens such as HIV and Ebola, as well as the

proliferation of insects near homes, have made the transmission of germs easier (Wolfe et al., 2005).
- The destruction of predators unbalances the food chain in favor of certain vectors carrying certain pathogens, let us quote the example of Lyme disease carried by the rodents present in large numbers of cornfields in the United States. In the end, the massive deforestation directly affects biodiversity by promoting genetic mutations and diversity.
- Factors related to pathogens: those factors include the ease of mutations acquisition, as is the case of RNA viruses.
- Factors related to vectors: precarious installations and lack of hygiene favor the proliferation of mosquitoes and flies that carry pathogenic germs.
- Host-related factors: certain host-related conditions predispose and contribute to the onset of the disease, for example, physiological or exogenous stress, immune depression, and so on (Morse, 1995a,b; Howard and Fletcher, 2012).

Thus human behavior was the main cause of the diseases' emergence, we mention some aspects: The sharing of habitat between humans and animals, displacement of populations, and the low rate of literacy were behind the nonapplication and nonassimilation of good hygienic practice associated with a deficient health policy.

Natural disasters and wars are often accompanied by the destruction of infrastructure and are still factors triggering emerging and reemerging diseases.

In the industrialized countries, other factors are added to the factors of emergence linked to the human activity which are different; we can quote the activities of leisure and the pets possession to the households. Also, the development of small farms around the cities, intensification of animal productions, and the centralization of food production chains using methods and techniques can promote the emergence of emerging and reemerging diseases (Morse, 1995a,b).

To standardize the international procedures and manage the emerging and reemerging infectious diseases, the international community has created several bodies to monitor the health of both humans and animals. In effect, in response to emergence and reemergence phenomena, the international scientific community through the tripartite alliance [OIE, WHO, Food and Agriculture Organization (FAO)] promotes collaborations for the prevention and control of health risks at the human–animal–environment interface through the adoption of strategies, concepts, and intervention plans, all according to the central dogma "One Health, One World."

"One Health, One World" is most promising. This alliance aims to formalize the sharing of responsibilities and the coordination of global actions to manage health risks at human–animal–environment interfaces (Belay et al., 2017).

The OIE is the largest organization in the animal field. It was created in 1924, to coordinate the actions of country members whose current membership is 178 states. This organization brings together several regional structures to optimize the epidemiosurveillance and the response. The main emerging and reemerging zoonoses are listed by periodic scientific papers from OIE experts and also contribute to the regulation of animal trade and the marketing of animal products by developing an international zoo-sanitary code (https://www.oie.int/doc/ged/D7650.PDF), which is systematically reviewed if necessary (OIE, 2015, http://www.oie.int/fileadmin/vademecum/fr/files/assets/basic-html/index.html#3).

The WHO represents the reference in human health. It was created in Geneva in 1948 to centralize and coordinate the actions of country members to fight against diseases of global interest and has numerous missions. It assists member states in the epidemiological surveillance of zoonotic and emerging diseases. At the same time, it encourages all research that aims to develop strategies or methods of prevention and regularly brings together experts to deal with certain emergency situations. The WHO regularly communicates, through reports, the observed phenomena of pathogens resistance to different drugs used which are considered as predisposing factors for reemergence (WHO, 1959).

The FAO is a key player in food production and agriculture. It participates in the prevention of emergencies related to disasters including situations of emergence and reemergence of diseases. The main forms of FAO intervention include needs assessments, rural rehabilitation, and technical assistance for planning and management of sustainable development (FAO, 2018).

Usually, three elements coexist and share the same biosphere: the animal, the man, and the ecosystem. To unify and standardize actions for the concept "one medicine" was initially proposed by the American researcher Calvin Schwabe (https://www.onehealthcommission.org/en/one_health_resources/whos_who_in_one_health/calvin_schwabe_one_health_project/). This concept proposes to unify the zoonotic efforts through close collaboration between human and veterinary medicine. The ultimate goal was the development of priorities and actions to be implemented. This concept is based on the fact that there is no difference between human and veterinary medicine and that there is a kind of complementarity (Zinsstag et al., 2011). Following the sudden onset of some emerging diseases, the concept of single medicine has evolved into the "one health" concept, with the incorporation of the single medicine concept into ecosystem health.

Therefore human and animal health depend on the health of the surrounding ecosystems and require coordination between health professionals and all the stakeholders in the concerned sectors, particularly educational, sociopolitical, economical and associative environments, etc.

CONCLUSION

The unexpected emergence and reemergence of viral diseases is currently a burning issue on the world stage. Despite the progress made, etiopathological knowledge of viral diseases, humanity would have a long way to go to unveil the unpredictable mechanisms that trigger the emergence and reemergence of highly pathogenic viral agents. This health risk is only the visible part of the iceberg, the immersed one corresponds to a multitude of complex factors of societal, environmental, and animal order, acting together in a synergistic way. The determinism of emergence and reemergence is an evolutionary process; the occurrence of emergence is acquired only after having exceeded the threshold of compensation or tolerance.

The list of emergence and reemergence factors is not exhaustive; other new factors related to climate and environmental change may arise and affect biodiversity in general and associated pathogens.

Only a holistic and concerted approach to the question of emergence and reemergence will make it easier to be apprehended. It will be based on the establishment and monitoring of specific parameters and on the continual identification of emergence risk factors within a framework of epidemiological and health surveillance.

To optimize the implementation of the fight against emergence and reemergence, the contribution of the legal and regulatory framework is essential in any integrative strategy, because it brings a mandatory dimension in the instructions application from the environment. Specialized scientists as well as guidelines from international bodies of guardians, namely WHO, OIE, and FAO.

On the other hand, the divergence between economically oriented policies and the need to preserve the environmental and health heritage imposes an additional burden and overhead which jeopardizes development. The dichotomous zone between economic and social development strategies and the protection of the environmental and health heritage is currently the basis of all sustainable development. The evolution of scientific and technical capacity to address the challenges of modernizing and integrating health and environmental education into the academic fabric can help sustain the balance of global health.

Acknowledgments

This chapter was supported by the financial support of the Ministry of Higher Education, University Hassan II of Casablanca, Faculty of Sciences and Techniques, Mohammedia and also Royal Gendarmerie of Morocco. The author thanks all the research staff at the Laboratory of Virology, Microbiology, Quality and Biotechnologies/Ecotoxicology & Biodiversity and Genetics Laboratory of Royal Gendarmeries for their support and technical help.

References

Ashford, R.W., 2003. When is a reservoir not a reservoir. Emerg. Infect. Dis. 9 (11), 1495–1496.

Belay, E.D., Kile, J.C., Hall, A.J., Barton-Behravesh, C., Parsons, M.B., Salyer, S., et al., 2017. Zoonotic disease programs for enhancing global health security. Emerg. Infect. Dis. 23 (1), S65–S70. Available from: https://doi.org/10.3201/eid2313.170544.

Belongia, E.A., 2002. Epidemiology and impact of coinfections acquired from Ixodes ticks. Vector Borne Zoonotic Dis. 2 (4), 265–273. Available from: http://www.ncbi.nlm.nih.gov/pubmed/12804168.

Bodian, D., 1955. Emerging concept of poliomyelitis infection. Science 122 (3159), 105–108.

Chastel, C., 2000. Emergential success: a new concept for a better appraisal of viral emergences and reemergences. Acta Virol. 44 (6), 375–376. Available from: https://www.cabdirect.org/cabdirect/abstract/20013052941.

Choffnes, E.R., Mack, A., 2015. Emerging Viral Diseases: The One Health Connection. Forum on Microbial Threats; Board on Global Health. Institute of Medicine. National Academies Press, Washington, DC, ISBN-13: 978-0-309-31397-1, ISBN-10: 0-309-31397-X. <https://www.ncbi.nlm.nih.gov/books/NBK280057/>.

Cicolella, A., 2010. Sante et environnement: la 2e révolution de santé publique. Santé Publique 3 (22), 343–351.

Cohen, M., 1998. Resurgent and emergent disease in a changing world. Br. Med. Bull. 54 (3), 523–532.

Domingo, E., Ruiz-Jarabo, C.M., Sierra, S., Arias, A., Pariente, N., Baranowski, E., et al., 2002. Emergence and selection of RNA virus variants: memory and extinction. Virus Res. 82 (1–2), 39–44.

Drevet, P., 2012. Approche géographique des risques d'émergence de maladies virales en Afrique forestière équatoriale: le VIH-1 au sud-est du Cameroun. <https://halshs.archives-ouvertes.fr/tel-00958735/>.

Drotman, P., 1998. Emerging infectious diseases: a brief biographical heritage. Emerg. Infect. Dis. 4 (3), 372–373.

FAO, 2018. Mission statement: FAO's role in emergencies. <http://www.fao.org/focus/e/disaster/mstate-e.htm>.

Fagherazzi-Pagel, H., 2008. Maladies émergentes et réémergentes chez l'homme, Concepts Facteurs d'émergence Alertes Riposte et coopération mondiale. <https://hal.archives-ouvertes.fr/hal-01456817v1>.

Gonzalo, L.C.R., 2017. Resurgence of almost eradicated paediatric diseases and the importance of childhood vaccination. Enferm. Infecc. Microbiol. Clin. 35 (9), 547–549. Available from: https://doi.org/10.1016/j.eimc.

Howard, C.R., Fletcher, N.F., 2012. Emerging virus diseases: can we ever expect the unexpected. Emerg. Microbes Infect. 1 (12), e46. Available from: https://doi.org/10.1038/emi.2012.47.

Hubalek, Z., 2003. Emerging human infectious diseases: anthroponoses, zoonoses, and sapronoses. Emerg. Infect. Dis. 9 (3), 403–404.

Kilbourne, E.D., 2006. Influenza pandemics of the 20th century. Emerg. Infect. Dis. 12 (1), 9–14. Available from: https://doi.org/10.3201/eid1201.051254.

Lederberg, J., Shope, R.E., Oaks, S.C., 1992. *Emerging Infections: Microbial Threats to Health in the United States. Institute of Medicine; Committee on Emerging Microbial Threats to Health*. National Academies Press, Washington, DC.

Maclachlan, J., Dubovi, E.J., Barthold, S.W., Swayne, D.F., Winton, J.R., 2016. Chapter 23—*Arenaviridae*. In: Fenner's Veterinary Virology, fifth ed. <https://www.sciencedirect.com/science/article/pii/B9780128009468000234>.

Morse, S.S., 1995a. Factors in the emergence of infectious diseases. Emerg. Infect. Dis. 1 (1), 7–15. Available from: https://doi.org/10.3201/eid0101.950102.

Morse, S.S., 1995b. Factors in the emergence of infectious diseases 1 (1), . Available from: ftp://ftp.cdc.gov/pub/EID/vol1no1/adobe/morse.vol1no1.pdfaccessed May 2018.

OIE, 2007. The 25th Conference of the Regional Commission for Asia, the Far East and Oceania of the World Organisation for Animal Health (OIE) was held in Queenstown (New Zealand) From 27 to 30 November. <http://www.oie.int/fr/pour-les-medias/communiques-de-presse/detail/article/25th-conference-of-the-oie-regional-commission-for-asia-the-far-east-and-oceania/>.

OIE, 2015. Les maladies de la faune sauvage. <http://www.oie.int/fileadmin/Home/fr/Media_Center/docs/pdf/Fact_sheets/WD_FR.pdf>.

Peigue-Lafeuille, H., Mirand, A., Archimbaud, C., Bailly, J.L., Henquel, C., 2014. Émergence et réémergence chez les entérovirus: de la poliomyélite à la maladie pieds-mains-bouche. Virologie 18 (2), 87–104.

Savey, M., Dufour, B., 2004. Diversité des zoonoses. Définitions et conséquences pour la surveillance et la lutte. Epidémiol santé animale. 46, 1–16.

Sharma, V.P., 1996. Re-emergence of malaria in India. Indian J. Med. Res. 103, 26–45.

Smolinski, M.S., Hamburg, M.A., Lederberg, J., 2003. Microbial Threats to Health: Emergence, Detection, and Response. Institute of Medicine. The *National Academies Press*, Washington, DC. Available from: http://dx.doi.org/10.17226/10636.

Stamm, L., Mudrak, B., 2013. Old foes, new challenges: syphilis, cholera and TB. Future Microbiol. 8 (2), 177–189. Available from: https://www.researchgate.net/publication/235394407_Old_foes_new_challenges_Syphilis_cholera_and_TB.

WHO, 1959. Activité de l'OMS en 1959: rapport annuel du Directeur général á l'Assemblée mondiale de la Santé et aux Nations Unies. <http://apps.who.int/iris/handle/10665/89736?locale = fr&null>.

Wolfe, N.D., Daszak, P., Kilpatrick, A., Burke, D.S., 2005. Bushmeat hunting, deforestation, and prediction of zoonotic disease. Emerg. Infect. Dis. 11 (12), 1822–1827. Available from: https://doi.org/10.3201/eid1112.040789.

Zinsstag, J., Schelling, E., Waltner-Toews, D., Tanner, M., 2011. From "one medicine" to "one health" and systemic approaches to health and well-being. Prev. Vet. Med. 101 (3–4), 148–156. Available from: https://doi.org/10.1016/j.prevetmed.2010.07.003.

Further Reading

OMS, 2017. Rapport de mission. <http://apps.who.int/iris/bitstream/handle/10665/259163/WHO-WHE-CPI-REP-2017.48-fre.pdf?sequence = 1&isAllowed = y&ua = 1>.

OMS, 2018. Constitution de l'OMS: ses principes. <http://www.who.int/about/mission/fr/>.

CHAPTER

2

Clustered Regularly Interspaced Short Palindromic Repeats/Cas9 Genome Editing Technology Against Emerging and Reemerging Virus

Aissam El-Aliani[1,2], *Yassine El Mallali*[2], *Mohammed El Mzibri*[2] *and Moulay Mustapha Ennaji*[1]

[1]Laboratory of Virology, Microbiology, Quality, Biotechnologies/Eco-Toxicology and Biodiversity, Faculty of Sciences and Techniques, Mohammedia, University Hassan II of Casablanca, Casablanca, Morocco

[2]Unit of Biology and Medical Research, National Center for Energy Sciences and Nuclear Techniques/Rabat (CNESTEN), Rabat, Morocco

INTRODUCTION

A decade ago since it burst onto the scene, clustered regularly interspaced short palindromic repeats (CRISPR)–Cas9 genes editing technology has shaken the field of genetics to its core, offering a revolutionist genome editing tool that is faster, cheaper, and more accurate than previous approaches such as transcription activator-like effector nuclease and Zinc-finger nucleases techniques (Urnov et al., 2010). It opens up an astonishing breadth of possible applications, from saving lives to changing the humanity forever.

Emerging and Reemerging Viral Pathogens
DOI: https://doi.org/10.1016/B978-0-12-814966-9.00002-0

Since the dawn of life, so-called bacteria and viruses have been fighting. Bacteriophages hunt bacteria in the ocean and kill 40% of them (Keen, 2015). Every single day phages do this by inserting their own genetic code into the bacteria and take them for use as factories (Kasman and Whitten, 2018). The bacteria try to resist but fail most the time because their protection tools are too weak, but sometimes bacteria survive by activating their most effective antivirus system (Rath et al., 2015). First, they save a part of the viral DNA in their own genetic code in a DNA archive called CRISPR (Fig. 2.1). Here it is stored safely until it is needed when the virus attacks again, the bacterium quickly makes an RNA copy from the DNA archive and arms a secret weapon, a protein called Cas9. This protein scans the bacterium's insides for signs of the virus invader by comparing every bit of DNA it finds to the sample from the archive. When it finds a 100% perfect match, it is activated and

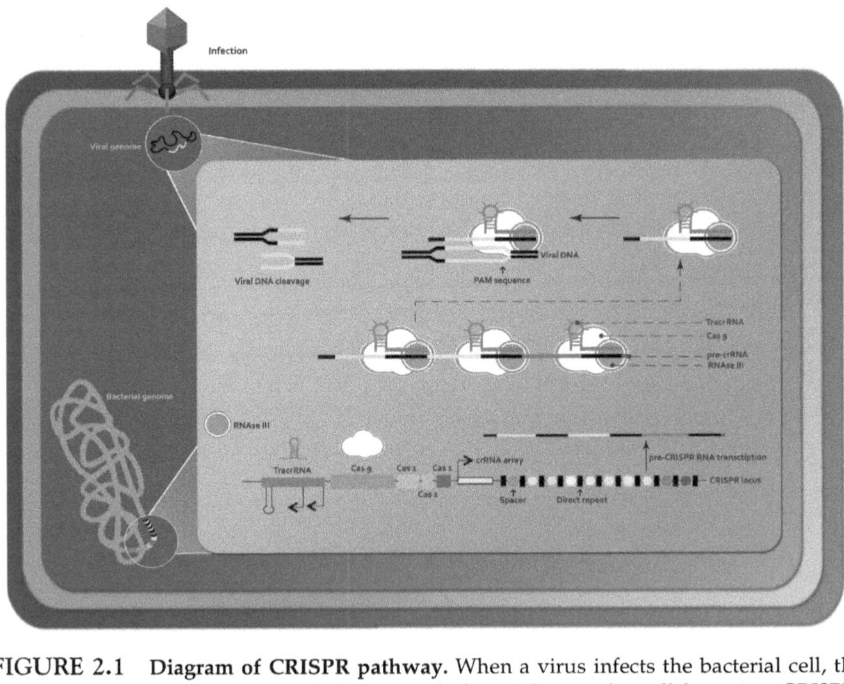

FIGURE 2.1 Diagram of CRISPR pathway. When a virus infects the bacterial cell, the CRISPR system can detect it and destroy it before it harms the cell by using CRISPR-associated proteins called Cas proteins that insert short viral DNA segments called spacers between the short palindromic repeats. The short palindromic repeats and spacers are transcribed and processed later into short segments called CRISPR RNAs (crRNAs). The crRNAs bind to tracrRNA and also a Cas protein. This complex then seeks out and cuts the viral sequences that are complementary to the crRNA and thus destroys the invading virus. *CRISPR*, Clustered regularly interspaced short palindromic repeats; *tracrRNA*, trans-activating crRNA.

cuts out the virus DNA making it useless and protecting the bacterium against the attack. Cas9 is very precise protein, almost like a DNA surgeon (Hille and Charpentier, 2016).

This adaptive immune defense mechanism used in bacteria began a revolution genome editing tool when scientists figured out that is programmable and we can just design a fragment of RNA called guide RNA (gRNA), which acts like a GPS system to recognize the genomic part that we want to change (Doudna and Charpentier, 2014).

CRISPR system has two components, a gRNA and the Cas9 endonuclease. When gRNA and Cas9 enzyme are expressed in a living cell, the gRNA/cas9 complex is recruited to the target sequence, which is directly upstream of the PAM (protospacer adjacent motif) sequence, through complementary base pairing of the gRNA to the genomic DNA (Fig. 2.1). Once the complex localized to the target DNA, Cas9 cuts the desire region with extreme accuracy resulting in a double-strand break (DSB). The DSB created by cas9 is then repaired by the cell's own repair mechanism such as HDR (homology-directed repair, pathway) or NHEJ (nonhomologous end joining, DNA repair pathway) (Fig. 2.2).

NHEJ DNA repair pathway used in the absence of a repaired template. With this pathway the ends of the DNA are simply ligated back together, which usually leads to the introduction of a small insertion or deletion mutations that disrupt the reading frame of the desired gene (Lieber et al., 2010). Alternatively, the HDR pathway can be utilized in the presence of a repaired template. This template will have the homology to the flanking region of the DSB. This method of repair is highly accurate and could be used to introduce specific nucleotide changes in the target gene (Davis and Chen, 2013). The NHEJ mechanism could be utilized to introduce random mutations, mostly in the form of insertion or deletion, and could be used to knock out the gene of interest (Su et al., 2016). On the other hand, HDR could be used for gene knockout, gene tagging, specific mutations, knock-ins and not for genes or promoter studies. To date, three variants of the cas9 endonuclease have been adopted in genome editing protocols (Zhang et al., 2017).

The wild-type Cas9 can site specifically cleave double-stranded DNA (Ceasar et al., 2016). The mutant form of cas9 is known as the cas9 Nickase. The cas9 Nickase has one of its molecular scissors disabled, resulting in the cleavage of only one DNA strand, and third, the cas9 Null mutant where both of the nucleus domains are inactivated. However, it still retains its ability to bind to DNA based on gRNA specificity (Doudna and Charpentier, 2014). Since a single-strand break, or nick is normally quickly repaired through the HDR pathway, using the intact complementary DNA strand as a repair template off-target, effects of the cas9 Nickase are minimized (Doudna and Charpentier, 2014). While the cas9 Null mutant does not introduce indel mutations or

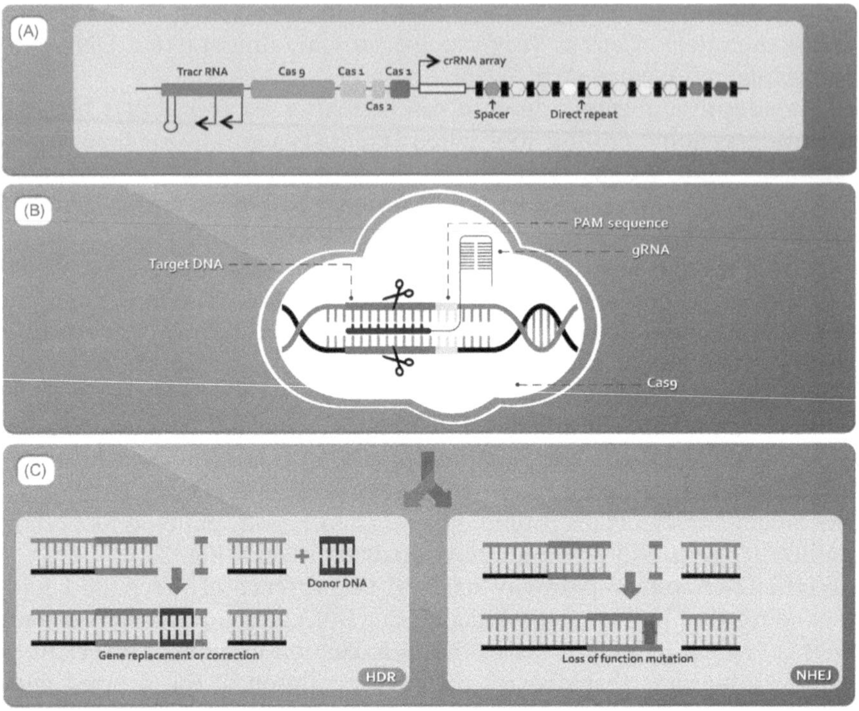

FIGURE 2.2 **Schematic representation of CRISPR system.** (A) Schematic representation of CRISPR loci: it contains a cluster of various genes, typically Cas9, Cas1, Cas2, as well as two noncoding RNA elements, tracrRNA and a repetitive sequences (direct repeats) interspaced by short nonrepetitive sequences called spacers. Each spacer is derived from foreign genetic material (protospacer), and each of them is associated with a PAM sequence whose recognition is exclusive to individual CRISPR systems. (B) Schematic of gRNA/Cas9 complex for genome editing: Cas9 is directed to its DNA target by base pairing between the gRNA and DNA. A PAM downstream of the gRNA-binding region is required for Cas9 recognition and cleavage. Cas9/gRNA cuts both strands of the target DNA, triggering endogenous DSB repair. (C) DSB and NHEJ DNA repair pathway: the resulting caused by the Cas9 cleavage is repaired by (NHEJ) pathway, disrupting the open reading frame of the targeted DNA sequence by causing an InDel frameshift or premature stop codon. On the other hand, HDR pathway exploited to introduce nucleotide modifications to genomic DNA by transfacting with the CRISPR/Cas9 system a DNA template containing the desired sequence that will replace the mutant sequence. The transfected template must have a high degree of homology to the sequence upstream and downstream of the DSB. *CRISPR*, Clustered regularly interspaced short palindromic repeats; *DSB*, double-strand break; *NHEJ*, nonhomologous end joining; *PAM*, protospacer adjacent motif.

directed recombination to the target genome, it offers great potential in genome targeting and can be used for the transcriptional activation by fusing the cas9 Null mutant or the transcriptional activator such as VP 64, transcriptional repression by the fusion of the cas9 Null mutant with transcriptional repressor or using a gRNA against a promoter region of the desired gene, DNA labeling by fusion of the cas9 Null mutant with florescent tags for genome imaging or for chromatin immunoprecipitation by fusion of the cas9 Null mutant with an antibody epitope tag to facilitate the pulldown of specific genomic loci (Beneke et al., 2017). CRISPR was first shown to work as a genome editing tool in human cell culture by 2012. Since then it has been used in a wide range of organisms including baker's yeast, zebrafish, fruit flies, nematodes plants, mice, and several other organisms. The CRISPR–cas9 system offers the first alternative to the current protein-based genome editing techniques such as zinc finger and TATEN. The simple and effective mechanism of CRISPR is considered the game changer in molecular genetics and has been applied to many scientific fields. CRISPR–cas9 system shows extensive applicability in our modern health-care system. It has the potential to become the platform in genetic therapeutics and personalized medicine.

HEPATITIS B VIRUS

Chronic hepatitis B is considered a major public health problem worldwide, with approximately 400 million chronic HBV (hepatitis B virus) carriers (Lau et al., 2007). HBV infection causes 1 million deaths per year due to cirrhosis and hepatocellular carcinoma (HCC) (Zamor, 2017). It has been five decades when Dane et al. identified the HBV virus (Gerlich, 2013). It is a DNA virus with a small and circular dsDNA genome, which contains 3173 bp (Fig. 2.3) and encoded for a few proteins such as HBV Protein X (HBx) and HBV surface antigen (HBsAg), which are responsible for multiple pleiotropic effects that are subsequently important in the transformation (Grimm et al., 2011).

The HBV genome may integrate into the host genome, and this can occur at many different sites and may cause disruption of key genes that regulate proliferative signaling, which is thought to be an important mechanism in the development of HCC in HBV-infected individuals (Ozkal-Baydin, 2014). HBx is oncogenic and can transform rodent hepatocytes and NIH 3T3 cells (Grimm et al., 2011). HBx may act by activating signal transduction pathways involved in the regulation of cell proliferation such as ERKs, SAPKs, and p38 protein kinase (Guan et al., 2004). It has also been reported to bind to p53 (Truant et al., 1995).

Using a gRNA prediction tool Li et al. found five candidate target sequences from the HBV genome by designing one gRNA (Table 2.1). The Cas9/gRNA coexpression vectors were transfected into A64 cells (Tang et al., 2017). The result showed that expression of the gRNAs induced the deletion of the entire HBV genome. To assess

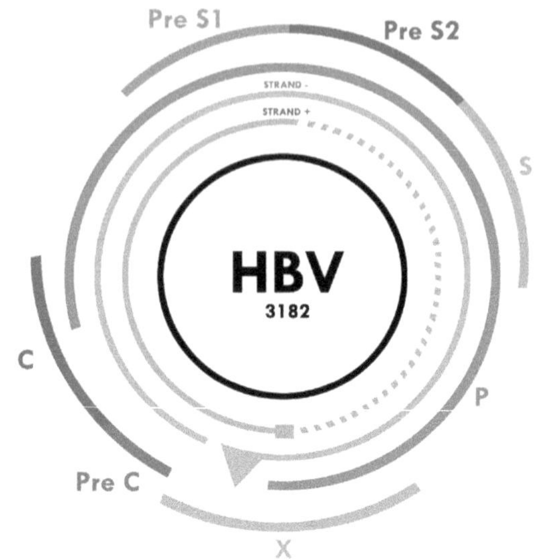

FIGURE 2.3 **Organization of the HBV genome.** The genome is represented in the center in black, and the ORFs are represented in different colors. Four overlapping ORFs are covering the entire genome. The four ORFs are ORF P, ORF S, ORF C, and ORF X. The ORF P covers 80% of the genome and codes for viral polymerase. ORF S overlaps the ORF P and codes for the three envelope proteins and has three codons of initiations for the PreS1 gene, the PreS2 gene, and the S gene. These genes will be used to code the HBs small proteins (S gene), HBs medium (gene preS2 and gene S), and HBs large (genes PreS1, PreS2, and gene S). This means that the envelope proteins all have the same C-terminus. ORF C that overlaps the 5′ end of the ORF P, the C gene codes for the capsid protein and for the precore, the translation of C and pre-C gives the precursor of the HBe protein. ORF X that overlaps the 3′ end of the ORF P code for the HBx protein. *HBV*, Hepatitis B virus; *HBx*, HBV Protein X; *ORFs*, open reading frames.

possible off-target effects of CRISPR–Cas9 and the integrated subgenomic HBV DNA fragments in the original HBV-containing cell line, whole-genome sequencing was performed (Tang et al., 2017). The raw data analysis showed six indels in 50 potential off-target regions, while four indels were identified in two cell lines, and only two indels were different after gRNA transfection, which suggests that gRNA did not have significant off-target indels. Recently Zhen et al. (2015) developed CRISPR/Cas9 system targeted the HBsAg and HBx-encoding region of HBV. The results showed in a cell culture system and in vivo that the HBV DNA levels and HBsAg were reduced as shown by qPCR and immunohistochemistry, respectively (Zhen et al., 2015). Kennedy et al. (2015) used lentiviral transduction of Cas9 and HBV-specific gRNAs. The results showed effective inhibition of HBV DNA production. Total HBV viral DNA levels were reduced by up to approximately 1000-fold while cccDNA levels were reduced by up to approximately 10-fold, and

TABLE 2.1 Specific gRNAs for Different Hepatitis B Virus (HBV) Target Gene in Various Location and the Obtained Results After the Gene Knockout

Target	gRNA sequences (5′→3′)	Target location	Cells ligne	Results	References
P1 and XCp	GGACTTCTCTAATTTTCTAGGG	261	Mouse liver in vivo	• Reduction in HBsAg level • Significant reduction on the production of HBV core and surface proteins	Lin et al. (2014)
	GGGCTTTCGCAAAATACCTATGG	621	Huh-7		
	GGGCCTCAGTCCGTTTCTCTTGG	648			
	GTTTTGCTCGCAGCCGGTCTGGG	1292			
	GGGGGAGGAGATTAGGTTAAAGG	1742			
	GCTGTGCCTTGGGTGGTTTGGG	1876			
	GTCGCAGAAGATCTCAATCTCGG	2421			
	GGAGTGGGAGCATTCGGGCCAGG	3028			
Conserved regions of different HBV genotypes	caccTACCGCAGAGTCTAGACTCG		Liver derived HepG2 cells	• Inhibit the replication of HBV of different genotypes	Liu et al. (2015)
	aaacCGAGTCTAGACTCTGCGGTA				
	caccCATTTGTTCAGTGGTTCGTA				
	aaacTACGAACCACTGAACAAATG				
	caccGTTGCCGGGCAACGGGGTAA				
	aaacTTACCCCGTTGCCCGGCAAC				
	caccAAACAAAGGACGTCCCGCGC				
	aaacGCGCGGGACGTCCTTTGTTT				

(Continued)

TABLE 2.1 (Continued)

Target	gRNA sequences (5′→3′)	Target location	Cells ligne	Results	References
	caccGGTCTCCATGCGACGTGCAG				
	aaacCTGCACGTCGCATGGAGACC				
	caccGTAGCTCCAAATTCTTTATA				
	aaacTATAAAGAATTTGGAGCTAC				
	caccGACCTTCGTCTGCGAGGCGA				
	aaacTCGCCTCGCAGACGAAGGTC				
	caccCCTTCCTGACTGGCGATTGG				
	aaacCCAATCGCCAGTCAGGAAGG				
HBsAg, core, and RT			HepAD38	• Reduction in viral DNA level	Kennedy et al. (2015)
HBsAg	GACCTTCGTCTGCGAGGCGA	2384	HepG2 cells	• Low decrease in pgRNA levels and HBsAg production	Ramanan et al. (2015)
HBsAg	TAAAGAATTTGGAGCTACTG	1989		• Decrease in pgRNA levels and HBsAg production	
HBsAg	GGGTTGCGTCAGCAAACACT	1179	Hep-NTCP	• Low decrease in pgRNA levels and HBsAg production	
HBsAg	TCCTCTGCCGATCCATACTG	1254		• Decrease in pgRNA levels and HBsAg production	

HBsAg, Hepatitis B virus surface antigen.

the majority of the residual viral DNA was mutationally inactivated (Kennedy et al., 2015).

In another study, Seeger and Sohn (2014) tested CRISPR/Cas9 HBV-specific gRNAs complex in HepG2 hepatoma cells and found inhibition of HBV infections up to eightfold, which was due to mutations and deletions in the virus genome caused by Cas9 cleavage and repaired by NHEJ.

In conclusion, studies from different laboratories around the world have shown the potential of CRISPR/Cas9 technology to treat HBV-associated diseases and could be a good way in eradicating persistent HBV infection.

ONCOGENIC HUMAN PAPILLOMAVIRUSES 16 AND 18

Human papillomaviruses (HPVs) are large family composed by more than 200 types (Schiffman et al., 2016). Some are spread through skin contact, while others are sexually transmitted and are responsible for the development of cervical cancer (CC) (Moody and Laimins, 2010). A small subgroup contains 40 types as HPV16, HPV18, HPV31, HPV33 has been considered high-risk HPVs (HR-HPVs) and found to be associated with more than 90% of CC, of which HPV16 and 18 accounts for approximately more than 90% (Braaten and Laufer, 2008). Worldwide, HPV infections account for more than half of all infection-linked cancers in females, and only less than 5% in males (Pytynia et al., 2014). At the molecular level the HPV genome is a linear double-stranded DNA of approximately 8 kbp with eight open reading frames (ORFs) and an upstream regulatory region, which regulates the transcription (Fig. 2.4) (Moody and Laimins, 2010).

The oncogenic properties of HR-HPVs such as 16 and 18 imply the expression of two proteins called E6 and E7, which are involved in cellular immortalization, transformation, and maintenance of the malignant phenotype (Hawley-Nelson et al., 1989).

The E7 protein binds the p105Rb protein with high affinity and facilitates its degradation via the proteasome. It is also able to sequester the p107 without influencing its stability. These interactions of the E7 protein with the pocket proteins make it possible to divert the cellular machinery by releasing the E2F/DP transcription factors, which stimulate the entry into the S phase of the cell cycle. This protein is also capable of activating the E/CDK2 cyclin complexes that are involved in the passage of the R cell cycle restriction point and the late G1 phase, and also the activation of the cyclin A/CDK2 complexes that are involved in the G1/S transition and progression in phase S (Martinez-Zapien et al., 2016; Yim and Park, 2005). Finally, E7 associates with HDAC (histones

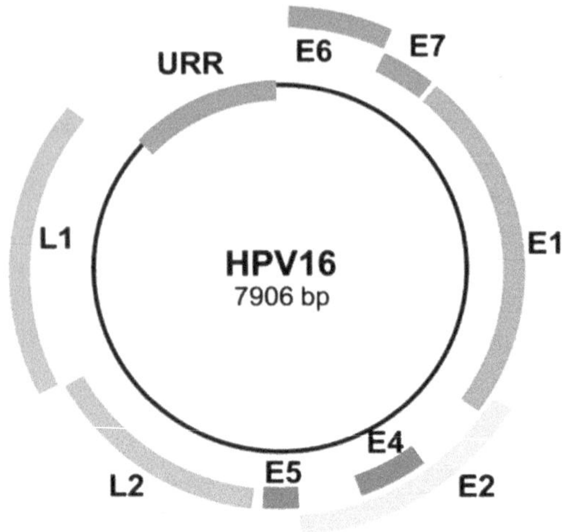

FIGURE 2.4 **Schematic representation of circular dsDNA genome of HPV showing the location of the early (E) and late genes (L) and of the URR region.** The viral genome is a circular double-stranded DNA molecule of about 8 kbp in length. The viral genome has an average of eight open reading frame organized into two regions; an early region that is approximately 4 kb in size and encodes nonstructural proteins involved in viral replication and cellular transformation (E6, E7, E1, E2, E4, E5); and a late region of about 3 kb that encodes two L1 and L2 capsid proteins that self-assemble to form the virion. There is a 1 kb LCR noncoding regulatory region that controls gene expression. *HPV*, Human papillomavirus; *LCR*, long control region; *URR*, upstream regulatory region.

deacetylase) proteins, which are normally related to pRb. This results in inactivation of the HDACs, which then allows histone acetylation, leading to the destabilization of the nucleosomal structure and chromatin remodeling favorable to gene transcription (Fig. 2.5).

Reactivation of cellular DNA replication in a differentiating cell is stressful for the cell that attempts to react by activating the expression of the p53 protein to induce apoptosis. But the HR-HPV E6 protein in particular will bind the p53 protein with high affinity and promote its hydrolysis in the proteasome (Thomas et al., 1999). The degradation of p53 has multiple consequences on the life of the cell (Fig. 2.5). This viral protein is also characterized by its ability to activate telomerase. Thus the E6 and E7 oncogenic genes are key targets for therapeutic intervention in HPV-associated cancer disease (Table 2.2).

In the hope of developing a gene-specific therapy for HPV-related cancer and effectively knocking down E6 and E7 expressions, Zhen and Li (2017) established in 2014 CRISPR/Cas9 complex targeting promoter regions of HPV16 E6/E7 and targeting E6, E7 transcripts. The transduction into cervical HPV-16-positive cell line SiHa showed that CRISPR/Cas9 targeting the promoter, as well as targeting E6 and E7 genes

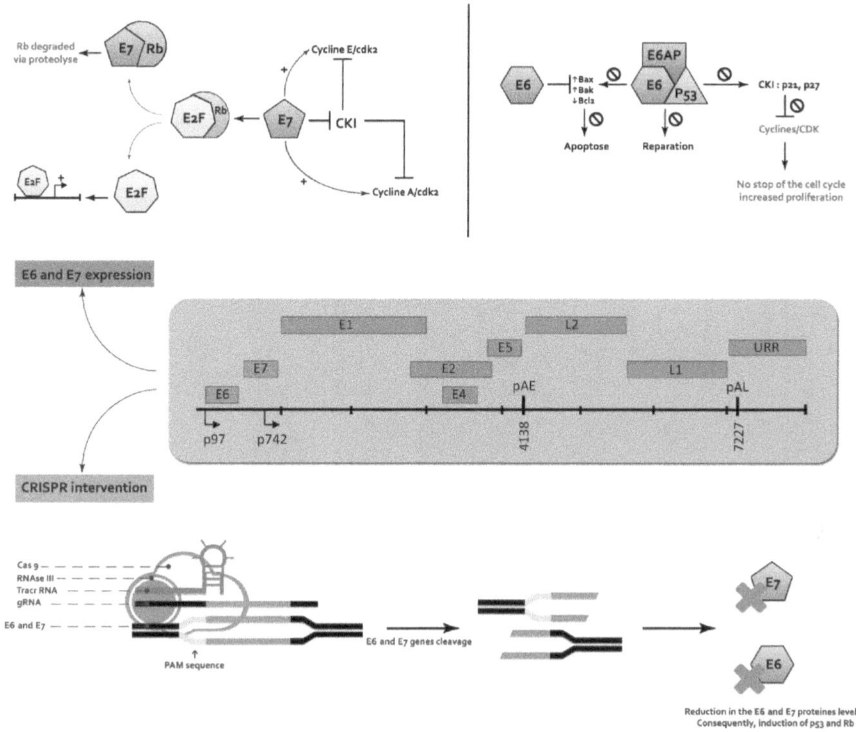

FIGURE 2.5 **Schematic representation of many targets of E6 and E7 oncoproteins of HPV.** Those proteins bind to different molecules in the cell cycle especially on the p53 and Rb leading to tumorigenic cell transformation and the role of CRISPR/Cas9 mechanism to stop the E6 and E7 expression. *CRISPR*, Clustered regularly interspaced short palindromic repeats; *HPV*, human papillomavirus.

resulted in the accumulation of p53 and p21 proteins. Then the cells inoculated subcutaneously into nude mice to establish the transplanted tumor animal models. The results showed dramatically the inhibition of tumorigenesis and growth of mice incubated by cells with CRISPR/Cas9 targeting (promoter + E6 + E7)-transcript (Zhen and Li, 2017). These results provide evidence for the application of CRISPR/Cas9 approach as a new treatment strategy.

Also, in another study, Kennedy et al. (2014) induced deletion and insertion of mutations in E6 or E7 by introducing CRISPR–Cas9 and E6- or E7-specific gRNAs into HeLa and SiHa cervical carcinoma cell lines, which contain integrated HPV18 or HPV16, respectively. This change will result in the activation of p53 and pRb, causing cell cycle arrest and subsequently cell death (Kennedy et al., 2014). The same results founded due to the work realized by Hu et al. They found that

TABLE 2.2 Specific gRNAs for E6 and E7 Human Papillomavirus (HPV) Oncogenes in Various Location and the Obtained Results After the Gene Knockout

Target	gRNA sequences (5′→3′)	Target location	Cells ligne	Results	References
Oncogenes E6 and E7 of HPV16	GCAACAGTTACTGCGACGTG	E6 (90)	SiHa	• Introduction of inactivating deletion and insertion mutations into the E6 or E7 gene • Induction of p53 and Rb • Cell cycle arrest and eventual cell death	Kennedy et al. (2014)
	GCCAGCTGGACAAGCAGAAC	E7 (567)			
Oncogenes E6 and E7 of HPV18	CGCGCTTTGAGGATCCAACA				
	CGAGCAATTAAGCGACTCAG	E6 (5)	HeLa		
		E7 (534)			
Promoter of E6 and E7 oncogenes of HPV16	CACCGACTAAGGGCGTAACCGAAAT	23	SiHa	• Inhibition of cell proliferation by the induction of p53 and p21	Zhen et al. (2014)
	AAACATTTCGGTTACGCCCTTAGT				
	CACCGGTTTCGGTTCAACCGATTT	36			
	AAACAAATCGGTTGAACCGAAACC				
Different region in E6/E7 oncogenes of HPV16	CACCGACTTTCTGGGTCGCTCCTGT	E6 (115)	SiHa	• Inhibition of cell proliferation by the induction of p53 and p21	Zhen et al. (2014)
	AAACACAGGAGCGACCCAGAAAGT				
	CACCGCAACAGTTACTGCGACGTG	E6 (205)			
	AAACCACGTCGCAGTAACTGTTGC				
	CACCGCTAATTAACAAATCACACAA	E6 (386)			
	AAACTTGTGTGATTTGTTAATTAG				
	CACCGTCCGGTTCTGCTTGTCCAGC	E7 (682)			
	AAACGCTGGACAAGCAGAACCGGA				
	CACCGACACGTAGACATTCGTACTT	E7 (777)			
	AAACAAGTACGAATGTCTACGTGT				

Oncogenes E7 of HPV16	AACCCAGCTGTAATCATGCA	E7 (564)	SiHa and CaSki And not in HPV negative C33A and HEK293 cells	• Inducing apoptosis and growth inhibition	Hu et al. (2014a,b)
	ACATTGCATGAATATATGTT	E7 (583)			
	GAGACAACTGATCTCTACTG	E7 (616)			
	GCTGGACAAGCAGAACCGGA	E7 (688)			
	GTCGATGTATGTCTTGTTGC	Not inducing DSB			

DSB, Double-strand break.

their HPV16-E7-specific gRNA designed could disrupt HPV16-E7 ORFs at specific sites, provoking the induction of apoptosis and growth inhibition in HPV-positive CaSki and SiHa cells and not in HPV negative HEK293 and C33A cells (Hu et al., 2014a,b).

Thereby the CRISPR/Cas9 technology has proven encouraging results when used on HPV-transformed cell lines in culture and has the potential to be developed as an effective strategy of genes therapy for HPV-associated CC in clinic scale.

EPSTEIN–BARR VIRUS

Epstein–Barr virus (EBV), an oncogenic virus widely spread in human populations, causes Burkitt's lymphoma and nasopharyngeal carcinoma (Chang et al., 2017). EBV infections are very common with more than 90% of people becoming infected by their 20s (Arvin et al., 2007). The Epstein–Barr genome is 165–175 kbp linear double-stranded DNA encodes approximately 85 genes (Fig. 2.6), 10 of which are expressed during the latency phase (Sivachandran et al., 2012). EBV has two subtypes, EBV-1 and EBV-2, which differ certain latency genes that have significant biological activity (EBNA-LP, EBNA-2, -3A, -38, and -3C). Sometimes, EBV can enter a latent state in B cells, where the episomal EBV genome expresses a low protein level involved in the events which lead to cellular transformation, particularly the EBNA-1 (EBV nuclear antigen 1) involved in the maintenance of the latent EBV episome, viral DNA replication, and cellular transformation. The second protein is EBNA-2, which regulates the expression of other latency genes such as LMP-1 and LMP-2 (Sivachandran et al., 2012). The LMP-1 protein is responsible for the activation of many pathways involved in cell transformation and induces invasion and metastasis factors. LMP-1 acts together with many latency-associated proteins and especially BARTs in Burkitt's lymphoma and nasopharyngeal carcinoma (Arvin et al., 2007).

In 2014 Wang and Quake derived cells from a patient with Burkitt's lymphoma with latent EBV infection and using a CRISPR/Cas9 vector targeted the EBV genome by developing a specific gRNA for many genes such as EBNA-3C, EBNA-1, and LMP-1 and they found a marked reduction in proliferation and a low level in viral load as well as restoration of the apoptosis pathway (Wang and Quake, 2014) In another study, Yuen et al. (2015, 2017) developed a new CRISPR/Cas9 vector using two specific gRNAs targeting the promoter region for BART gene to make deletion of 558 bp in the promoter region for BART resulting the loss expression of BART miRNA.

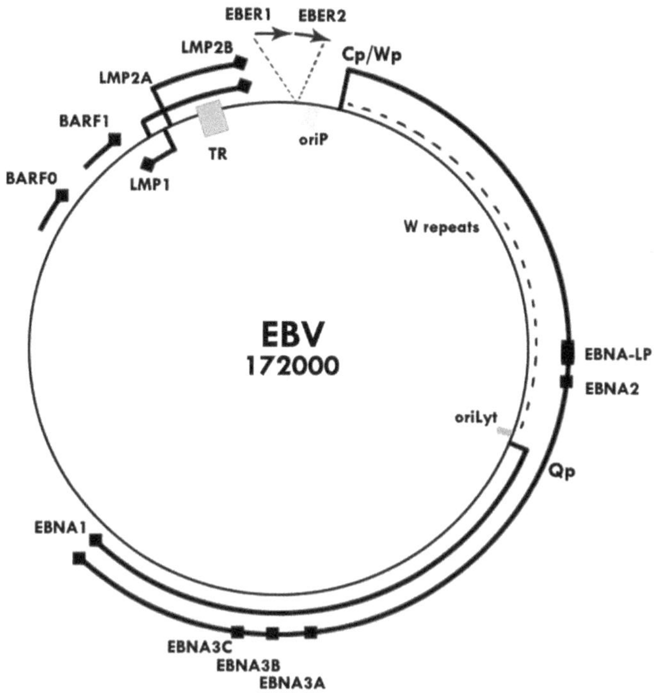

FIGURE 2.6 **The EBV genome consists of a 172 kbp double-stranded DNA. It is organized in a manner similar to that of other lymphocryptoviruses.** Hundred genes have been described, of which at least 11 are expressed during the latency phase. These genes are grouped into three main families: six nuclear genes, two LMP genes, genes encoding small RNAs (EBER). EBNA-1 is required for episomal replication and maintenance of the viral genome. EBNA-2 is essential for the process of B cell immortalization and the expression of EBNA-1 and EBNA-3 proteins. EBNA-3 consists of a family of three high molecular weight genes located in tandem on the Epstein–Barr virus genome: EBNA-3A (or EBNA-3), EBNA-3B (or EBNA-4), and EBNA-3C (or EBNA-6). It negatively regulates the transactivation of genes by EBNA-2. EBNA-LP (or EBNA-5) encodes a set of extremely polymorphic proteins. Its function is poorly known. LMP-1 is expressed in the absence of EBNA-2 during activation of the lytic cycle of B lymphocytes. LMP-2 encodes a membrane protein containing 12 hydrophobic transmembrane domains. It is colocalized with LMP-1 at the plasma membrane of infected lymphocytes. EBER-1 and EBER-2 are the most abundant EBV RNAs during the latency phase of infected B cells. These small RNAs are encoded in nonpolyadenylated form. The majority of EBERs are localized in the nucleus where they form a complex with the La protein. The La protein is a ubiquitous protein in eukaryotic cells and is associated with the three terminal end of the newly synthesized RNA.

HUMAN IMMUNODEFICIENCY VIRUS

About 48 years after AIDS was first recognized (Centers for Disease Control, 1981), it remains a major public health problem worldwide due to its permanent integration into the host genome and the absence of inhibition by antiviral drugs (Moss, 2013; Ruelas and Greene, 2013). New therapy called highly active antiretroviral therapy (HAART) intended to control human immunodeficiency virus (HIV-1) infection and impeding AIDS development (Broder, 2010). HAART includes a mixture of compounds that act on various steps of the viral life cycle and powerfully inhibits viral replication in cells that support HIV-1 infection and robustly reduces plasma viremia. HAART profoundly reduces viral replication in cells that support HIV-1 infection, reduces plasma viremia to a minimal level but does not suppress the replication and expression of the viral genome in tissues nor target infected cells that serve as a reservoir for the virus, including macrophages, microglia, astrocytes, and lymphoid cells (Broder, 2010). Despite this, AIDS remains incurable due to the permanent integration of HIV-1 into the cell genome.

To eliminate the integrated HIV-1 genome Hu et al. (2014a,b) have developed a Cas9/LTR-gRNA system targeting the HIV-1 LTR U3 region. The result showed the inactivating viral gene expression and replication in latently infected promonocytic, microglial, and T cells. Also, the deep sequencing showed the absence of off-target cleavage, which means that the Cas9/gRNAs system do not cause genotoxicity to the host cells (Hu et al., 2014a,b).

In another study, Zhu et al. tested various sites in HIV-1 DNA for CRISPR/Cas9 in cells that are latently infected by HIV-1. Sequencing results showed that all the target sites in HIV-1 DNA were efficiently mutated, which was also proved by the increase of HIV-1 gene expression and the reduction of virus production by 20-fold (Zhu et al., 2015).

In a recent study published in 2018, Ophinni et al. designed RNA-guided CRISPR/Cas9 targeting the two regulatory genes *tat* and *rev* (Fig. 2.7) with specific gRNAs across six major HIV-1 subtypes. The designed CRISPR/Cas9 complex was introduced into 293 T and HeLa cells abolished the expression of Tat and Rev proteins. The target gene was mutated at the Cas9 cleavage site with high frequency and various indel mutations, which is demonstrated by deep sequencing. Conversely, no mutations were detected at off-target sites (Ophinni et al., 2018).

Thus the CRISPR/Cas9 genome editing tool has the potential to provide a specific and efficacious approach against HIV-1/AIDS and can be employed both prophylactically and therapeutically.

CONCLUSION

For a long time, bacteria and archaea utilize sequence-specific DNA nucleases to interfere with viral replication. Inspired by CRISPR's evolutionary origins, many researchers around the globe developed many complex systems of CRISPR/Cas9 to fight the emergence and

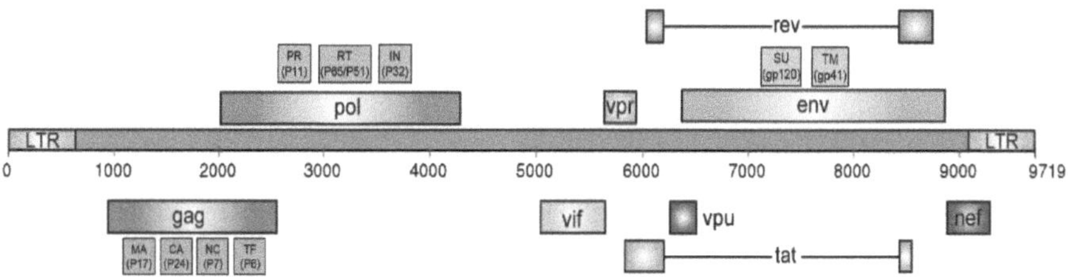

FIGURE 2.7 **Genomic organization of HIV-1.** The genome of HIV-1 is about 9700 nucleotides and consists of three main genes called gag "group specific antigen," pol "polymerase," and env "envelope." They define the structure of the virus and are common to all retroviruses. The gag gene encodes the structural proteins (MA, CA, NC, and TF), while the pol gene encodes the proteins necessary for viral replication (PR, TI, and IN) and the env gene encodes the envelope glycoproteins (gp120 and gp41). Six other genes, located between pol and env, code for viral protein expression-regulating proteins that are tat, rev, alive, vpr, vpu, and nef. These so-called auxiliary proteins are essential for viral replication in vivo. *HIV-1*, Human immunodeficiency virus.

reemergence virus. Although CRISPR/Cas9 has only been available to be used in eukaryotic cells for only about 2 years, it has already found many potential applications to human diseases including genetic disorders, cancer, and viruses. So far, CRISPR/Cas9 system has been applied to many emerging and reemerging viruses as discussed earlier. The only remaining challenge for CRISPR/Cas technology is the problem of off-target that reduces its specificity.

References

Arvin, A., Campadelli-Fiume, G., Mocarski, E., Moore, P.S., Roizman, B., Whitley, R., et al., (Eds.), 2007. Human Herpesviruses: Biology, Therapy, and Immunoprophylaxis. Cambridge University Press.

Beneke, T., Madden, R., Makin, L., Valli, J., Sunter, J., Gluenz, E., 2017. A CRISPR Cas9 high-throughput genome editing toolkit for kinetoplastids. R. Soc. Open Sci. 4 (5), 170095.

Braaten, K.P., Laufer, M.R., 2008. Human papillomavirus (HPV), HPV-related disease, and the HPV vaccine. Rev. Obstet. Gynecol. 1 (1), 2.

Broder, S., 2010. The development of antiretroviral therapy and its impact on the HIV-1/AIDS pandemic. Antiviral Res. 85 (1), 1–18.

Ceasar, S.A., Rajan, V., Prykhozhij, S.V., Berman, J.N., Ignacimuthu, S., 2016. Insert, remove or replace: a highly advanced genome editing system using CRISPR/Cas9. Biochim. Biophys. Acta, Mol. Cell. Res. 1863 (9), 2333–2344.

Centers for Disease Control, 1981. Pneumocystis pneumonia Los Angeles. MMWR 30, 250–252.

Chang, Y., Moore, P.S., Weiss, R.A., 2017. Human oncogenic viruses: nature and discovery. Philos Trans R Soc Lond B Biol Sci. 372 (1732). Available from: https://doi.org/10.1098/rstb.2016.0264.

Davis, A.J., Chen, D.J., 2013. DNA double strand break repair via non-homologous end-joining. Transl. Cancer Res. 2 (3), 130.

Doudna, J.A., Charpentier, E., 2014. The new frontier of genome engineering with CRISPR-Cas9. Science 346 (6213), 1258096.

Gerlich, W.H., 2013. Medical virology of hepatitis B: how it began and where we are now. Virol. J. 10 (1), 239.
Grimm, D., Thimme, R., Blum, H.E., 2011. HBV life cycle and novel drug targets. Hepatol. Int. 5 (2), 644–653.
Guan, J., Chen, X.P., Zhu, H., Luo, S.F., Cao, B., Ding, L., 2004. Involvement of extracellular signal-regulated kinase/mitogen-activated protein kinase pathway in multidrug resistance induced by HBx in hepatoma cell line. World J. Gastroenterol.: WJG 10 (23), 3522.
Hawley-Nelson, P., Vousden, K.H., Hubbert, N.L., Lowy, D.R., Schiller, J.T., 1989. HPV16 E6 and E7 proteins cooperate to immortalize human foreskin keratinocytes. EMBO J. 8 (12), 3905–3910.
Hille, F., Charpentier, E., 2016. CRISPR-Cas: biology, mechanisms and relevance. Philos. Trans. R. Soc. Lond. B: Biol. Sci. 371 (1707), 20150496.
Hu, W., Kaminski, R., Yang, F., Zhang, Y., Cosentino, L., Li, F., et al., 2014a. RNA-directed gene editing specifically eradicates latent and prevents new HIV-1 infection. Proc. Natl. Acad. Sci. U.S.A. 111 (31), 11461–11466.
Hu, Z., Yu, L., Zhu, D., Ding, W., Wang, X., Zhang, C., et al., 2014b. Disruption of HPV16-E7 by CRISPR/Cas system induces apoptosis and growth inhibition in HPV16 positive human cervical cancer cells. Biomed. Res. Int. 2014, 612823.
Kasman, L.M., & Whitten, R.A. (2018). Bacteriophages.
Keen, E.C., 2015. A century of phage research: bacteriophages and the shaping of modern biology. Bioessays 37 (1), 6–9.
Kennedy, E.M., Kornepati, A.V., Goldstein, M., Bogerd, H.P., Poling, B.C., Whisnant, A.W., et al., 2014. Inactivation of the human papillomavirus E6 or E7 gene in cervical carcinoma cells using a bacterial CRISPR/Cas RNA-guided endonuclease. J. Virol. 88, 11965–11972.
Kennedy, E.M., Bassit, L.C., Mueller, H., Kornepati, A.V., Bogerd, H.P., Nie, T., et al., 2015. Suppression of hepatitis B virus DNA accumulation in chronically infected cells using a bacterial CRISPR/Cas RNA-guided DNA endonuclease. Virology 476, 196–205.
Lau, G., Marcellin, P., Peters, M., 2007. Chronic hepatitis B: a global health problem requiring coherent worldwide treatment strategies. Hepatol. Int. 1 (2), 316–325.
Lieber, M.R., Gu, J., Lu, H., Shimazaki, N., Tsai, A.G., 2010. Nonhomologous DNA end joining (NHEJ) and chromosomal translocations in humans. Genome Stability and Human Diseases. Springer, Dordrecht, pp. 279–296.
Lin, S.R., Yang, H.C., Kuo, Y.T., Liu, C.J., Yang, T.Y., Sung, K.C., et al., 2014. The CRISPR/Cas9 system facilitates clearance of the intrahepatic HBV templates in vivo. Mol. Ther. Nucleic Acids 19 (3), e186.
Liu, X., Hao, R., Chen, S., Guo, D., Chen, Y., 2015. Inhibition of hepatitis B virus by the CRISPR/Cas9 system via targeting the conserved regions of the viral genome. J. Gen. Virol. 96 (8), 2252–2261.
Martinez-Zapien, D., Ruiz, F.X., Poirson, J., Mitschler, A., Ramirez, J., Forster, A., et al., 2016. Structure of the E6/E6AP/p53 complex required for HPV-mediated degradation of p53. Nature 529 (7587), 541.
Moody, C.A., Laimins, L.A., 2010. Human papillomavirus oncoproteins: pathways to transformation. Nat. Rev. Cancer 10 (8), 550.
Moss, J.A., 2013. HIV/AIDS review. Radiol. Technol. 84 (3), 247–267.
Ophinni, Y., Inoue, M., Kotaki, T., Kameoka, M., 2018. CRISPR/Cas9 system targeting regulatory genes of HIV-1 inhibits viral replication in infected T-cell cultures. Sci. Rep. 8 (1), 7784.
Ozkal-Baydin, P., 2014. How did hepatitis B virus effect the host genome in the last decade? World J. Hepatol. 6 (12), 851.
Pytynia, K.B., Dahlstrom, K.R., Sturgis, E.M., 2014. Epidemiology of HPV-associated oropharyngeal cancer. Oral. Oncol. 50 (5), 380–386.

Ramanan, V., Shlomai, A., Cox, D.B., Schwartz, R.E., Michailidis, E., Bhatta, A., et al., 2015. CRISPR/Cas9 cleavage of viral DNA efficiently suppresses hepatitis B virus. Sci. Rep. 5, 10833.

Rath, D., Amlinger, L., Rath, A., Lundgren, M., 2015. The CRISPR-Cas immune system: biology, mechanisms and applications. Biochimie 117, 119–128.

Ruelas, D.S., Greene, W.C., 2013. An integrated overview of HIV-1 latency. Cell 155 (3), 519–529.

Schiffman, M., Doorbar, J., Wentzensen, N., De Sanjosé, S., Fakhry, C., Monk, B.J., et al., 2016. Carcinogenic human papillomavirus infection. Nat. Rev. Dis. Primers 2, 16086.

Seeger, C., Sohn, J.A., 2014. Targeting hepatitis B virus with CRISPR/Cas9. Mol. Ther. Nucleic Acids 3, e216.

Sivachandran, N., Wang, X., Frappier, L., 2012. Functions of the Epstein-Barr virus EBNA1 protein in viral reactivation and lytic infection. J. Virol. 86, 6146–6158.

Su, T., Liu, F., Gu, P., Jin, H., Chang, Y., Wang, Q., et al., 2016. A CRISPR-Cas9 assisted non-homologous end-joining strategy for one-step engineering of bacterial genome. Sci. Rep. 6, 37895.

Tang, L., Mao, B., Li, Y., Lv, Q., Zhang, L., Chen, C., et al., 2017. Knockout of OsNramp5 using the CRISPR/Cas9 system produces low Cd-accumulating indica rice without compromising yield. Sci. Rep. 7 (1), 14438.

Thomas, M., Pim, D., Banks, L., 1999. The role of the E6-p53 interaction in the molecular pathogenesis of HPV. Oncogene 18 (53), 7690.

Truant, R., Antunovic, J., Greenblatt, J., Prives, C., Cromlish, J.A., 1995. Direct interaction of the hepatitis B virus HBx protein with p53 leads to inhibition by HBx of p53 response element-directed transactivation. J. Virol. 69 (3), 1851–1859.

Urnov, F.D., Rebar, E.J., Holmes, M.C., Zhang, H.S., Gregory, P.D., 2010. Genome editing with engineered zinc finger nucleases. Nat. Rev. Genet. 11 (9), 636.

Wang, J., Quake, S.R., 2014. RNA-guided endonuclease provides a therapeutic strategy to cure latent herpesviridae infection. Proc. Natl. Acad. Sci. U.S.A. 111 (36), 13157–13162.

Yim, E.K., Park, J.S., 2005. The role of HPV E6 and E7 oncoproteins in HPV-associated cervical carcinogenesis. Cancer Res. Treat. 37 (6), 319.

Yuen, K.S., Chan, C.P., Wong, N.H.M., Ho, C.H., Ho, T.H., Lei, T., et al., 2015. CRISPR/Cas9-mediated genome editing of Epstein–Barr virus in human cells. J. Gen. Virol. 96 (3), 626–636.

Yuen, K.S., Chan, C.P., Kok, K.H., Jin, D.Y., 2017. Mutagenesis and genome engineering of Epstein–Barr virus in cultured human cells by CRISPR/Cas9. In Vitro Mutagenesis. Humana Press, New York, pp. 23–31.

Zamor, P.J., 2017. Viral hepatitis and hepatocellular carcinoma: etiology and management. J. Gastrointest. Oncol. 8 (2), 229.

Zhang, J.P., Li, X.L., Li, G.H., Chen, W., Arakaki, C., Botimer, G.D., et al., 2017. Efficient precise knockin with a double cut HDR donor after CRISPR/Cas9-mediated double-stranded DNA cleavage. Genome. Biol. 18 (1), 35.

Zhen, S., Li, X., 2017. Oncogenic human papillomavirus: application of CRISPR/Cas9 therapeutic strategies for cervical cancer. Cell. Physiol. Biochem. 44 (6), 2455–2466.

Zhen, S., Hua, L., Takahashi, Y., Narita, S., Liu, Y.H., Li, Y., 2014. In vitro and in vivo growth suppression of human papillomavirus 16-positive cervical cancer cells by CRISPR/Cas9. Biochem. Biophys. Res. Commun. 450 (4), 1422–1426.

Zhen, S., Hua, L., Liu, Y.H., Gao, L.C., Fu, J., Wan, D.Y., et al., 2015. Harnessing the clustered regularly interspaced short palindromic repeat (CRISPR)/CRISPR-associated Cas9 system to disrupt the hepatitis B virus. Gene Ther. 22 (5), 404–412.

Zhu, W., Lei, R., Le Duff, Y., Li, J., Guo, F., Wainberg, M.A., et al., 2015. The CRISPR/Cas9 system inactivates latent HIV-1 proviral DNA. Retrovirology 12 (1), 22.

CHAPTER

3

Infectious Bronchitis Virus in Poultry: Molecular Epidemiology and Factors Leading to the Emergence and Reemergence of Novel Strains of Infectious Bronchitis Virus

Youssef Ennaji, Khadija Khataby and Moulay Mustapha Ennaji

Laboratory of Virology, Microbiology, Quality, Biotechnologies/Eco-Toxicology and Biodiversity, Faculty of Sciences and Techniques, Mohammedia, University Hassan II of Casablanca, Casablanca, Morocco

INTRODUCTION

Infectious bronchitis virus (IBV) is a coronavirus that causes an acute and highly contagious disease in chickens. The virus can cause substantial economic losses throughout the poultry industry worldwide. It can affect the upper respiratory tract and the reproductive tract, and some strains can cause nephritis (Cavanagh, 1997). IBV is the prototype species of the Coronavirus family, Gamma coronavirus genus, classified in the order Nidovirales, and it is the type species of the genus coronavirus of the domestic chicken (*Gallus gallus*) (Cavanagh and Naqi, 2003; Cook et al., 2012). Infectious bronchitis (IB) was first observed by

Emerging and Reemerging Viral Pathogens
DOI: https://doi.org/10.1016/B978-0-12-814966-9.00003-2

Schalk and Hawn (1931) in North Dakota in the United States in 2- to 3-week-old chickens. However, the nature of the infectious agent was not determined at that time and was assumed that IB was mainly a disease of young chickens. Hence, the disease was named "infectious bronchitis of young chicks." Five years later, it was demonstrated that the causative agent of this disease is a virus, which was named IBV (Beach and Schalm, 1936). Since that initial discovery, many different serotypes, defined by neutralizing antibodies, and genetic types, based on the deduced amino acid sequence (from the nucleic acid sequence) of the spike gene, have been described around the world (El-Houadfi et al., 1986; Jackwood et al., 1997).

EPIDEMIOLOGY OF INFECTIOUS BRONCHITIS VIRUS

The disease was first reported in the early 1930s, and since then has been documented in all countries with an intensive poultry industry (Ignjatovic and Sapats, 2000). It has a wide geographical distribution and it was found in regions of Africa, Asia, Australia, Europe, and the United States (Khataby et al., 2016). There are several widely distributed classic and variant IBV genotypes (de Wit et al., 2011). Some IBV genotypes and serotypes are closely related to the vaccine strains while others are variants that are unique to their geographical regions (Bande et al., 2017). Generally, IBV serotypes show variations in approximately 20%–25% in their S1 glycoprotein sequences. However, the variation can sometimes be as high as 50%, which affects the cross-protection toward virus strains (Cavanagh et al., 1992). Recently, a S1-gene-based phylogenetic classification of IBV identified 6 different viral genotypes, 32 distinct lineages, and several unassigned recombinants with interlineage origin. Interestingly, the distribution and diversity of these IBV genotypes differ with geographical location (de Wit et al., 2011; Valastro et al., 2016). The global distributions of major IBV serotypes such as Mass-type, 4/91 (793B or CR88)-like, D274-like (D207, D212 or D1466, D3896), and D3128, QX-like, and Italy-02 are shown in Fig. 3.1.

Infectious Bronchitis Virus Types in the United States

In the United States, the first case of IB was reported in the early 1930s (Schalk and Hawn, 1931). Since then, many IBV strains have been identified. Mainly Connecticut, SE17, Delaware strains, and the most commonly isolated type of IBV is Arkansas (Ark) (Jackwood et al., 2005), and the presence of Ark-like isolates indicates that this virus continues to change (Nix et al., 2000).

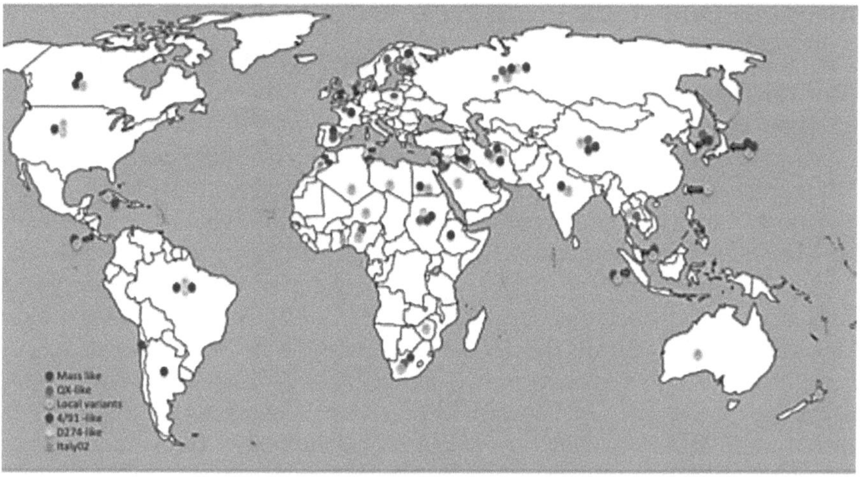

FIGURE 3.1 Distribution of major IBV serotypes around the world. *IBV*, Infectious bronchitis virus (Bande et al., 2017).

The Delaware IBV variant, nominated DE072 (Gelb et al., 1997), was first reported in 1992 and found to be distributed across the Northeastern United States. Based on S1 sequence, this variant resembles the Dutch D1466 variant (Lee and Jackwood, 2001). It is not known how the D1466 variant entered the country. The variant was later found to be prevalent in Georgia. The DE072-specific vaccine was then used to control the infection with little or no success. However, the use of the DE072 vaccine probably leads to the emergence of Georgia 98 (GA98) and GA08 variants (Lee and Jackwood, 2001).

California-type viruses were first isolated in the 1990s and were designated California variant (Moore et al., 1998). In 1999 another related but unique virus designated CAL99 was reported (Mondal and Cardona, 2007; Schikora et al., 2003). Since those reports, several other unique California viruses including CA/557/03 and CA/1737/04 (Jackwood et al., 2007) have been reported, indicating that the California-type viruses continue to evolve.

In 2007 and 2008, two new IBV variants were detected Georgia and South Carolina broilers with respiratory disease, wherein Mass and Ark vaccines do not protect. The viruses were distinct from each other and designated GA07 and GA08. The molecular analysis of the glycoprotein S1 has shown that the two variants have a unique sequence of the gene S1 (de Wit et al., 2011; Jackwood, 2012). Also, the sequence analysis showed the GA07 virus to be similar to CA1737/04 and the GA08 virus to be somewhat similar to CA/557/03, suggesting possible origins (Jackwood, 2012).

Infectious Bronchitis Virus Types in South America

In South America, Mass was first isolated in Brazil (Hipolito, 1957). Later in 1986, the variant Ark emerged, causing devastations to Brazilian poultry (De Wit et al., 2011). Subsequently, 12 new Brazilian isolates were identified based on S1-gene-specific reverse transcription polymerase chain reaction (RT-PCR) and restriction fragment length polymorphism (RFLP). Interestingly, the IBVPR07 isolate, belonging to the Mass serotype, was found to have high tropism for the gonads and trachea (Montassier et al., 2008). In Mexico, several different genotypes have now been isolated which include Conn, Mass, and Ark type (Jackwood, 2012). Similarly, in 2001, new variants were identified. Of these, Max/1765/99 variant was isolated from 64% of chickens showing respiratory problems in Mexico, three new isolates were found to be similar with BL-56 earlier reported in 1996, whereas two other indigenous isolates were antigenically similar to Conn genotypes (Gelb et al., 2001). In recent years, IBVQ1 type, originally isolated in China, has been detected in Chile, Peru, Argentina, and Colombia (Sesti et al., 2014).

Infectious Bronchitis Virus Types in Europe

Until the late 1970s, it was believed that only IBVs of the Mass serotype were important causes of disease in Europe (Jones, 2010; de Wit et al., 2011). In 1980s the Doorn Institute of The Netherlands isolated four serotypes designated as D207 (also known as D274), D212 (also known as D1466), D3896, and D3128, from Mass isolate-vaccinated flocks (Davelaar et al., 1984). Many IBV variants were isolated, from other European countries including IBV PL-84084 in France (Picault et al., 1987), B1648 in Belgium (Meulemans et al., 1987), 624/I, Fa 6881/97, AZ 27/98, AZ 20/97, and BS 216/01 in Italy (Capua et al., 1994; Zanella et al., 2003), and Spanish strains of 97/314, 98/313, 00/337, and 00/338 (Dolz et al., 2006).

In the United Kingdom, 793/B (also known as 4/91 and/or CR88) was identified as the predominant serotype (Cavanagh et al., 1999). It was shown to have a nucleotide sequence in the hypervariable regions of the S1 spike gene quite distinct from Mass and Dutch variant viruses (Cavanagh et al., 2005).

In Italy, an early variant designated Italy/624I/94 was reported (Capua et al., 1994) as well as a more recent type designated Italy-02 (Italy/Italy-02/497/02). Italy-02 was found in almost all European countries but in 2004, it was reported to be declining in prevalence in all countries except Spain (Worthington et al., 2008). Because of its pathogenicity, the most significant IBV type to become widespread in Europe in recent years is the QX IBV type and so-called QX-like types (Monne et al., 2008).

Infectious Bronchitis Virus Types in Africa

In many African countries, the Mass IBV serotypes cause sporadic IB outbreaks in the commercial poultry industry. A number of local variants are reported in Africa in addition to the widely known vaccine serotypes such as Mass and 4/91 strains (de Wit et al., 2011).

In North Africa, an unusual variant with tropism to gastrointestinal system, known as IBV G strain, was isolated in Morocco in the early 1980s (El-Houadfi et al., 1986). However, recent studies identified several other local nonvaccine types, including the QX-like strains and Italy-02, originally localized in China and Europe, respectively (Bande et al., 2017).

In Tunisia, three isolates were identified by molecular test and virus neutralization (VN) test and designated TN20/00, TN200/01, and TN335/01. These isolates were found closely related to European strains including D274 and 793B (Bourogâa et al., 2009).

In Eastern Libya, recent studies showed the presence of 12 IBV strains that are phylogenetically classified into two distinctive clusters which were detected by RT-PCR in broiler flocks (Awad et al., 2014).

In Egypt, the variants of the IBV have been identified since the 1950s, with the isolation and identification by tests of sero-neutralization of a variant closely linked to the variant of Dutch3128 (Sheble et al., 1986). Subsequently, several related variants either to the types Mass (Egypt /Masse/F/03) or to other European variants [D274 (Egypt/D274/D/89)] or of the Variants Israelis have been identified by the genetic analysis of the IB in this country (Abdel-Moneim et al., 2006, 2012).

In Southern Africa, IBV was isolated in 1980 by Morley and Thomson (1984) and was associated with swollen head syndrome causing severe problems. It was confirmed as a variant that showed to be poorly protected by Massachusetts vaccines (Cook et al., 1999). The only other study done on the detection of the variants of IBV in sub-Saharan Africa is the report of Ducatez et al. (2009), which has characterized a new variant named "IBADAN" in Nigeria and in Niger, which was antigenically different from the other genotypes known.

FACTORS LEADING TO THE EMERGENCE OF NEW STRAINS OF INFECTIOUS BRONCHITIS VIRUS

The emergence and evolution of viral pathogens cause a major problem in the poultry industry. Mutation and recombination processes are involved in the genetic and phenotypic variations of IBV in chickens, leading to the emergence of new variant strains, and give rise to virus population diversity to be modeled by the host, particularly by the immune system. The consequence is a continuous emergence of new IBV variants with regard to pathotypes, serotypes, and protectotypes.

IBV, like many other RNA viruses, as well as the coronaviruses, has a high error rate during the transcription of its genomes (Lai and Cavanagh, 1997), it creates genetic diversity through rapid replication and large population sizes coupled with a high mutation rate and recombination. Mutations include substitutions, which are the result of a high error rate and limited proofreading capability of the viral RNA-dependent RNA polymerase (RdRp), and insertions and deletions, caused by recombination events or by RdRp stuttering or slippage.

The viral genes encoding the spike S, replicase, and nucleocapsid proteins can be considered the main genomic regions, which indicate the evolution processes of IBV.

Investigations carried out to date have highlighted the role of three factors: (1) lack of RNA polymerase proofreading, leading to replication errors in RNA genomes with mutation at the order of 10^{-4}; (2) interference of continuous use of live and often multiple attenuated vaccines formulated with different IBV strains; and (3) immune pressure exerted on circulating viruses by the constant presence of partially immune bird populations (Umar et al., 2016).

However, molecular studies have shown that a new serotype or variant can emerge as a result of only a few changes in the amino acid composition in the S1 part of the virus spike protein, while most of the virus genomes remain unchanged (Cavanagh, 2007). This could be due to immunological pressure caused by the widespread use of vaccines, to recombination as a consequence of mixed infections, or to a reduction of dominant serotypes as a result of vaccination, allowing other field strains to emerge (Lee, 2002; Liu et al., 2006).

Due to the high variability and important biological properties of the S1 glycoprotein, antigenic evolution in IBV has been primarily associated with changes in the sequence of the S1 glycoprotein, which contains regions associated with virus attachment to cell receptors and relevant epitopes that induce the production of neutralizing antibodies (Cavanagh et al., 1988). Therefore different serotypes, subtypes, and antigenic variants of IBV are thought to be generated by nucleotide point mutations, insertions, deletions (Kusters et al., 1987), or RNA recombination of S1 gene (Wang and Huang, 2000), resulting in IB outbreaks even in vaccinated chicken flocks.

Actually, serotypic determinants have been identified in the first 395 amino acid region of the S1 subunit, which contains three major hypervariable regions. Amino acid changes within the three S1 glycoprotein hypervariable regions determine the most relevant phenotypic changes, resulting in new serotypes and the induction of non-cross-reacting

VN antibodies (non-cross-protecting). Variants may attain increased virulence, efficient receptor binding, rapid transmission, and persistence in host system causing significant disease in vaccinated flocks of all ages (Wang and Huang, 2000; Dhama et al., 2014).

DIAGNOSIS TECHNIQUES OF THE EMERGING STRAINS OF INFECTIOUS BRONCHITIS VIRUS

IBV affects chickens of all ages, involving the respiratory system, also renal and reproductive systems. And because the clinical signs are not specific, the need for differential diagnostic methods was very important to realize. These methods focus on either isolating or detecting the virus itself as well as detecting serum antibodies to it.

Sampling

Samples appropriate to the form of IB observed must be obtained as soon as signs of clinical disease are apparent. Laryngotracheal swabs from live birds or tracheal and lung tissues from fresh carcasses from diseased birds should be collected. Also, kidney, oviduct, or proventriculus samples from birds with nephritis could be used for laboratory diagnosis of IBV. All samples should be placed in virus transport medium containing penicillin (10,000 IU/mL) and streptomycin (10 mg/mL) and kept in ice and then frozen (OIE, 2013).

Virus Isolation

Specific pathogen-free embryonated chicken egg (SPF-ECE) is recommended for primary isolation of IBV. Those embryonated eggs used for virus isolation should originate preferably from SPF chickens or from breeder sources that have been neither infected nor vaccinated with IBV. Suspensions of tissues (10%–20% w/v) are prepared in sterile phosphate buffered saline. After being clarified by low-speed centrifugation and filtration through bacteriological filters, 0.1–0.2 mL of sample supernatant is inoculated into the allantoic cavity of 9- to 11-day-old embryos (Delaplane, 1947). Eggs are candled daily for 7 days with mortality within the first 24 hours being considered nonspecific death. Normally, the allantoic fluids of all eggs are pooled after harvesting 3–6 days after infection; this pool is diluted 1/5 or 1/10 in antibiotic broth. Blind passage into another set of eggs for up to a total of three to four passages is conducted. The last passage is left for 7 days to screen the presence of pathognomonic embryonic changes consisting of stunted and curled embryos with feather dystrophy (clubbing) and urate deposits in the mesonephros on the second to fourth passage. Isolation of IBV must be confirmed by serum neutralization or reverse transcription polymerase chain reaction (RT-PCR) (OIE, 2013; Delaplane, 1947).

Methods for Identification

The initial tests performed on IBV isolates are directed at eliminating other viruses from diagnostic consideration. Chorioallantoic membranes from infected eggs are collected, homogenized, and tested for avian adenovirus group 1 by immunodiffusion or PCR (OIE, 2013). Group 1 avian adenovirus infections of commercial chickens are common, and the virus often produces stunted embryos indistinguishable from IBV-infected embryos. Furthermore, IBV exerts hemagglutination (HA) activity only after phospholipase C treatment of concentrated virus-infected allanto-aminiotic fluids (Bingham et al., 1975).

Genetic-based tests (RT-PCR or RT-PCR–RFLP) are used commonly to identify an isolate as IBV. Also, in situ hybridization can be used to detect viral nucleic acid (Collisson et al., 1990).

IBV can also be detected using immunofluorescence or immunoperoxidase on the tracheal or kidney section from the field isolates or on the chorioallantoic membrane from the inoculated embryos (Abdel-Moneim et al., 2009). Other techniques may be used, for example, cells present in the allantoic fluid of infected eggs may be tested for IBV antigen using fluorescent antibody tests (Clarke et al., 1972). Furthermore, serotyping of IBV isolates and strains has been done using HA inhibition (King and Hopkins, 1984). However, nonspecific reactions or lower sensitivity especially in field samples may occur (Benyeda et al., 2010), for that we needed a reliable technique as and VN tests in chick embryos (Dawson and Gough, 1971) and also as enzyme-linked immunosorbent assay. This technique is quick, inexpensive, and sensitive, which is suitable for screening a large number of samples, IBV diagnosis, and serotype identification as well (Karaca and Syed, 1993). It has been proved useful in grouping and differentiating strains of IBV (Ignjatovic et al., 1991).

Treatment and Vaccination

There is no specific antiviral therapy available to control IBV field infection, but we can reduce the effect of the complicating bacterial infections by using antimicrobial therapy. Also reduced mortalities in nephrogenic strains can be achieved by reducing the protein concentrations in ration, providing electrolytes in drinking water, and using diuretics (Abdel-Moneim, 2017). So, the only solution we have left is vaccination.

Live Attenuated Vaccine

Live attenuated vaccines are the first-generation IBV vaccines used to control IBV infection in the field. Currently, live IBV vaccines are normally attenuated by multiple repeat passage of a virulent virus in SPF-ECEs (Cavanagh, 2003). However, extensive passage should be avoided to prevent the reduction in immunogenicity. There is an evidence that some attenuated vaccines showed increase in virulence after back passage in chickens (Hopkins and Yoder, 1986). These vaccines are commercially available for application via drinking water or by coarse spray at 1 day or within the first week of age. Live vaccination of 1-day-old chicks induced a rapid decline in maternally derived antibodies due to binding and partial neutralization of vaccine viruses (Mondal and Naqi, 2001). Since the duration of immunity following live attenuated vaccines is short, booster vaccination is carried out with the same or combinations of other strains, 2–3 weeks after prime vaccination (Cavanagh, 2003). Given the nature of these live attenuated vaccines, further passage beyond the master seed stock must be kept to a minimum to prevent potential loss of immunogenicity. The stocks must be grown in SPF chicken eggs to prevent the introduction of other potential pathogens (OIE, 2013).

Most of the commercially available live attenuated vaccines are derived from virulent strains such as Massachusetts-based M41 serotype and the Dutch H52 and H120 strains, although some strains with regional or local impact have been used in different parts of the world (Sasipreeyajan et al., 2012).

For logistics and economic reasons, some commercially available live attenuated IBV vaccines have been combined with other virus vaccines such as those against Newcastle disease virus, Marek's disease virus, and infectious bursal disease virus. However, it is not clear whether the combination may influence immune response to the combined antigen (Vagnozzi et al., 2010), and if excess IBV component is present, IBV may interfere with other virus response such as NDV response (Thornton and Muskett, 1975).

Some of the limitations of live attenuated viral vaccines include reversion to virulence, tissue damage, and interference by MDA. Tissue damage due to live vaccines may lead to pathological disorders or secondary bacterial infections, especially in day-old chick (Tarpey et al., 2006). And it has been found that H52 and H120 IBV vaccines induce considerable pathology in the trachea (Zhang et al., 2010). And potential recombination between vaccine strains and virulent field strains may lead to the emergence of new IBV serotypes (Lee et al., 2010; McKinley et al., 2008).

Inactivated or Killed Vaccines

Inactivated or killed vaccines have been used either alone or in combination with live attenuated IBV vaccines (Finney et al., 1990). These vaccines are administered by injection to layers and breeders at point of lay (13–18 weeks of age). Inactivated IBV vaccines have poor efficacy unless the chickens have previously been primed by vaccination with a live virus vaccine (OIE, 2013). And since inactivated vaccines do not replicate, they are unlikely to revert and cause pathological effects. Being injectable, administration of killed vaccines is either difficult or impracticable in large poultry setting. Likewise, issues of injection-site reactions may also lead to carcass rejection or reduction in value (Cook et al., 2012).

Recombinant Vaccines

Recombinant DNA vaccines have been enhanced to target multiple serotypes and their efficacy has been improved using delivery vectors, nano-adjuvants, and in ovo vaccination approaches. Although most recombinant IB DNA vaccines are yet to be licensed, it is expected that these types of vaccines may hold sway as future vaccines for inducing a cross-protection against multiple IBV serotypes (Jackwood, 1999).

CONCLUSION

Even with spending huge amounts of money to control IB, outbreaks involving classical and newly emerging virus serotypes are constantly reported. Though live attenuated vaccines are still common in the field, these vaccines provide only a little or partial cross-protection occurs between vaccine strains and new field which require the development of new vaccines as the recombinant vaccines. There is no doubt that newer generation vaccines such as the recombinant vector DNA vaccines, plasmid DNA vaccines, and multiepitope vaccines may stand as future alternatives as these vaccines have potential to deliver numerous antigens, thus producing broad-based antibody and cell-mediated immune response against numerous serotypes and also, the requirement of new strategies for vaccination.

References

Abdel-Moneim, A.S., 2017. *Coronaviridae*: infectious bronchitis virus. In: Bayry, J. (Ed.), Emerging and Re-Emerging Infectious Diseases of Livestock, pp. 133–166.

Abdel-Moneim, A.S., El-Kady, M.F., Ladman, B.S., Gelb, J., 2006. S1 gene sequence analysis of a nephropathogenic strain of avian infectious Bronchitis virus in Egypt. Virol. J. 3, 78.

Abdel-Moneim, A.S., Zlotowski, P., Veits, J., Keil, G.M., Teifke, J.P., 2009. Immunohistochemistry for detection of avian infectious bronchitis virus strain M41 in the proventriculus and nervous system of experimentally infected chicken embryos. Virol. J. 6, 15.

Abdel-Moneim, A.S., Afifi, M.A., El-Kady, M.F., 2012. Emergence of a novel genotype of avian infectious bronchitis virus in Egypt. Arch. Virol. 157, 2453–2457.

Awad, F., Baylis, M., Ganapathy, K., 2014. Detection of variant infectious bronchitis viruses in broiler flocks in Libya. Int. J. Vet. Med. Sci 1, 78–82.

Bande, F., Arshad, S.S., Omar, A.R., Hair-Bejo, M., Mahmuda, A., Nair, N., 2017. Global distributions and strain diversity of avian infectious bronchitis virus: a review. Anim. Health Res. Rev 18 (1), 70–83.

Beach, J.R., Schalm, O.W., 1936. A filterable virus, distinct from that of laryngotracheitis, the cause of a respiratory disease of chicks. Poult. Sci. 1, 199–206.

Benyeda, Z., Szeredi, L., Mató, T., Süveges, T., Balka, G., Abonyi-Tóth, Z., et al., 2010. Comparative histopathology and immunohistochemistry of QX-like, Massachusetts and 793/B serotypes of infectious bronchitis virus infection in chickens. J. Comp. Pathol 143 (4), 276–283.

Bingham, R.W., Madge, M.H., Tyrrell, D.A., 1975. Hemagglutination by avian infectious bronchitis virus—a coronavirus. J. Gen. Virol 28 (3), 381–390.

Bourogâa, H., Miled, K., Gribâa, L., El Behi, I., Ghram, A., 2009. Characterization of new variants of avian infectious bronchitis virus in Tunisia. Avian Dis. 53, 426–433.

Capua, I., Gough, R.E., Mancini, M., Casaccia, C., Weiss, C., 1994. A novel infectious bronchitis strain infecting broiler-chickens in Italy. J. Vet. Med. B 41 (2), 83–89.

Cavanagh, S., 1997. Content analysis: concepts, methods and applications. Nurse Researcher 4, 5–16.

Cavanagh, D., 2003. Severe acute respiratory syndrome vaccine development: experiences of vaccination against avian infectious bronchitis coronavirus. Avian Pathol. 32 (6), 567–582.

Cavanagh, D., 2007. Coronavirus avian infectious bronchitis virus. Vet. Res. 38, 281–297.

Cavanagh, D., Naqi, S.A., 2003. Infectious bronchitis. In: Saif, Y.M. (Ed.), Diseases of Poultry., eleventh ed. University Press, Ames, IA, pp. 101–119.

Cavanagh, D., Davis, P.J., Cook, J.K., Li, D., Kant, A., Koch, G., 1992. Location of the amino acid differences in the S1 spike glycoprotein subunit of closely related serotypes of infectious bronchitis virus. Avian Pathol. 21, 33–43.

Cavanagh, D., Davis, P.J., Mockett, A.P.A., 1988. Amino acids within hypervariable region 1 of avian coronavirus IBV (Massachusetts serotype) spike glycoprotein are associated with neutralization epitopes. Virus Res. 11, 141–150.

Cavanagh, D., Mawditt, K., Britton, P., Naylor, C.J., 1999. Longitudinal field studies of infectious bronchitis virus and avian pneumovirus in broilers using type-specific polymerase chain reactions. Avian Pathol. 28 (6), 593–605.

Cavanagh, D., Picault, J.P., Gough, R., Hess, M., Mawditt, K., Britton, P., 2005. Variation in the spike protein of the 793/B type of infectious bronchitis virus, in the field and during alternate passage in chickens and embryonated eggs. Avian Pathol. 34, 20–25.

Clarke, J.K., MC Ferran, J.B., Gay, F.W., 1972. Use of allantoic cells for the detection of avian infectious bronchitis virus. Arch. Gesamte Virusforsch 36, 62–70.

Collisson, E.W., Li, J.Z., Sneed, L.W., Peters, M.L., Wang, L., 1990. Detection of avian infectious bronchitis viral infection using in situ hybridization and recombinant DNA. Vet. Microbiol. 24 (3–4), 261–271.

Cook, J.K.A., Orbell, S.J., Woods, M.A., Huggins, M.B., 1999. Breadth of protection of the respiratory tract provided by different live-attenuated infectious bronchitis vaccines against challenge with infectious bronchitis viruses of heterologous serotypes. Avian Pathol. 28, 477–485.

Cook, J.K.A., Jackwood, M., Jones, R.C., 2012. The long view: 40 years of infectious bronchitis research. Avian Pathol. 41, 239–250.

Davelaar, F.G., Kouwenhoven, B., Burger, A.G., 1984. Occurrence and significance of infectious bronchitis virus variant strains in egg and broiler production in the Netherlands. Vet. Q. 6 (3), 114–120.

Dawson, P.S., Gough, R.E., 1971. Antigenic variation in strains of avian infectious bronchitis virus. Arch. Gesamte Virusforsch. 34 (1), 32–39.

Delaplane, J., 1947. Technique for the isolation of infectious bronchitis or Newcastle virus including observations on the use of streptomycin in overcoming bacterial contaminants. In: Proceedings of Northeast Conference of Laboratory Workers in Pullorum Disease Control, vol. 19, pp. 11–13.

De Wit, J.J., Cook, J.K., Van der Heijden, H.M., 2011. Infectious bronchitis virus variants: a review of the history, current situation and control measures. Avian Pathol. 40, 223–235.

Dhama, K., Singh, S.D., Barathidasan, R., Desingu, P.A., Chakraborty, S., Tiwari, R., et al., 2014. Emergence of avian infectious bronchitis virus and its variants need better diagnosis, prevention and control strategies: a global perspective. Pak. J. Biol. Sci 17, 751–767.

Dolz, R., Pujols, J., Ordóñez, G., Porta, R., Majó, N., 2006. Antigenic and molecular characterization of isolates of the Italy 02 infectious bronchitis virus genotype. Avian Pathol. 35 (2), 77–85.

Ducatez, M.F., Martin, A.M., Owoade, A.A., Olatoye, I.O., Alkali, B.R., Maikano, I., et al., 2009. Characterization of a new genotype and serotype of infectious bronchitis virus in Western Africa. J. Gen. Virol 90, 2679–2685.

El-Houadfi, M., Jones, R.C., Cook, J.K.A., Ambali, A.G., 1986. The isolation and characterisation of six avian infectious bronchitis viruses isolated in Morocco. Avian Pathol. 15, 93–105.

Finney, P., Box, P., Holmes, H., 1990. Studies with a bivalent infectious bronchitis killed virus vaccine. Avian Pathol. 19.3 435–450.

Gelb, J., Keeler Jr., C.L., Nix, W.A., Rosenberger, J.K., Cloud, S.S., 1997. Antigenic and S-1 genomic characterization of the Delaware variant serotype of infectious bronchitis virus. Avian Dis. 41, 661–669.

Gelb, J., Ladman, B.S., Tamayo, M., Gonzalez, M., Sivanandan, V., 2001. Novel infectious bronchitis virus S1 genotypes in Mexico 1998–1999. Avian Dis. 45, 1060–1063.

Hipolito, O., 1957. Isolamento e identificacao do virus da bronquite infecciosa das galinhas no Brasil. Arquivo Escuela Veterinaria Universidade de Minas Gerais 10, 131–151.

Hopkins, S.R., Yoder Jr., H.W., 1986. Reversion to virulence of chicken-passaged infectious bronchitis vaccine virus. Avian Dis. 30 (1), 221–223.

Ignjatovic, J., Sapats, S., 2000. Avian infectious bronchitis virus. Rev. Sci. Tech. Off. Int. Epiz. 19 (2), 493–508.

Ignjatovic, J., Mcwaters, P., Galli L., 1991. Antigenic relationship of Australian infectious bronchitis viruses: analysis using polyclonal and monoclonal antibodies. In: Proceedings of the Second International Symposium on Infectious Bronchitis. Rauischholzhausen, Germany, pp. 161–167.

Jackwood, M.W., 1999. Current and future recombinant viral vaccines for poultry. Adv. Vet. Med 41, 517–522.

Jackwood, M.W., 2012. Review of infectious bronchitis virus around the world. Avian Dis. 56, 634–641.

Jackwood, M.W., Yousef, N.M., Hilt, D.A., 1997. Further development and use of a molecular serotype identification test for infectious bronchitis virus. Avian Dis. 41 (1), 105–110.

Jackwood, M.W., Hilt, D.A., Lee, C.W., Kwon, H.M., Callison, S.A., Moore, K.M., et al., 2005. Data from 11 years of molecular typing infectious bronchitis virus field isolates. Avian Dis. 49, 614–618.

Jackwood, M.W., Hilt, D.A., Williams, S.M., Woolcock, P., Cardona, C., O'Connor, R., 2007. Molecular and serologic characterization, pathogenicity, and protection studies with infectious bronchitis virus field isolates from California. Avian Dis. 51, 527–533.

Jones, R.C., 2010. Europe: history, current situation and control measures for infectious bronchitis. Braz. J. Poultry Sci 12 (2), 125–128.

Karaca, K., Syed, N., 1993. A monoclonal antibody blocking ELISA to detect serotype-specific infectious bronchitis virus antibodies. Vet. Microbiol. 34 (3), 249–257.

Khataby, K., Fellahi, S., Loutfi, C., Ennaji, M.M., 2016. Avian infectious bronchitis virus in Africa: a review. Vet. Quart. 34, 1–5.

King, D.J., Hopkins, S.R., 1984. Rapid serotyping of infectious bronchitis virus isolates with the hemagglutination inhibition test. Avian Dis. 28, 727–733.

Kusters, J.G., Niesters, H.G.M., Bleumink-Pluym, N.M.C., 1987. Molecular epidemiology of infectious bronchitis virus in the Netherlands. J. Gen. Virol 68, 343–352.

Lai, M.M., Cavanagh, D., 1997. The molecular biology of coronaviruses. Adv. Virus Res 48, 1–100.

Lee, C.W., 2002. Evolution of avian infectious bronchitis virus: genetic drift and recombination. Korean J. Vet. Ser 25, 97–103.

Lee, C.W., Jackwood, M.W., 2001. Origin and evolution of Georgia 98 (GA98), a new serotype of avian infectious bronchitis virus. Virus Res. 80, 33–39.

Lee, H.J., Youn, H.N., Kwon, J.S., Lee, Y.J., Kim, J.H., Lee, J.B., et al., 2010. Characterization of a novel live attenuated infectious bronchitis virus vaccine candidate derived from a Korean nephropathogenic strain. Vaccine. 28 (16), 2887–2894.

Liu, S.W., Zhang, Q.X., Chen, J.D., Han, Z.X., Liu, X., Feng, L., et al., 2006. Genetic diversity of avian infectious bronchitis coronavirus strains isolated in China between 1995 and 2004. Arch. Virol. 151, 1133–1148.

McKinley, E.T., Hilt, D.A., Jackwood, M.W., 2008. Avian coronavirus infectious bronchitis attenuated live vaccines undergo selection of subpopulations and mutations following vaccination. Vaccine. 26 (10), 1274–1284.

Meulemans, G., Carlier, M.C., Gonze, M., Petit, P., Vandenbroeck, M., 1987. Incidence, characterization and prophylaxis of nephropathogenic avian infectious bronchitis viruses. Vet. Rec. 120 (9), 205–206.

Mondal, S.P., Cardona, C.J., 2007. Genetic and phenotypic characterization of the California 99 (Cal99) variant of IBV. Virus Genes. 34, 327–341.

Mondal, S.P., Naqi, S.A., 2001. Maternal antibody to infectious bronchitis virus: its role in protection against infection and development of active immunity to vaccine. Vet. Immunol. Immunopathol 79 (1–2), 31–40.

Monne, I., Cattoli, G., Jones, R., Worthington, K., Wijmenga, W., 2008. QX genotypes of infectious bronchitis virus circulating in Europe. Vet. Rec. 163, 606–607.

Montassier, M.F.S., Brentano, L., Montassier, H.J., Richtzenhain, L.J., 2008. Genetic grouping of avian infectious bronchitis virus isolated in Brazil based on RT-PCR/RFLP analysis of the S1 gene. Pesq. Vet. Bras. 28, 190–194.

Moore, K.M., Bennett, J.D., Seal, B.S., Jackwood, M.W., 1998. Sequence comparison of avian infectious bronchitis virus S1 glycoproteins of the Florida serotype and five variant isolates from Georgia and California. Virus Genes. 17, 63–83.

Morley, A.J., Thomson, D.K., 1984. Swollen-head syndrome in broiler chickens. Avian Dis. 28, 238–243.

Nix, W.A., Troeber, D.S., Kingham, B.F., Keeler Jr., C.L., Gelb Jr., J., 2000. Emergence of subtype strains of the Arkansas serotype of infectious bronchitis virus in Delmarva broiler chickens. Avian Dis. 44, 568–581.

OIE, 2013. Chapter 2.3.2: Avian infectious bronchitis virus. Terrestrial Manual, pp. 1–15.

Picault, J.P., Giraud, P., Drouin, P., Guittet, M., Bennejean, G., Lamande, J., et al., 1987. Isolation of a TRTV-like virus from chickens with swollen-head syndrome. Vet Rec. 121 (6), 135.

Sasipreeyajan, J., Pohuang, T., Sirikobkul, N., 2012. Efficacy of different vaccination programs against Thai QX-like infectious bronchitis virus. Thai J. Vet. Med 42 (1), 73–79.
Schalk, A.F., Hawn, M.C., 1931. An apparently new respiratory disease of baby chicks. J. Am. Vet. Med. Assoc 78, 413–422.
Schikora, B.M., Shih, L.M., Hietala, S.K., 2003. Genetic diversity of avian infectious bronchitis virus California variants isolated between 1988 and 2001 based on the S1 subunit of the spike glycoprotein. Arch. Virol. 148, 115–136.
Sesti, L., Sara, L., Alvarado, L., Coregana, J., Orosco, R., Romero, J.C., et al., 2014. Diagnosis, epidemiology and control of the Q1 variant strain in Peru, Colombia, Argentina and Chile. In: Leirz, M., Huffels-Redman, U., Kaleta, E.F., Heckman, J. (Eds.), 8th International Symposium on Avian Corona and Pneumovirus Infections/ 2nd Meeting Cost Action, Rauischholzhausen, Germany.
Sheble, A., Sabry, M.Z., Davelaar, F.G., Burger, A.G., Khafagy, A.K., Moustafa, F., et al., 1986. Present status of infectious bronchitis in Egypt. Egypt. Vet. Med. Assoc. 4, 393–411.
Tarpey, I., Orbell, S.J., Britton, P., Casais, R., Hodgson, T., Lin, F., et al., 2006. Safety and efficacy of an infectious bronchitis virus used for chicken embryo vaccination. Vaccine. 24, 6830–6838.
Thornton, D.H., Muskett, J.C., 1975. Effect of infectious bronchitis vaccination on the performance of live Newcastle disease vaccine. Vet. Rec. 96 (21), 467–468.
Umar, S., Shah, M.A.A., Munir, M.T., Ahsan, U., Kaboudi, K., 2016. Infectious bronchitis virus: evolution and vaccination. World's Poult. Sci. J. 72, 49–60.
Vagnozzi, A., Garcia, M., Riblet, S.M., Zavala, G., 2010. Protection induced by infectious laryngotracheitis virus vaccines alone and combined with Newcastle disease virus and/or infectious bronchitis virus vaccines. Avian Dis. 54 (4), 1210–1219.
Valastro, V., Holmes, E.C., Britton, P., Fusaro, A., Jackwood, M.W., Cattoli, G., et al., 2016. S1 gene-based phylogeny of infectious bronchitis virus: an attempt to harmonize virus classification. Infect. Gen. Evol 39, 349–364.
Wang, C.H., Huang, Y.C., 2000. Relationship between serotypes and genotypes based on the hypervariable region of the S1 gene of infectious bronchitis virus. Arch. Virol 145, 291–300.
Worthington, K.J., Currie, R.J., Jones, R.C., 2008. A reverse transcriptase–polymerase chain reaction survey of infectious bronchitis virus genotypes in Western Europe from 2002 to 2006. Avian Pathol. 37, 247–257.
Zanella, A., Lavazza, A., Marchi, R., Martin, A.M., Paganelli, F., 2003. Avian infectious bronchitis: characterization of new isolates from Italy. Avian Dis. 47 (1), 180–185.
Zhang, Y., Wang, H.N., Wang, T., Fan, W.Q., Zhang, A.Y., Wei, K., et al., 2010. Complete genome sequence and recombination analysis of infectious bronchitis virus attenuated vaccine strain H120. Virus Genes. 41 (3), 377–388.

Further Reading

Yu, L., et al., 2001. Characterization of three infectious bronchitis virus isolates from china associated with proventriculus in vaccinated chickens. Avian Dis. 45, 416–424.

CHAPTER

4

Molecular Modeling of Major Structural Protein Genes of Avian Coronavirus: Infectious Bronchitis Virus Mass H120 and Italy02 Strains

Khadija Khataby[1,2], *Yassine Kasmi*[1], *Amal Souiri*[1], *Chafiqa Loutfi*[2] *and Moulay Mustapha Ennaji*[1]

[1]Laboratory of Virology, Microbiology, Quality, Biotechnologies/ Eco-Toxicology and Biodiversity, Faculty of Sciences and Techniques, Mohammedia, University Hassan II of Casablanca, Casablanca, Morocco

[2]Society Biopharma, Rabat, Morocco

INTRODUCTION

Avian infectious bronchitis (IB) is a highly contagious respiratory infectious disease hazardous to the poultry industry. It can infect chickens at all ages and replicates in many tissues, causing respiratory symptoms, diarrhea, decline of egg production and quality, etc. (Cavanagh, 2007a,b; Abd et al., 2009). Prevention of IB is of economic importance to the poultry industry due to the high morbidity and production losses associated with the disease (Cavanagh, 2005). Although vaccines are now being used widely and extensively, outbreaks of IB still occur frequently due to the epidemic IB virus (IBV) strains (Zou et al., 2010). It is well known that little or no cross-protection occurs between different

Emerging and Reemerging Viral Pathogens
DOI: https://doi.org/10.1016/B978-0-12-814966-9.00004-4

serotypes of IBV, and new serotypes may appear in the future, complicating the prevention and control of IB.

In Morocco, the epidemiological situation of IBV is very complex due to the antigenic diversity associated with the emergence of new serotypes/genotypes and variants, vaccination failures linked to a possible maladjustment of the vaccine strain used and/or poor vaccination practices, and inadequate biosecurity measures by livestock keepers. The avian IBV strains in circulation are serotypes/genotypes Italy02 and Mass H120 identified since 2010 (Fellahi et al., 2015).

The etiologic agent of IB is IBV, a prototype of the *Coronaviridae* family, which is an enveloped, positive sense, single-stranded RNA virus (Boursnell et al., 1987).

The viral genome is around 27.6 kb in length and encodes four structural proteins, nucleocapsid protein (N), membrane glycoprotein (M), spike glycoprotein (S), and small envelope protein (E) (Lai et al., 1981). The S glycoprotein is posttranslationally cleaved at protease cleavage recognition motifs into the animal-terminal S1 and carboxyl-terminal S2 subunits by cellular protease (Jackwood et al., 2001; Cavanagh et al., 1986). The S1 glycoprotein contains epitopes that induce virus-neutralizing, serotype-specific antibodies, hemagglutination inhibition antibodies, and cross-reactivity enzyme-linked immunosorbent assay (ELISA) antibodies (Niesters et al., 1987). It also plays an important role in tissue tropism and the degree of virulence of the virus (Casais et al., 2003). The appearance of these variants hinders the prophylactic strategy carried out by the breeders of the Moroccan poultry farm. In order to solve this problem, we have opted to study the structure of the hypervariable region of the S1 protein of serotype Italy02 and Mass in silico by molecular modeling, where the largest number of epitopes identified by neutralizing antibodies is observed (Koch et al., 1992).

Structural bioinformatics is a branch of bioinformatics that focuses on the prediction of macromolecular structures, such as the structure of three-dimensional (3D) proteins (Zhang et al., 2005).One of the main questions in the problem of protein structure prediction is the challenge of understanding how the primary protein structure information is translated into a 3D structure and how to use this information for the development of prediction of the 3D structure (Creighton, 1990).

Experimental methods for determining the 3D structure of proteins are cumbersome and costly in terms of time and resources. The predictive methods in silico propose a fast and efficient alternative, based on a set of physical, statistical, and biological laws. Generally, there are two main classes of methods: the first class is called "comparative modeling" and the second is called "ab initio" (Piuzzi, 2010). The first method depends on the existence of homologous proteins, whose structures are determined experimentally. The second method is only on physical and statistical laws. The algorithms used by the latter are very greedy in computing time, and the results obtained progress with advances in computer science.

Actually, despite the immense progress of ab initio methods, comparative methods are still those which offer the best predictions of antigenic sites of proteins. Therefore the objective of the present study, which is reported for the first time in Morocco, aims to compare the structural conformation of the S1 protein in 3D form, and to predict the common neutralizing epitopes between the vaccine strain Mass H120, which is the most dominant serotype in Morocco, and the serotype Italy02 to better understand their pathogenic and immunogenicity.

MATERIAL AND METHODS

Sequence and Structural Data

The viral strains used in this study were Italy02 and Mass (H120). The amino acid sequences for the proteins to be modeled were obtained from Genbank NCBI-USA, and their access numbers are (KM594188: Italy02) and (M21970: H120). The evolutionary characterization of IBV is essentially based on the analysis of the three hypervariable regions of the S1 gene (HVR1, HVR2, and HVR3), located in the following positions: 114–201nt, 297–423nt, and 822–1161nt, respectively, corresponding to amino acid residues 38–67, 91–141, and 274–38 (Bourogâa et al., 2009).

MODELING OF THE HYPERVARIABLE REGION OF S1 SPICULE PROTEINS

This paper focuses on the molecular modeling of the structure of the hypervariable regions of the S1 protein of the Mass H120 and Italy02 serotypes circulating in Morocco. To meet the stated objective, an alignment of the protein sequences of these two strains was carried out in order to detect the homologous and common regions, so that it could be applied to the model described by Piuzzi (2010). All these manipulations were developed by The CHIMERA V.01 software.

Homology modeling allows replacing the missing structural information, provided that it has the structure of a protein with strong homology between the two sequences "Italy02 and H120." It is estimated that two structures can be considered identical when their root, mean-square, deviation (RMSD) (obtained by the superposition of the atoms of their respective main chains) is less than 2 Å.

In this study, modeling was done using the I-TASSER server, then the COACH, and another Meta server to determine and predict the common immunogenic active sites, between these two IBV strains. The 3D modeling was carried first, with I-TASSER, which is a server offering a service of prediction of the structure and function of the protein studied. It makes it possible to produce high-quality 3D models from the amino acid sequences. The results provided by the I-TASSER server are in the form of several 3D models, classified according to a score called "TM-score." If the TM-score is greater than 0.5, it indicates that the model generated a valid topology. However, a score less than 0.17 indicates a random topology (Piuzzi, 2010).

After the 3D modeling of the hypervariable regions, the next step was to calculate and determine the active site of the modeled regions, the site where the ligand interaction takes place, which results in activation or deactivation of the biological function of the protein.

The calculation of the active site was carried out with COACH which is a meta server for the prediction of the "ligand binding domain." The 3D models generated by I-TASSER are taken into account by the COACH server to predict all active sites with their ligands. The active sites obtained by COACH are coordinates in this form of "X/Y/Z/Xs/Ys/Zs" or "X, Y, and Z" represent the position of the active site in 3D space and for "Ys and Zs" show the size of the box containing the immunogenic active site.

Another server was used, named "RAMACHANDRAN," giving information on the conformation of the protein in 3D. Thanks to the diagrams generated by this server, the potential secondary structures can be identified according to the torsion angles. There are two types of torsion angles, the angle phi (φ) and the angle psi (ψ). The angle psi (Ψ) represents the angle of rotation around the Cα–C bond (of C = O) of the plane 1 and the angle phi (φ) represents the angle of rotation around the Cα–N bond (of NH) of the plane 2.

EVALUATION AND REFINEMENT OF THE THREE-DIMENSIONAL MODEL

The quality and precision of the models obtained were evaluated by the geometry of the different regions of the model and the identification of possible errors. The evaluation of the quality of the 3D modeled protein was carried out by the "PROSESS: Protein Structure Evaluation Suite & Server." PROSESS is a web server designed to evaluate and validate protein structures and allows us to integrate a variety of analyzes:

- covalent and geometric quality
- noncovalent bond quality
- quality of the torsion angle
- chemical shift quality

PROSESS produces detailed tables with explanations, images, and graphs that summarize the results by comparing them with values observed in high-quality protein structures. This server is used to coordinate the location of hydrogen bonds, secondary structure, and geometric analysis, which can then be used for computation of aliasing and solvent energy, and chemical shift correlations, to correlate the mobility of the structure with chemical shift, as well as for the calculation of torsional angle and chemical changes (Berjanskii et al., 2010).

RESULTS

The study of the spatial conformation of the structure of the hypervariable region of the S1 protein in 3D and the prediction of the neutralizing epitopes of the virus were carried out using tools of molecular modeling.

MODELING OF THE HYPERVARIABLE REGION OF S1 SPICULE PROTEINS

Spatial Conformation of the S1 Structure in Three-Dimensional

Homology modeling between the two protein sequences (Italy02 and Mass) showed a similarity percentage of 81%. This homology was evaluated by the I-TASSER server, which allows us to generate 3D models from the protein sequence. These models were then ranked in specific order and defined by the TM-score which measures the deviation distance (Angstrom) between the residual position of the model and the native structure. The score obtained by I-TASSER revealed that the two modeled sequences used in this study have a more significant TM-score that exceeds the value of 0.5 (TM-Score = 0.63), confirming that the model is biologically significant and has a correct structural topology (Fig. 4.1, Table 4.1).

Projection and disposition of the 3D structure of the Italy02 strain on that of the Mass strain was validated by the RMSD factor. This factor evaluates the degree of deviation between the two 3D structures.

The RMSD is equal to 0.3 Å, whereas most hypervariable regions have an RMSD equal to zero, indicating that both structures are

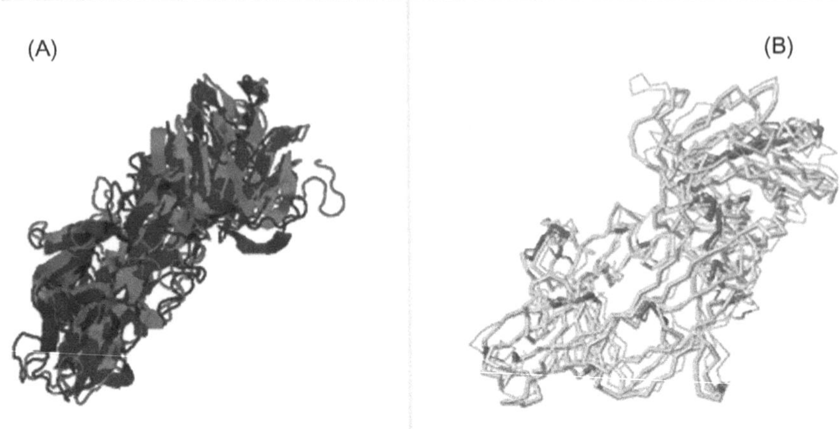

FIGURE 4.1 (A) 3D structure predicted by I-TASSER; Blue color: strain H120. Red color: strain Italy02. (B) The common regions between the two structures (red). *3D*, Three-dimensional. (*Note: For interpretation of the references to color in this figure legend, the reader is referred to the web version of this chapter*).

TABLE 4.1 Scores TM and C As Well As Predicted Active Sites and Exogenous Molecules

Sequences	C score	TM-score	RMSD	Active sites	Ligand
Italy02	−0.91	0.60 ± 0.14	9.6 ± 4.6 Å	169, 171, 179, 208, 224, 229, 232, 233, 234, 235, 236, 237	Mg^{2+}
H120	−0.91	0.60 ± 0.14	9.6 ± 4.6 Å	225, 228, 229, 230, 232, 233, 235, 338, 341, 433, 435, 440, 476, 485	Mg^{2+}

RMSD, Root, mean-square, deviation.

identical and share homologous and common regions (Fig. 4.1B). In addition, both strains share common active sites in the S1 spike protein and are located at residues 229, 230, 232, 233, and 235 (Table 4.1). This study also revealed the presence of a molecule of magnesium associated with the structure of the amino acids common between the two sequences of the strains studied.

STABILITY OF THE STRUCTURE OF THE S1 PROTEIN IN THREE-DIMENSIONAL

In order to confirm the stability of the 3D structure, several Beta and Alpha sheets were demonstrated by RAMACHANDRAN test. The analysis of these results showed a variability in the stability of the sequences, depending on the number of residues outside the stability zones. The amino acids are distributed between Beta sheets and Alpha helices (Fig. 4.2, Table 4.2).

The amino acids distributed in the upper left quadrant indicate those found in the Beta leaflets. They have an angle Phi less than −30 and an angle psi greater than 90. The set of proteins in the white space is of suspended structure and of unknown nature (Fig. 4.2).

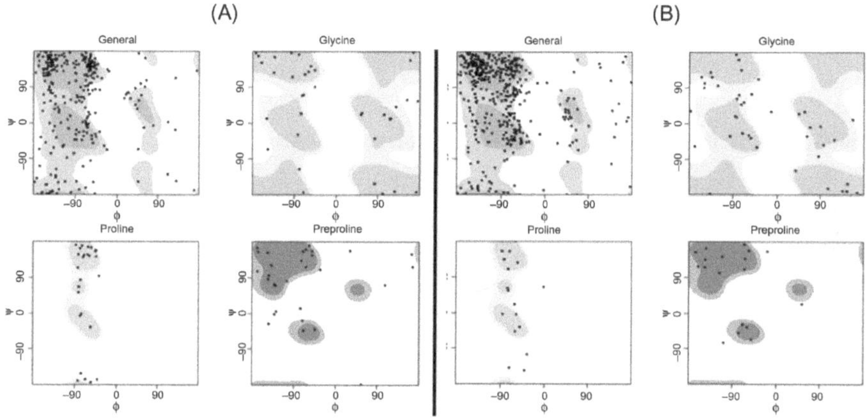

FIGURE 4.2 **RAMACHANDRAN plot (A) H120 and (B) Italy02.**

TABLE 4.2 Results of RAMACHANDRAN Analysis

Residues types	Percentage of outliers	Number of percentage of outliers	Total number
ITALY02			
Générale (non-Gly, non-Pro, non-pre-Pro)	12.58	39	310
Glycine	10.26	4	39
Proline	19.35	6	31
Preproline	22.58	7	31
H120			
General (non-Gly, non-Pro, non-pre-Pro)	10.43	48	460
Glycine	13.33	6	45
Proline	31.58	6	19
Preproline	15.79	3	19

The arrangement of the amino acids in the lower left part coinciding on the one hand, with the right-handed alpha-helix conformation, and on the other hand, a small number of amino acids is located in the upper right quadrant, By showing alpha helices rotating to the left through their conformation angles. They also have an average stability of between 10 and 12 (Fig. 4.2).

EVALUATION OF THE QUALITY OF THE THREE-DIMENSIONAL MODEL AND PREDICTION OF ANTIGENIC SITES

Evaluation of the Quality of the Three-Dimensional Model

The analysis of the evaluation of the structural quality of 3D sequences by the PROSESS server showed that they have an overall quality of 2.5. All residues, within the range of $20 < R < 60$ are characterized only by noncovalent bonds with a value of two anomalies, while residues included in $74 < R < 500$ are indicated by the packing and noncovalent bonds, whose average quality is equal to 3.5, then for the other covalent bonds, it reaches a value of 4.5 (Fig. 4.3).

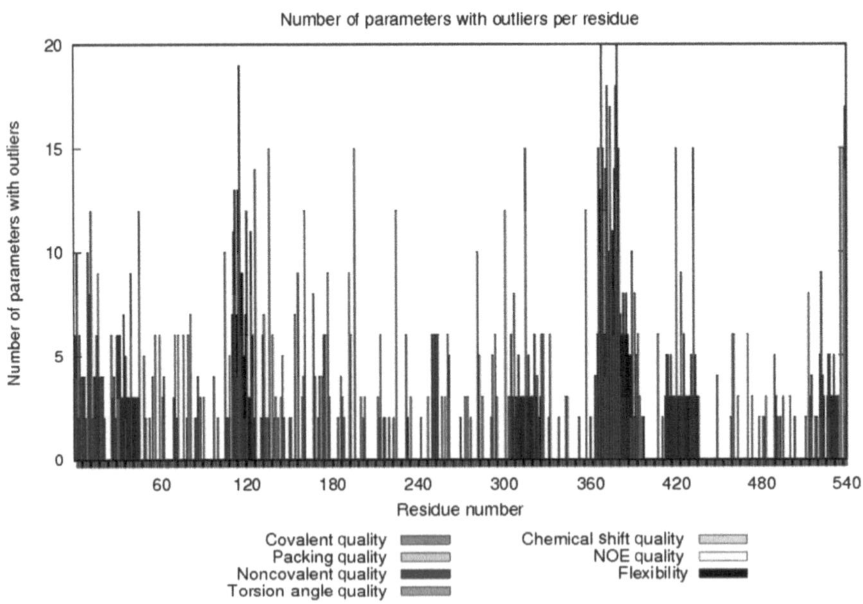

FIGURE 4.3 **Aggregated results of sequence residue problems.**

PREDICTION OF ANTIGENIC SITES

The results of the prediction of neutralizing epitopes at the level of the S1 protein in 3D showed that the serotype Italy02 and the vaccine strain H120 of serotype Mass share common epitopes in the hypervariable regions of the S1 spicule protein, which may have antigenic and immunogenic role. These epitopes are at residues 38–67, 91–141, and 274–387

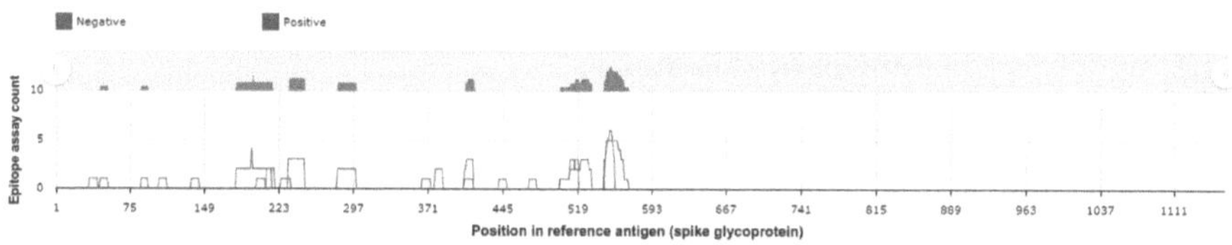

Prediction of epitopes at the spicule protein structure S1.

DISCUSSION

This study would focus on the in silico prediction of peptides of the S1 spicule protein from two IBV strains, Italy02 and Mass H120. The choice of the S1 subunit was not made by chance but was chosen by its ability to undergo mutations in the hypervariable regions, giving new strains of IBV (Cavanagh, 2007a,b). The S1 subunit anchors to the outer surface of the viral particle, making it the more easily recognized antigen, by the IB-specific antibody, compared to other IBV antigens.

The S1 gene is now commonly used as a marker of the IBV classification. Although it is highly variable, it remains the first choice for the development of subunits of vaccines against IB (Zou et al., 2015). Furthermore, since there are still relatively conserved regions or epitopes in the S1 subunit, S1 could also be used as a targeted antigen in the development of diagnostic agents (Zou et al., 2015). However, there is little information on the structure of the S1 gene protein in 3D, which is carried out to predict epitopes on this gene, hence the objective of this paper, which aimed to predict and identify the most immunogenic antigenic sites are critical for vaccine development.

The results of the homology modeling showed that the two studied serotypes had almost the same spatial conformation of the hypervariable region of the S1 protein in 3D and shared homologous and common regions with a similarity percentage of 81%. These results are in agreement with the data reported by Chothia and Lesk (1986). These authors have shown that for RMSD values below 2 Å, the two structures can be considered as similar. Thus from 60% homology, homology modeling allows correct prediction in 70% of cases (Chothia and Lesk, 1986).

Analysis of the results of structural stability showed that there were residual stability and variability between the two protein sequences (Italy02 and Mass) in 3D. Jones and Jordan (1972) reported that the serotype Mass H120 is more stable than the serotype Italy02, whose ratio of the predicted strains is 1.5. These data could be explained by the evolutionary power of this virus as a function of time, where mutation occurs at a speed faster than normal in the hypervariable sequence of the S1 gene, which is the subject of this study.

The study of the prediction of epitopes revealed the presence of common active residues at the level of the hypervariable region of the spike protein S1, which can exercise a common function by intervening in the juice of internalization of the virus of the cell, thus the role of cathepsins.

In addition, detection of a magnesium molecule was detected associating with the structure of amino acids around Aln280, a common predicted region and considered to be one of the most immunogenic regions in both IBV strains.

The presence of the magnesium molecule around this site stimulates immunogenicity, which has been researched because of its functionality in the body (Tam et al., 2003). These authors have shown that this combination site antigen–magnesium has a strong relationship with the immune system, both in the nonspecific and specific immune response, also called innate and acquired immune response (Tam et al., 2003). That is, as a cofactor for the synthesis of immunoglobulins, C′3 convertase, immune cell adhesion, antibody-dependent cytolysis, IgM lymphocyte binding, macrophage response to lymphokines, and the adhesion of helper T lymphocytes (Tam et al., 2003). These data are in accordance with the results described above (Zou et al., 2015). These authors have demonstrated that this molecule promotes more antigenic and immunogenic power around this active site, giving an immune response of 100% neutralizing antibodies. The reason might be that despite significant differences in the S1 protein, much of the virus genome remains unchanged, and there are common epitopes among different strains of IBV, which play a major role in protective immunity (Cavanagh, 1997).

Based on the results presented here, the two protein sequences studied have a 3D spatial conformation and common predicted neutralizing epitopes, where it seems that both strains have the same pathogenicity and tissue tropism.

Thus it is highly probable that the H120 vaccine strain confers cross-protection against a challenge with new strains Italy02 circulating in Morocco. So far, Mass strains have been mainly used as live vaccines because of their epizootic distributions and cross-protective capacity (Ignjatovic and Sapats, 2000, Bijlenga et al., 2004).

CONCLUSION

The in silico study presented here shows that the two serotypes Italy02 and Mass H120 circulating in Morocco share an identical structure in 3D, with a similarity percentage of 81%, as well as common predicted neutralizing epitopes. To realize this data, experimental research on the cross-protection between the two serotypes detected in this country is necessary.

COMPETING INTERESTS

Authors declare they have no competing interests and had access to all generated data and that they contributed to the analyses and interpretation.

References

Abd, E.R.S., El-Kenawy, A.A., Neumann, U., Herrler, G., Winter, C., 2009. Comparative analysis of the sialic acid binding activity and the tropism for the respiratory epithelium of four different strains of avian infectious bronchitis virus. Avian Pathol. 38 (1), 41–45.

Berjanskii, M., Liang, Y., Zhou, J., Tang, P., Stothard, P., Zhou, Y., et al., 2010. PROSESS: a protein structure evaluation suite and server. Nucleic Acids Res. 38, W633–W640.

Bijlenga, G., Cook, J.K.A., Gelb, J., de Wit, J.J., 2004. Development and use of the H strain of avian infectious bronchitis virus from the Netherlands as a vaccine: a review. Avian Pathol 33, 550–557.

Boursnell, M.E., Brown, T.D., Foulds, I.J., Green, P.F., Tomley, F.M., Binns, M.M., 1987. Completion of the sequence of the genome of the coronavirus avian infectious bronchitis virus. J. Gen. Virol. 68, 57–77.

Casais, R., Dove, B., Cavanagh, D., Britton, P., 2003. Recombinant avian infectious bronchitis virus expressing a heterologous spike gene demonstrates that the spike protein is a determinant of cell tropism. J. Virol. 77 (16), 9084–9089.

Cavanagh, D., 1997. *Nidovirales:* a new order comprising *Coronaviridae* and *Arterividae.* Arch. Virol 142 (3), 629–633.

Cavanagh, D., 2005. Coronaviruses in poultry and other birds. Avian Pathol. 34 (6), 439–448.

Cavanagh, D., 2007a. Coronavirus avian infectious bronchitis virus. Vet. Res 38, 281–297.

Cavanagh, D., 2007b. Coronavirus avian infectious bronchitis virus. Vet. Res. 38 (2), 281–297.

Cavanagh, D., Davis, P.J., Darbyshire, J.H., Peters, R.W., 1986. Coronavirus IBV: virus retaining spike glycopolypeptide S2 but not S1 is unable to induce virus-neutralizing or haemagglutination-inhibiting antibody, or induce chicken tracheal protection. J. Gen. Virol. 67 (Pt 7), 1435–1442.

Chothia, C., Lesk, A.M., 1986. The relation between the divergence of sequence and structure in proteins. EMBO J 4, 823–826.

Fellahi, S., Ducatez, M., El Harrak, M., Guerin, J.L., Touil, N., Sebbar, G., et al., 2015. Prevalence and molecular characterization of avian infection bronchitis virus in poultry flocks in Morocco from 2010–2014 and the first report of Italy02 genotype in Africa. Avian Pathol. 44, 287–295.

Ignjatovic, J., Sapats, S., 2000. Avianinfectiousbronchitis virus. Rev. Sci. Tech 19, 493–508.

Jackwood, M.W., Hilt, D.A., Callison, S.A., Lee, C.W., Plaza, H., Wade, E., 2001. Spike glycoprotein cleavage recognition site analysis of infectious bronchitis virus. Avian Dis. 45 (2), 366–372.

Jones, R.C., Jordan, F.T., 1972. Persistence of virus in the tissues and development of the oviduct in the fowl following infection at day old with infectious bronchitis vims. Res. Vet. Sci. 13, 52–60.

Koch, A.E., Polverini, P.J., Kunkel, S.L., Harlow, L.A., DiPietro, L.A., Elner, V.M., Strieter, R.M., 1992. Interleukin-8 as a macrophage-derived mediator of angiogenesis. Science 258 (5089), 1798–1801.

Lai, M.M., Brayton, P.R., Armen, R.C., Patton, C.D., Pugh, C., Stohlman, S.A., 1981. Mouse hepatitis virus A59: mRNA structure and genetic localization of the sequence divergence from hepatotropic strain MHV-3. J. Virol. 39 (3), 823–834.

Niesters, H.G., Kusters, J.G., Lenstra, J.A., Spaan, W.J., Horzined, M.C., van der Zeijst, B. A., 1987. The neutralization epitopes on the spike protein of infectious bronchitis virus and their antigenic variation. Adv. Exp. Med. Biol. 218, 483–492.

Piuzzi, M., 2010. Détermination de la structure de protéines à l'aide de données faiblement résolues (Thèse Biochimie). Université Pierre et Marie Curie, Paris VI, Français.

Tam, M., Gómez, S., González-Gross, M., Marcos, A., 2003. Possible roles of magnesium on the immune system. Eur. J. Clin. Nutr 57, 1193–1197.

Zhang, Y., et al., 2005. Cloning, expression and characterization of the human NOB1 gene. Mol Biol Rep. 32 (3), 185–189.

Zou, N.L., Zhao, F.F., Wang, Y.P., Liu, P., Cao, S.J., Wen, X.T., et al., 2010. Genetic analysis revealed LX4 genotype strains of avian infectious bronchitis virus became predominant in recent years in Sichuan area, China. Virus Genes 41 (2), 202–209.

Zou, N.L., Wang, F.F., Duan, Z., Xia, J., Wen, X., Yan, Q., et al., 2015. Development and characterization of neutralizing monoclonal antibodies against the S1 subunit protein of QX-like avian infectious bronchitis virus strain Sczy3. Monoclon. Antib. Immunodiagn. Immunother. 34, 17–24.

CHAPTER

5

Biological Databases in Virology

Pramodkumar Pyarelal Gupta[1], *Virupaksha Ajit Bastikar*[2], *Santosh Subhash Chhajed*[3] *and Shanker Lal Kothari*[4]

[1]School of Biotechnology and Bioinformatics, DY Patil Deemed to be University, Navi Mumbai, India [2]Amity Institute of Biotechnology, Amity University, Mumbai, India [3]Department of Pharmaceutical Chemistry, METS Institute of Pharmacy, Bhujbal Knowledge City, Nashik, India [4]Amity Institute of Biotechnology, Amity University, Jaipur, India

INTRODUCTION

Viruses are most likely the most abundant biological entities on the planet, and the total number of virus particles exceeding by 10 times the total number of cells present on this planet (Suttle, 2007). Viral disease outbreaks and timely epidemics to different parts of the world cause severe clinical manifestations and congenital malformations (Centers for Disease Control and Prevention (CDC), 1999, 2003; Quang Ha, 2000; Trifonov et al., 2009). Identification of the causative agents using comparative genomic and proteomic analysis can be significant in preventing the spread of the virus and discovering suitable targets in prevention and treatment options (Pickett et al., 2012). The study and analysis of infectious disease have played an important role in unwinding the cause of many disease and death. The devotion of scientific community toward the fight against the virus and their infectious disaster has generated a vast amount of information regarding genomic, proteomic whole and partial sequence and structural data including, virus–host, virus drug target–drug interaction. Among the millions of different types of viruses, only about 5000 different types are well

Emerging and Reemerging Viral Pathogens
DOI: https://doi.org/10.1016/B978-0-12-814966-9.00005-6

TABLE 5.1 Genomic Diversity Among Viruses

Property	Parameters
Nucleic acid	DNA
	RNA
	Both DNA and RNA (at different stages in the life cycle)
Shape	Linear
	Circular
	Segmented
Strandedness	Single-stranded
	Double-stranded
	Double-stranded with regions of single-strandedness
Sense	Positive sense (+)
	Negative sense (−)
	Ambisense (±)

defined (Breitbart and Rohwer, 2005; Dimmock et al., 2007). The genomic diversity of the viruses can be found in Table 5.1.

Bioinformatics resources that provide the genomic structure and phenotypic characteristics of known viruses make the process more resourceful by supporting data mining in the design and development of hypotheses for the emerging new strains (Yan, 2008). Bioinformatics database that supplies genomic, proteomic, functional information related to protein−protein and drug interaction has become necessary in virological-based studies. Timely review, data-curation, and upgradation of these bioinformatics-related databases would be not only helpful for scientific end users but also important for the researchers to make further progress in the field (Yan, 2008).

RELATIONAL DATABASE MODEL

In 1970 EF Codd from IBM has published "A Relational Model of Data for Large Shared Data Banks" (Codd, 1970). In the paper, the author had explained the term "relation" in an accepted mathematical sense. For a given sets (S1, S2, . . ., Sn), where R is a relation on these n sets. Relation is a mathematical assembly; users find it much uncomplicated, while thinking about relation as a table. Table is recognizes as a two–dimensional (2D) arrangement that is made of rows and columns. There are three degrees of relationship: one to one (1:1), one to many (1:*M*), and many to many (*M*:*N*) (College of Computer Technology, n.d.).

CHARACTERISTICS OF A RELATIONAL TABLE

1. A table is recognized as a –2D array composed of rows and column.
2. Each table row denotes as a single entity.
3. Each table column characterizes attributes and has a distinct name.
4. Each row/column from table represents a single data value.
5. All values in a column should be in common data format.
6. Each column has a specific range of values known as attributes.
7. Each table must have an attribute or amalgamation of attributes that uniquely identify each row (College of Computer Technology, n.d.).

KEYS

In the relational architecture model, keys are prime important and are used to confirm that in a given table, each row is uniquely identifiable. Keys are also used to associate relationship among tables and guarantee the integrity of data. Generally, key are composed of one or more attributes that determines the other attributes. The role of key is primarily based on a notion known as determination, such as if you know the value *X* attribute then you can pool out the value of attribute *Y*, for example, in a biological database where accession numbers identify all of the data attributes, such as name of particular sequence name, description, sequence length, evidence, and other related listed information in the database (Table 5.2).

The principle of relational database in data connection and information retrieval of digital libraries are important for understanding biological database system. Biological database includes nucleotide, protein sequence, attributes, ontology, annotation data, citations, tabular, and image files. Accession numbers are the common key components in

TABLE 5.2 Relational Data Attributes

Accession no.	Name	Description	Sequence length	Evidence
ACC1234	Capsid protein 1	Viral protein	120	Protein
ACC1235	Capsid protein 2	Viral protein	165	Protein
ACC1236	Capsid protein 3	Viral protein	101	Protein

cross-referencing two databases with common dataset into it (Toomula et al., 2012). In many public and private databanks the degree of relations are very high. Consider an example of biochemical database, where a single gene/protein/metabolite takes part in multiple reactions.

NATIONAL CENTRE FOR BIOTECHNOLOGY INFORMATION VIRAL GENOME RESOURCES

The NCBI (National Centre for Biotechnology Information) Viral Genomes Project (https://www.ncbi.nlm.nih.gov/genome/viruses/) (Fig. 5.1) was launched in response to the rising necessitate for a public, virus-specific, reference sequence resource (Bao et al., 2004; Brister et al., 2015). The Viral Genome project encompasses all complete genomes deposited in INSDC (Karsch-Mizrachi et al., 2012) databases (Benson et al., 2014), DDBJ (Kosuge et al., 2014), and EMBL (Brooksbank et al., 2014), and creates RefSeq (Pruitt et al., 2007) records for each viral species. Each record of RefSeq is derived from INSDC database sequence record but may include supportive annotation and/or information from other databases too. Accessions for RefSeq genome records are prefixed with "NC," which marks them differently from INSDC records. For example, RefSeq genome record for Enterobacteria phage T4 had accession NC_000866 but was derived from INSDC record AF158101 (Brister et al., 2015).

The first genome submitted for a specific species is selected as a RefSeq. Following the created RefSeq, other validated genomes for that

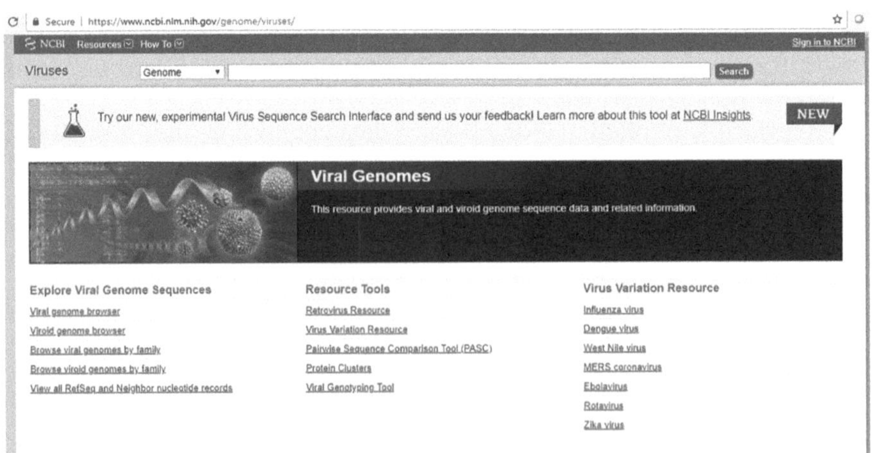

FIGURE 5.1 Home page of NCBI viral genomes. *NCBI,* National Centre for Biotechnology Information.

species are indexed as genome neighbors. Hence, the viral RefSeq data model is more taxonomic centric or species-centric, and all the RefSeq records and genome neighbors are indexed at the species level (Brister et al., 2015).

Reference viral RefSeq records are curated by biologists and field of experts, using in-house designed annotation tools and scientific literature (Brister et al., 2015). Expert guidance is provided from outside of NCBI as a team of Viral Genome Advisor that supports in curation for specific RefSeq records. Mostly, it is used for the maintenance of adenovirus and herpes virus RefSeq records (Davison, 2010) and could be extended to other virus genomes (Rodney Brister et al., 2012).

VIRALZONE

Viruses constitute a diverse molecular architecture that complicates in the interpretation of viral genomic and proteomic data (Hulo et al., 2011). Most viruses have a comparatively minute genome encoding for only a few numbers of proteins: Circovirus is one of the smallest viruses with a 1.7-kb genome coding has only two proteins (Finsterbusch and Mankertz, 2009). Currently, there are 83 virus families, where each varies in its replication plan which calls for distinctive proteins and distinctive enzyme (Macnaughton and Lai, 2006). For example, the replication cycles of ebolavirus and human herpes virus 1 have nothing in common (Hulo et al., 2011).

The Swiss-Prot virus annotation team has built up a website committed to viruses and related information known as ViralZone (www.expasy.org/viralzone) (Fig. 5.2). The perception of this website is to connect specific knowledge for each virus family with its viral genomic and protein sequences (Hulo et al., 2011). All the annotated information is offered in a brief and accessible virus fact sheet. These facts are composed of condensed information about the genome, replication cycle, taxonomy, and epidemiology as well as graphics describing virion organization, genome transcription, and translation strategies (Hulo et al., 2011).

Until March 2017 the whole site comprises 653 fact sheets covering the whole known virosphere: 101 families, 543 genera, and 9 individual species, including 216 viral molecular biology pages. The datasets are linked to 1712 reference proteomes, 16,700 manually reviewed proteins, and 2,918,867 unreviewed viral proteins from UniProt release 2017_02 (ViralZone, 2017).

ViralZone is regularly updated with new annotated and curated information extracted from publications and scientific meetings abstract. Users can forward their contribution by sending minor changes or feedback or improvised ideas to viralzone@isb-sib.ch.

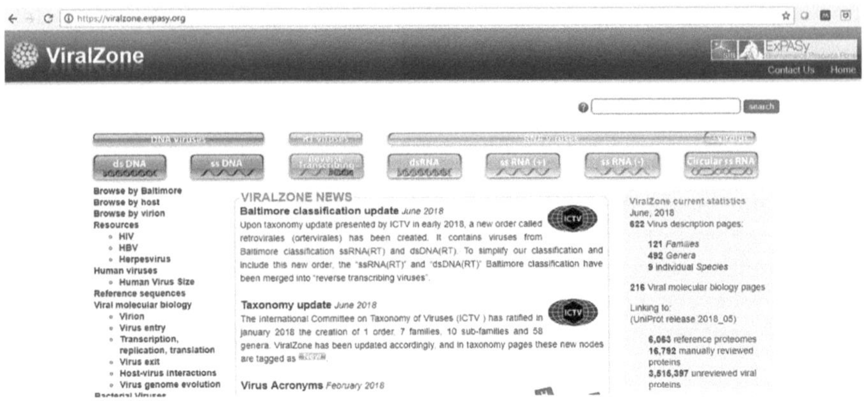

FIGURE 5.2 **ViralZone webpage.**

VIRUS PATHOGEN DATABASE AND ANALYSIS RESOURCE

The Virus Pathogen Database and Analysis Resource (ViPR) (www.ViPRbrc.org) is supported by Bioinformatics Resource Centers program and National Institute of Allergy and Infectious Diseases (Greene et al., 2007; Lefkowitz, 2004; Pickett et al., 2012). ViPR is an integrated repository for virus data, annotation, and analysis tools for multiple virus families (Pickett et al., 2012) and in-houses for human pathogenic viruses belonging to the following families listed in Table 5.3 until April 10, 2017, to support experimental virology researchers focusing on the development of diagnostics, prophylactics, vaccines, and treatments for these pathogens and further plans to support additional virus families to be incorporated in the future (Pickett et al., 2012). (Virus Pathogen Resource, 2017).

The database holds various types of information such as sequence records, gene and protein annotations, genomes, three-dimensional (3D) protein structures, immune epitope locations, host factor data, antiviral drug, clinical and surveillance metadata, and novel data derived from comparative genomics analysis listed in Table 5.4.

VIRUS PATHOGEN DATABASE AND ANALYSIS RESOURCE ANALYSIS AND VISUALIZATION TOOL

The users can find tools for metadata-driven statistical sequence analysis, multiple sequence alignment, phylogenetic tree construction,

TABLE 5.3 Virus Family and Data Lists

Family	Data
Arenaviridae	3 Genera—969 complete segments
Bunyaviridae	5 Genera—6417 complete segments
Caliciviridae	6 Genera—1221 complete genomes
Coronaviridae	2 Subfamilies—1857 complete genomes
Filoviridae	3 Genera—509 complete genomes
Flaviviridae	4 Genera—10416 complete genomes
Hepeviridae	2 Genera—321 complete genomes
Herpesviridae	3 Subfamilies—820 complete genomes
Paramyxoviridae	2 Subfamilies—2625 complete genomes
Picornaviridae	31 Genera—4027 complete genomes
Poxviridae	2 Subfamilies—393 complete genomes
Reoviridae	2 Subfamilies—44420 complete segments
Rhabdoviridae	15 Genera—1138 complete genomes
Togaviridae	2 Genera—1374 complete genomes

Virus Pathogen Resource, 2017. Retrieved April 16, 2017, from: <https://www.viprbrc.org/brc/home.spg?decorator = vipr>.

TABLE 5.4 Data Aggregated by Virus Pathogen Database and Analysis Resource (ViPR)

Sr no.	Data aggregated by ViPR	Data amount
1	Families	14
2	Genera	125
3	Species	4872
4	Strains (GenBank)	512,760
5	Sequences (GenBank)	684,236
6	Proteins (GenBank and UniProt)	1,533,677
7	3D Protein Structures (PDB)	13,405
8	Experimentally determined epitopes (IEDB)	1,079,089
9	Genomes with clinical metadata (NIAID GSCID, manual curation)	3394
10	Host Factor Experiments (NIAID Systems Biology, ViPR DBPs)	25

3D, Three-dimensional; *NIAID*, National Institute of Allergy and Infectious Diseases; *PDB*, Protein Data Bank.
Virus Pathogen Resource, 2017. Retrieved April 16, 2017, from: <https://www.viprbrc.org/brc/home.spg?decorator = vipr>.

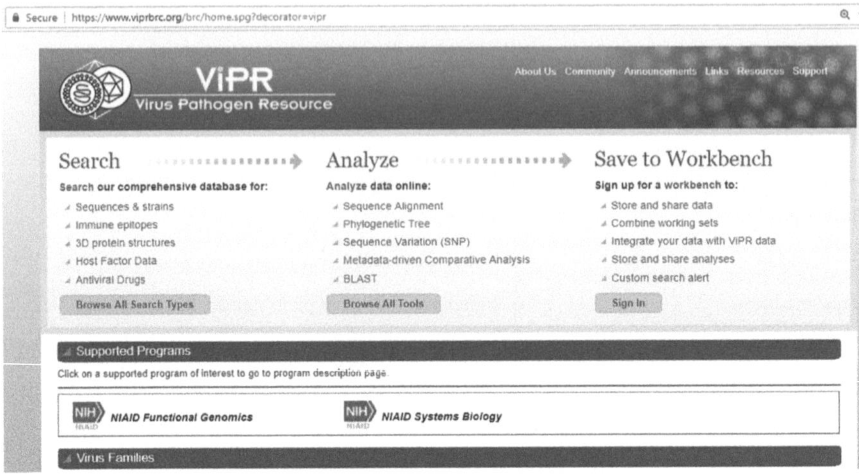

FIGURE 5.3 **ViPR database home page.** *ViPR*, Virus Pathogen Database and Analysis Resource.

BLAST comparison and sequence variation determination, sequence polymorphism analysis, metadata-driven comparative genomics statistical analysis, genome annotator, GBrowse genome viewer, sequence format conversion, BLAST sequence similarity search, 3D protein structure visualization and movie generation, sequence feature variant type analysis (Fig. 5.3) (Virus Pathogen Resource, 2017).

VIRAL PROTEIN STRUCTURE RESOURCE

Viral Protein Structure Resource (ViPs) (http://196.1.114.46:1800/virus/home.html) is a relational viral database that in-houses all known protein structures of the virus, determined by experimental methods such as X-ray crystallography, NMR, or predicted by homology protein modeling concept (Viral Protein Structure Resource, 2017). ViPs is funded by Ministry of Communications and Information Technology, Government of India, and has been developed and maintained by Bioinformatics Centre, University of Pune, Maharashtra, India (Viral Protein Structure Resource, 2017).

The database has two aspects:

1. In the first aspect the structural data of viral protein from Protein Data Bank (PDB) archive is classified according to taxonomic hierarchy. Detailed information on primary sequence, secondary structure, 3D-structure, domains, GO-terms, taxonomy, host,

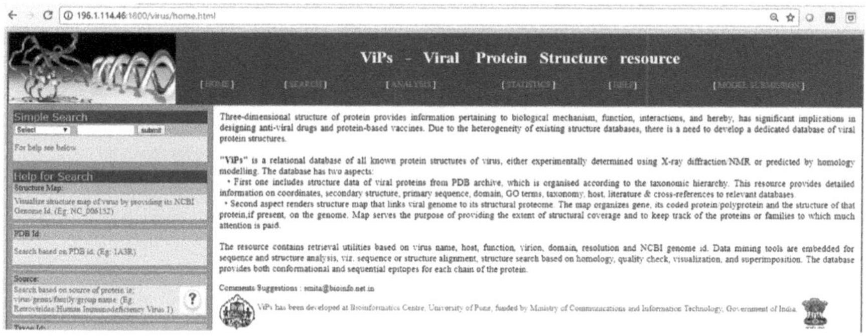

FIGURE 5.4 **Simple search with virus name.**

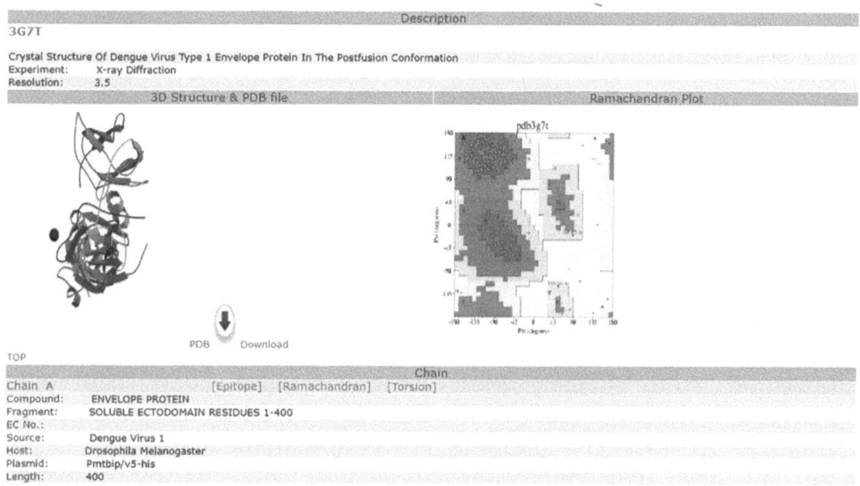

FIGURE 5.5 **Structure quality check and detail information.**

literature, and cross-reference to the relevant database is provided (Viral Protein Structure Resource, 2017).

2. In the second a structural map links to the viral genome to its structural proteome. The map systematizes gene, protein/polyprotein, and protein structure, if available on the genome map (Viral Protein Structure Resource, 2017).

The users can retrieve the information using a query of virus name, host, function, virion, domain, resolution NCBI genome id and PDB id. Users can perform sequence and structure analysis and alignment, homology search, structure quality check, visualization, and superimposition, Figs. 5.4 and 5.5 (Viral Protein Structure Resource, 2017).

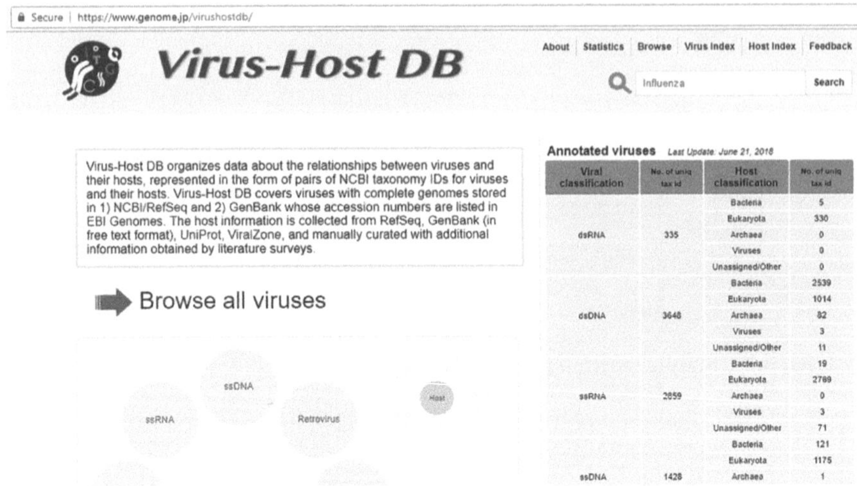

FIGURE 5.6 **Virus–Host Database home page.**

VIRUS–HOST DATABASE

The GenomeNet Virus–Host Database (http://www.genome.jp/virushostdb/note.html) (Fig. 5.6) derives a TaxID-based link between viruses and their hosts (Mihara et al., 2016; Virus–Host Database, 2017). Initially, the data were extracted from natural host and laboratory host information from RefSeq viral genome entries and UniProtKB (Apweiler, 2004; Pruitt et al., 2007). Datasets were manually curated for RefSeq free text to TaxID mapping. For viral genome entries that lacked significant host data, the information was collected by surveying through literature sources. The database was connected to different biological database and finally it provides a link to external reference sources such as ViralZone (Hulo et al., 2011), the NCBI taxonomy database (Federhen, 2012), the Kyoto Encyclopedia of Genes and Genomes database (Kanehisa and Goto, 2000), and the International Committee on Taxonomy of Viruses database (Andrew et al., 2011).

About 38% of the total viral entries are manually curated in the Virus–Host Database. Manually curated entries were linked with literature or other types of evidence, whereas RefSeq or UniProt is added to automatically created entries (Mihara et al., 2016).

DESCRIPTIONS OF PLANT VIRUSES

Descriptions of plant viruses (DPVweb) (http://www.dpvweb.net/) acts as a central source of information about viruses, viroid, and satellites of plants, fungi, and protozoa (Adams, 2006). Initially the project is developed as an electronic version of the Association of Applied Biologists (AAB) description of plant virus is a standalone program for personal computer launched in 1998, holds detail descriptions of selected plant viruses collectively with information on taxonomy and sequences (Adams, 2006; Adams et al., 1998). DPV served as a collective resource widely used for teaching, disease management, and research activities. DPV web is simple and user-friendly accessible biological database; the website has five major sites: Home, DPV, Notes, Sequences, and Analysis.

Home: It provides the overview and introduction of the site (Adams, 2006).

DPV: It holds the detailed information and descriptions about individual viruses added in DPV. More than 400 viruses or virus groups' data are included in DPV. Numbers 1–354 were originally published in paper form by the AAB between 1970 and 1989, while later additional descriptions have been added since 1998 (Adams et al., 1998). The description provides a list of detail information on virus diseases, geographical distribution, host range, symptoms, transmission, vectors, serology, relationships with other viruses, purification protocols, properties of virus particles, particle structure and composition, molecular structure, genome properties, cytopathology, ecology, and control procedures as well as references to the scientific literature, images of symptoms, electron micrographs of virus particles, genome maps, and so on (Adams, 2006).

Notes: It comes with a brief description of each family and genus included within the project (Adams, 2006).

Sequences: It gives the list of accession numbers used by Genbank/EMBL/DDBJ databases for the sequences of viruses, satellites, and viroids included in the DPV (Adams, 2006).

Analysis: It provides links to client software that accesses the data from the user's PC to perform multitask such as the download of nucleotide or protein sequences in FASTA format, calculating codon usage, prediction of the transmembrane domain (Adams, 2006).

Other sources: List of the virus and related information database is added in Table 5.5.

TABLE 5.5 Another Biological Database That Stores Virus and Related Information

Sr no.	Database	Web link	Description
1	All the virology on the WWW	http://www.virology.net/	Links to various sources of viral information on the Internet
2	euHCVdb	http://euhcvdb.ibcp.fr/euHCVdb/	HCV information and links to analysis tools
3	GenBank	http://www.ncbi.nlm.nih.gov/gquery/gquery.fcgi	Entrez query system
4	HCV database	https://hcv.lanl.gov/content/hcv-db/index	HCV information
5	HIV databases	http://www.hiv.lanl.gov/content/index	HIV information
6	ICTV	http://www.ictvdb.org/	Virus classification and nomenclature taxonomy
7	Influenza virus resource	https://www.ncbi.nlm.nih.gov/genome/viruses/variation/flu/	Influenza sequence database and tools
8	Viral bioinformatics resource center	http://athena.bioc.uvic.ca/	Viral data sources
9	PDB	http://www.rcsb.org/pdb/	3D biological macromolecular structure data
10	Retroviruses	http://www.ncbi.nih.gov/retroviruses/	Complete reference genomes, a genotyping tool, and alignment tools
11	VBRC	http://athena.bioc.uvic.ca/index.php	Databases of viral genomes and tools for comparative genomic analyses
12	VIDA	http://www.biochem.ucl.ac.uk/bsm/virus_database/VIDA.html	Homologous protein families
13	VIPERdb	http://viperdb.scripps.edu/main.php	Icosahedral virus capsid structures

3D, Three-dimensional; *euHCVdb*, European Hepatitis C Virus Database; *HCV*, hepatitis C virus; *PDB*, Protein Data Bank.

References

Adams, M.J., 2006. DPVweb: a comprehensive database of plant and fungal virus genes and genomes. Nucleic Acids Res. 34 (90001), D382–D385. Available from: https://doi.org/10.1093/nar/gkj023.

Adams, M.J., Antoniw, J.F., Barker, H., Jones, A.T., Murant, A.F., Robinson, D., 1998. Descriptions of Plant Viruses on CD-ROM. Warwick.

Andrew, M.Q., King, E.L., Adams, M.J., Carstens, E.B., 2011. Virus Taxonomy Ninth Report of the International Committee on Taxonomy of Viruses. Elsevier, Retrieved from: <https://www.sciencedirect.com/science/book/9780123846846>.

Apweiler, R., 2004. UniProt: the universal protein knowledgebase. Nucleic Acids Res. 32 (90001), 115D–119D. Available from: https://doi.org/10.1093/nar/gkh131.

Bao, Y., Federhen, S., Leipe, D., Pham, V., Resenchuk, S., Rozanov, M., et al., 2004. National Center for Biotechnology Information viral genomes project. J. Virol. 78 (14), 7291–7298. Available from: https://doi.org/10.1128/JVI.78.14.7291-7298.2004.

Benson, D.A., Clark, K., Karsch-Mizrachi, I., Lipman, D.J., Ostell, J., Sayers, E.W., 2014. GenBank. Nucleic Acids Res. 42 (D1), D32–D37. Available from: https://doi.org/10.1093/nar/gkt1030.

Breitbart, M., Rohwer, F., 2005. Here a virus, there a virus, everywhere the same virus? Trends Microbiol. 13 (6), 278–284. Available from: https://doi.org/10.1016/j.tim.2005.04.003.

Brister, J.R., Ako-adjei, D., Bao, Y., Blinkova, O., 2015. NCBI viral genomes resource. Nucleic Acids Res. 43 (D1), D571–D577. Available from: https://doi.org/10.1093/nar/gku1207.

Brooksbank, C., Bergman, M.T., Apweiler, R., Birney, E., Thornton, J., 2014. The European Bioinformatics Institute's data resources 2014. Nucleic Acids Res. 42 (D1), D18–D25. Available from: https://doi.org/10.1093/nar/gkt1206.

Centers for Disease Control and Prevention (CDC), 1999. Outbreak of West Nile-like viral encephalitis—New York, 1999. Morb. Mortal. Wkly. Rep. 48 (38), 845–849. Retrieved from: <http://www.ncbi.nlm.nih.gov/pubmed/10563521>.

Centers for Disease Control and Prevention (CDC), 2003. Severe acute respiratory syndrome—Taiwan. JAMA 289 (22), 2930. Available from: https://doi.org/10.1001/jama.289.22.2930.

Codd, E.F., 1970. A relational model of data for large shared data banks. Commun. ACM 13 (6), 377–387.

College of Computer Technology, n.d. Creating a Relational Database Design. Retrieved June 5, 2018, from: <https://www.sqa.org.uk/e-learning/SoftDevRDS02CD/page_44.htm>.

Davison, A.J., 2010. Herpesvirus systematics. Vet. Microbiol. 143 (1), 52–69. Available from: https://doi.org/10.1016/j.vetmic.2010.02.014.

Dimmock, N., Easton, A., Leppard, K., 2007. Introduction to Modern Virology (Sixth). Blackwell Publishing.

Federhen, S., 2012. The NCBI taxonomy database. Nucleic Acids Res. 40 (D1), D136–D143. Available from: https://doi.org/10.1093/nar/gkr1178.

Finsterbusch, T., Mankertz, A., 2009. Porcine circoviruses—small but powerful. Virus Res. 143 (2), 177–183. Available from: https://doi.org/10.1016/j.virusres.2009.02.009.

Greene, J.M., Collins, F., Lefkowitz, E.J., Roos, D., Scheuermann, R.H., Sobral, B., et al., 2007. National Institute of Allergy and Infectious Diseases Bioinformatics Resource Centers: new assets for pathogen informatics. Infect. Immun. 75 (7), 3212–3219. Available from: https://doi.org/10.1128/IAI.00105-07.

Hulo, C., de Castro, E., Masson, P., Bougueleret, L., Bairoch, A., Xenarios, I., et al., 2011. ViralZone: a knowledge resource to understand virus diversity. Nucleic Acids Res. 39 (Suppl. 1), D576–D582. Available from: https://doi.org/10.1093/nar/gkq901.

Kanehisa, M., Goto, S., 2000. KEGG: Kyoto encyclopedia of genes and genomes. Nucleic Acids Res. 28 (1), 27–30. Retrieved from: <http://www.ncbi.nlm.nih.gov/pubmed/10592173>.

Karsch-Mizrachi, I., Nakamura, Y., Cochrane, G., 2012. The International Nucleotide Sequence Database Collaboration. Nucleic Acids Res. 40 (D1), D33–D37. Available from: https://doi.org/10.1093/nar/gkr1006.

Kosuge, T., Mashima, J., Kodama, Y., Fujisawa, T., Kaminuma, E., Ogasawara, O., et al., 2014. DDBJ progress report: a new submission system for leading to a correct annotation. Nucleic Acids Res. 42 (D1), D44–D49. Available from: https://doi.org/10.1093/nar/gkt1066.

Lefkowitz, E.J., 2004. Poxvirus Bioinformatics Resource Center: a comprehensive Poxviridae informational and analytical resource. Nucleic Acids Res. 33 (Database issue), D311–D316. Available from: https://doi.org/10.1093/nar/gki110.

Macnaughton, T.B., Lai, M.M.C., 2006. HDV RNA replication: ancient relic or primer? Curr. Top. Microbiol. Immunol. 307, 25–45. Retrieved from: <http://www.ncbi.nlm.nih.gov/pubmed/16903219>.

Mihara, T., Nishimura, Y., Shimizu, Y., Nishiyama, H., Yoshikawa, G., Uehara, H., et al., 2016. Linking virus genomes with host taxonomy. Viruses 8 (12), 66. Available from: https://doi.org/10.3390/v8030066.

Pickett, B.E., Sadat, E.L., Zhang, Y., Noronha, J.M., Squires, R.B., Hunt, V., et al., 2012. ViPR: an open bioinformatics database and analysis resource for virology research. Nucleic Acids Res. 40 (D1), D593–D598. Available from: https://doi.org/10.1093/nar/gkr859.

Pruitt, K.D., Tatusova, T., Maglott, D.R., 2007. NCBI reference sequences (RefSeq): a curated non-redundant sequence database of genomes, transcripts and proteins. Nucleic Acids Res. 35 (Database), D61–D65. Available from: https://doi.org/10.1093/nar/gkl842.

Quang Ha, D., 2000. Dengue epidemic in Southern Vietnam, 1998. Emerg. Infect. Dis. 6 (4), 422–425. Available from: https://doi.org/10.3201/eid0604.000421.

Rodney Brister, J., Le Mercier, P., Hu, J.C., 2012. Microbial virus genome annotation—mustering the troops to fight the sequence onslaught. Virology 434 (2), 175–180. Available from: https://doi.org/10.1016/j.virol.2012.09.027.

Suttle, C.A., 2007. Marine viruses—major players in the global ecosystem. Nat. Rev. Microbiol. 5 (10), 801–812. Available from: https://doi.org/10.1038/nrmicro1750.

Toomula, N., Kumar, A., Sathish Kumar, D., Bheemidi, V.S., 2012. Biological databases—integration of life science data. J. Comput. Sci. Syst. Biol. 04 (05). Available from: https://doi.org/10.4172/jcsb.1000081.

Trifonov, V., Khiabanian, H., Rabadan, R., 2009. Geographic dependence, surveillance, and origins of the 2009 influenza A (H1N1) virus. N. Engl. J. Med. 361 (2), 115–119. Available from: https://doi.org/10.1056/NEJMp0904572.

Viral Protein Structure Resource, 2017. Retrieved February 11, 2017, from: <http://196.1.114.46:1800/virus/home.html>.

ViralZone, 2017. Retrieved April 15, 2017, from: <http://viralzone.expasy.org/>.

Virus Pathogen Resource, 2017. Retrieved April 16, 2017, from: <https://www.viprbrc.org/brc/home.spg?decorator = vipr>.

Virus–Host Database, 2017. Retrieved April 10, 2017, from: <http://www.genome.jp/virushostdb/>.

Yan, Q., 2008. Bioinformatics databases and tools in virology research: an overview. In Silico Biol. 8 (2), 71–85. Retrieved from: <https://content.iospress.com/articles/in-silico-biology/isb00345>.

CHAPTER

6

Involvement and Roles of Long Noncoding RNAs in the Molecular Mechanisms of Emerging and Reemerging Viral Infections

Maryame Lamsisi and Moulay Mustapha Ennaji

Laboratory of Virology, Microbiology, Quality, Biotechnologies/Eco-Toxicology and Biodiversity, Faculty of Sciences and Techniques, Mohammedia, University Hassan II of Casablanca, Casablanca, Morocco

ABBREVIATIONS

AFP alpha-fetoprotein
ATF2 activating transcription factor 2
CHC chronic HCV infection
EV71 enterovirus 71
HBV hepatitis B virus
HCC hepatocellular carcinoma
HDV hepatitis D virus
HEV hepatitis R
HIV human immunodeficiency virus
IAV influenza A virus

Emerging and Reemerging Viral Pathogens
DOI: https://doi.org/10.1016/B978-0-12-814966-9.00006-8

IL interleukine
IRG immunity-related GTPase
ISGs interferon-stimulated genes
JEV Japanese encephalitis virus
LncRNAs long noncoding RNAs
MDM monocyte-derived macrophages
MiRNAs microRNAs
NcRNAs noncoding RNAs
NGS next-generation sequencing
NI nonidentified
PCR polymerase chain reaction
PiRNAs Piwi RNAs
PRC2 polycomb repressive complex 2
qPCR quantitative PCR
qRT-PCR quantitative real-time PCR
RNA-Seq RNA sequencing
SARS-CoV severe acute respiratory syndrome coronavirus
siRNAs small interfering RNAs

INTRODUCTION

Genome analysis, especially using the next-generation sequencing approaches, has revealed that only 2% of the human genome is composed of protein-coding genes, while the largest part that was considered dark matter transcribed into diverse RNAs with no protein-coding capacity (Takayama and Inoue, 2015). Noncoding RNAs (ncRNAs) are classified based on their length; there are short ncRNAs of less than 200 nucleotides and long ncRNAs (lncRNA) larger than 200 nucleotides. Further division of short differentiate Piwi-interacting RNAs, small interfering RNAs, and microRNAs (miRNAs) (Wang et al., 2013).

Transcription of lncRNAs generally involves RNA polymerase II or III. Notably, these transcripts present similarities with mRNAs such as the 5′cap and the polyadenylated tail in some cases at the 3′end. Nevertheless, LncRNAs are less abundant in the cell compared to mRNAs and their expression is more specific to the cell, tissue, organ type, and to the developmental stage (Nguyen and Carninci, 2016). According to their position relative to the neighboring protein-coding gene, lncRNAs are classified as sense, antisense, bidirectional, intronic, or intergenic. In most cases, lncRNAs regulate expression of neighboring genes with whom they probably share the same enhancers, in *Cis*, or in *Trans*, if the genes are distant (Laurent et al., 2015).

Recent studies have revealed different roles of lncRNAs, and they are now known for their important functions within the cell, as they are involved in the regulation of cell physiology through epigenetic regulation of gene expression, either by modifying the chromatin state or regulating transcriptional and posttranscriptional events. Thus lncRNAs are incriminated in several pathologies, such as Alzheimer, cancer, and hepatocellular carcinoma (HCC) as reviewed recently by DiStefano (Hasegawa and Nakagawa, 2015; Liu and Ding, 2017; DiStefano, 2018; Josipovic et al., 2018).

Comparing to available data on lncRNAs in malignant diseases, discoveries on the involvement of these transcripts in emerging viral pathologies are in their infancy. Nevertheless, accumulating evidence from transcriptomic studies showed different expression patterns of lncRNAs during viral infections and suggests regulatory roles of these molecules (Peng et al., 2010; Zhang et al., 2016a,b; Fan et al., 2017, Wang et al., 2017). To assess the mechanism underlying this regulation, coexpression and pathway analysis studies have been carried. Actually, they revealed the correlation of lncRNAs with mRNAs and identified lncRNAs–mRNAs partners that may interfere together in many signaling pathways, related to viral invasion and antiviral immune response (Ma et al., 2017b).

In fact, lncRNAs play roles in controlling the virus–host interface. After viral entry, lncRNAs regulate genes of immune response responsible for interferon (INF) production in order to transduce signals to the immune cells for activation (Ouyang et al., 2014; Tang et al., 2017). However, viruses hijack these mechanisms by interfering with cellular lncRNAs, along with coding genes and proteins. Moreover, viral pathogens dispose of their own lncRNAs within their genome and they use them to create favorable environment to their replication and progeny (Saayman et al., 2014). In this chapter, we discuss the molecular mechanisms involving lncRNAs in viral–host interface, with an emphasis on examples of emerging viral pathogens.

LONG NONCODING RNAS IN VIRUS BIOLOGY: EXAMPLES OF EMERGING VIRAL PATHOGENS

Viral pathogens initiate their infectious process when sufficient viral particles are available in contact with accessible, susceptible, and permissive cells. Giving that, viruses are obligate parasites that dispose of small genomes, they need to hijack and manipulate their host molecular resources to synthesize the essential proteins required for completing their life cycle and replicate their genome to produce new viral particles. Using noncoding transcripts such as miRNAs and lncRNAs is among the newly discovered strategies of viruses that are currently the subject of extensive studies. However, there has been few published data to understand the link between lncRNAs and emerging viral infections. The main recent findings are summarized in Table 6.1.

TABLE 6.1 Main Studies on Long Noncoding RNAs (lncRNAs) in Emerging Viral Pathogens

Virus	LncRNAs	Author/year	Detection and quantification technique	Main findings
Respiratory viruses	NI	Peng et al. (2010)	• NGS • Microarray • qRT-PCR	• Discovery of the differential expression of lncRNAs during viral infections in SARS-CoV and IAV-infected cells • LncRNAs are possibly involved in regulating the host response
	NI	Josset et al. (2014)	RNA-Seq	• Differentially expressed lncRNAs were coexpressed with differentially expressed mRNAs • *Cis*-Regulatory lncRNAs were correlated with neighboring coding genes • 5 lncRNAs were predicted to be ISGs
	VIN	Winterling et al. (2014)	• Microarray • qRT-PCR	• Different IAV strains (H1N1, H3N2, and H7N7) induce the expression of VIN • VIN expression constitutes a specific response to some viral infections and contributes to the virulence of the responsible viral pathogens

(*Continued*)

TABLE 6.1 (Continued)

Virus	LncRNAs	Author/ year	Detection and quantification technique	Main findings
HIV	NEAT1	Zhang et al. (2013)	• qPCR array • qRT-PCR	• NEAT1 expression levels are enhanced following HIV-1 infection • HIV-1 production is mediated by depletion of NEAT1 • NEAT1 induces the production of paraspeckle bodies as a mechanism of HIV-1 viral replication
	HIV-encoded lncRNA	Saayman et al. (2014)	qRT-PCR	HIV gene expression is regulated by HIV-encoded lncRNA through an epigenetic mechanism
	NRON	Imam et al. (2015)	• PCR array • NGS • Semiquantitative and qRT-PCR	NRON is regulated by HIV-1 accessory proteins Nef and Vpu and modulates HIV-1 replication through NFAT-mediated pathway
		Li et al. (2016)	qRT-PCR	• NRON participates in maintaining HIV-1 latency • NRON represses HIV-1 replication and transcription through Tat protein degradation
	NI	Trypsteen et al. (2016)	• Microarray • qRT-PCR	Five differentially expressed lncRNAs–mRNAs were involved in HIV pathogenesis

(Continued)

TABLE 6.1 (Continued)

Virus	LncRNAs	Author/ year	Detection and quantification technique	Main findings
	ASP RNA	Zapata et al. (2017)	• qRT-PCR	• Expression of ASP RNA suppresses viral replication • Expression of ASP RNA promotes the establishment and maintenance of HIV-1 latency through its interaction with the PRC2 complex
	LNC 173	Postler et al. (2017)	• In silico analysis • qRT-PCR	LINC00173 is upregulated during infection with HIV
Herpesviruses	PAN	Borah et al. (2011)	Immunofluorescence	PAN RNA was also found to interact in the nucleus with PABPC1
			• RNA-protein pulldown assays • IP	• PABPC1 suppresses PAN RNA expression • ORF57 targets PABPC1 to enhance PAN RNA accumulation
			• ChIRP-Seq • ELISA	• PAN can bind promoters of viral genes and activate their expression to induce the removal of the H3K27me3 mark for activation • PAN acts as a repressor by interacting with components of the repressive PRC2 complex

(*Continued*)

TABLE 6.1 (Continued)

Virus	LncRNAs	Author/ year	Detection and quantification technique	Main findings
		Campbell et al. (2014)		• PAN RNA promotes LANA-episome disassociation during KSHV reactivation
	ALT		• RT-PCR • RNA-Seq	ALT RNA is identified as a bona fide lncRNA of KSHV
	NI	O'Grady et al. (2013)	RNA-Seq	• During reactivation of EBV infection, late antisense transcripts are overexpressed in the nucleus • The process of viral production is highly regulated by noncoding transcripts
	NAG7/ LINC00312	Zhang et al. (2013)	Microarray	• NAG 7 is downregulated in NPC tissues compared to normal controls • NAG 7 expression is negatively correlated to tumor size and positively correlated to lymph node metastasis
	BART lncRNAs	Marquitz et al. (2015)	RNA-Seq	First report of the function of spliced and polyadenylated transcripts of BARTs genes in regulating cellular growth of EBV-induced epithelial cancer

(*Continued*)

TABLE 6.1 (Continued)

Virus	LncRNAs	Author/ year	Detection and quantification technique	Main findings
	-oriPtL-oriPtR	Cao et al. (2015)	RNA-Seq	The latency origin of replication oriP encodes two late noncoding transcripts that interact with paraspeckles in the innate immune response to viruses and promote lytic gene expression of EBV
	LOC553103	He et al. (2016)	qRT-PCR	• LOC553103 promotes EBV-associated cancer metastasis and invasion • EBV-miR-BART6-3p suppresses metastasis through regulation of LOC553103
	H19		• PCR Array • qRT-PCR	• H19 is highly expressed in EBV-transformed LCL • H19 is present in LCL exosomes
	SNHG8	Huang et al. (2016)	• RT-PCR • RNA-Seq	SNHG8 regulates genes and EBV proteins, which modulate gastric cancer progression
EV71	NI	Yin et al. (2013)	• Microarray • qRT-PCR	• LncRNAs are aberrantly expressed in EV71-infected cells • LncRNAs regulate host–EV71 interactions

(Continued)

TABLE 6.1 (Continued)

Virus	LncRNAs	Author/ year	Detection and quantification technique	Main findings
HEV	BISPR	Paliwal et al. (2017)	• RNA-Seq • qRT-PCR	BISPR regulates viral replication through modulating BST2 expression levels

EBV, Epstein–Barr virus; *EV*, enterovirus; *HEV*, hepatitis E virus; *HIV*, human immunodeficiency virus; *IAV*, influenza A virus; *IP*, immunoprecipitation; *ISGs*, interferon-stimulated genes; *KSHV*, Kaposi's sarcoma-associated herpesvirus; *LANA*, latency-associated nuclear antigen; *LCL*, lymphoblastoid cell line; *NFAT*, nuclear factor of activated T cells; *NGS*, next-generation sequencing; *NI*, nonidentified; *NPC*, nasopharyngeal carcinoma; *NRON*, noncoding repressor of NFAT; *PABPC1*, poly(A)-binding protein C1; *PCR*, polymerase chain reaction; *qPCR*, quantitative PCR; *qRT-PCR*, quantitative real-time PCR; *RNA-Seq*, RNA sequencing; *SARS-CoV*, severe acute respiratory syndrome coronavirus; *VIN*, viral-induced noncoding.

LONG NONCODING RNAS CONTRIBUTE TO VIRAL PATHOGENICITY IN RESPIRATORY DISEASES

Previously, Peng et al. (2010) investigated lncRNAs as a response signature to viral infections using whole-transcriptome analysis of the host response to severe acute respiratory syndrome coronavirus (SARS-CoV). This study showed a differential expression of lncRNAs during SARS-CoV infection, which suggests an important role as new regulators of viral infections (Peng et al., 2010). Later, Josset et al. performed a transcriptional analysis of mouse SARS-CoV and influenza A virus (IAV) infected lungs and found 5329 differentially expressed lncRNAs. Importantly, some of these lncRNAs were found to be associated with nearby coding genes. In addition, these lncRNAs were predicted and validated to be stimulated by INF production, and therefore, they may play a key role in viral pathogenesis (Josset et al., 2014). Using microarray systems, Winterling et al. identified 42 differentially expressed lncRNAs in response to IAV infection. Among studied lncRNAs, vesicular stomatitis virus and H1N1, H3N2, H7N7 strains of IAV, induced viral-induced noncoding RNA (VIN RNA). Moreover, RNA interference knockdown of VIN RNA repressed the protein synthesis and replication of IAV (Winterling et al. 2014). These findings suggest a common mechanism of respiratory viruses' pathogenicity involving lncRNAs.

LONG NONCODING RNAS REGULATE HUMAN IMMUNODEFICIENCY VIRUS REPLICATION AND MAINTAIN ITS LATENCY

In human immunodeficiency virus (HIV), Zhang et al. (2013) characterized for the first time the lncRNA NEAT1 for its role in HIV-1 biology. Among 83 studied lncRNAs, NEAT1 was one of many lncRNAs that have been enhanced in expression upon HIV-1 infection (Zhang et al., 2013). Furthermore, NEAT1 has been shown to posttranscriptionally regulate HIV-1 replication through inducing paraspeckle bodies expression and the regulation of nucleus-to-cytoplasm export of Rev-dependent instability element (INS)–containing HIV-1 mRNAs (Zhang et al., 2013). In addition, Saayman et al. (2014) studied an HIV-encoded lncRNA that has been previously discovered and showed that this lncRNA binds to gene promoters and suggests its function as a negative epigenetic regulator of viral transcription and a promoter of HIV latency. NRON [noncoding repressor of NFAT (nuclear factor of activated T cells)] is another lncRNA that has been aberrantly expressed following the infection with HIV-1 and that it regulates HIV-1 replication through modulating NFAT activity (Imam et al., 2015). Moreover, HIV accessory protein, including Nef, which is an early HIV-1 protein, reduced NRON levels, resulting to viral replication, while the late protein Vpu increased its levels and thus promote HIV viral latency (Imam et al., 2015). Another mechanism of NRON-mediated latency of HIV was proposed by Li et al. (2016); they suggest that NRON is involved in inhibiting viral transcription through directing tat protein to the ubiquitin/proteasome components (CUL4B and PSMD11) resulting in its degradation. Several other lncRNAs have been reported for their putative roles in regulating HIV viral life cycle. To study the different expression patterns of lncRNA during viral life cycle, Trypsteen et al. (2016) performed a transcriptome profiling during HIV integration, reverse transcription, and HIV particle production. They also investigated the roles of these lncRNAs and found their involvement in different pathways of cell cycle regulation including responses to DNA damage, apoptosis, and proteasomal and ubiquitination pathways (Trypsteen et al., 2016). In addition, several lncRNAs have been reported to regulate gene expression through their interaction with chromatin modifying complexes, especially the polycomb repressor complex 2 (PRC2) that directs methylation marks to the promoters of targeted genes. Similarly, lncRNA ASP RNA was reported to interact with PRC2 at the HIV-1 5′LTR region and results in H3K27me3 mark accumulation (Zapata et al., 2017). Along with these findings, ASP RNA reduced RNA polymerase II expression (Zapata et al., 2017). As a result, HIV-1 replication is repressed and its latency is established and maintained (Zapata et al., 2017). More recently, Postlera et al. used a metaanalytic approach to investigate differentially expressed lncRNAs during HIV-1 infection. One of the identified lncRNAs was proposed to be a regulator of cytokines regulation in T cells (Postler et al., 2017).

LONG NONCODING RNAS IN HERPESVIRUSES LATENT TO LYTIC CYCLE TRANSITION

Herpesviruses infect a wide range of organisms including plants and mammals. Mammalian Herpesviruses undergo two life cycles, latency is the phase where the virus does not produce any viral particles while lytic phases is when the virus reactivates and replicate its genome to produce new progeny. The most studies on herpesviruses and lncRNAs have been focused on Kaposi's sarcoma-associated herpesvirus (KSHV) and Epstein–Bar virus (EBV).

LONG NONCODING RNAS REGULATE THE TRANSITION FROM LATENT TO LYTIC PHASE IN KAPOSI'S SARCOMA-ASSOCIATED HERPESVIRUS INFECTION

KSHV is associated with several diseases such as AIDS-related malignancy named Kaposi's sarcoma, KSHV inflammatory cytokine syndrome, and lymphoproliferative disorders primary effusion lymphoma (PEL), Multicentric Castleman's Disease and PEL (Borah et al., 2011; Conrad, 2016). During KSHV infection, a 1.1 kb long ncRNA PAN is produced in the early stages and accumulates in the nucleus. In fact, PAN RNA expression represents 80% of the total polyadenylated transcripts produced in the lytic phase by both infected cells and KSHV, which suggest an important role in latent to lytic phase transition of the virus (Borah et al., 2012). KSHV recruit several molecules in order to activate the expression of its own transcripts and the shutdown of host RNAs expression; these include ORF50 and ORF57 proteins that modulate viral gene expression and RNA processing. ORF57 interacts with PAN RNA and enhances its accumulation. ORF57 is also found to maintain PAN RNA stability in the absence of ENE elements through binding to a 9 nt motif contained in an element called MRE (Mta responsive element) located at the 5′ region of PAN. In addition, MRE element interacts with other proteins as E1B-AP5 and poly(A)-binding protein C1 (PABPC1) (Massimelli et al., 2011). Coherent with these findings, PAN RNA was also found to interact in the nucleus with PABPC1. PABPC1 is a protein located in the cytoplasm of the infected host cells and responsible for modulating mRNA translation, maintaining their stability and processing. During lytic phase of KSHV infection, SOX protein induces PAN RNA expression and PABPC1 relocation to the nucleus which is required for the expression of late KSHV gene, such as vIL-6 and k8.1 (Borah et al., 2011). Later, Massimelli further investigated the interaction relating PAN RNA to PABPC1 and ORF57 and found that PABPC1 actually suppresses PAN RNA expression and that ORF57 targets PABPC1 to enhance PAN RNA accumulation (Massimelli et al., 2013).

In fact, PAN RNA promotes the transition of KSHV infection from latent to lytic state through its interaction with the latency-associated nuclear antigen (LANA). LANA is important to maintain viral latency through inhibiting the expression of lytic genes. During latency, LANA is associate with a large portion of the KSHV episomes, while during reactivation, high levels of PAN RNA are transcribed to block LANAs interaction sites and induce its disassociation of LANA from KSHV episome, thus promoting viral gene expression (Campbell et al., 2014). The mechanism underlying PAN RNA regulation of viral gene expression appears to involve epigenetic pathways as well. PAN can bind promoters of viral genes and activate their expression such as the lytic regulator Rta, the demethylases JMJD3 (Jumonji domain containing 3) and UTX (ubiquitously transcribed tetratricopeptide repeat, X chromosome) and histone methyltransferase MLL2 (mixed-lineage leukemia protein 2) to induce the removal of the H3K27me3 mark and mark it to be activated. PAN RNA can also act as a repressor by interacting with SUZ12 (suppressor of zeste 12) and EZH2 (enhancer of zeste 2) which are components of the repressive PRC2 complex (polycomb repressive complex 2) and mediate the trimethylation of H3K27. Consequently, the elimination of repressive marks and activation of chromatin marks.

In addition to PAN RNAs, other lncRNAs have been reported during KSHV infection such as T3.0, T1.2, T1.5, T6.1, and ALT. ALT is identified in a recent study as a bona fide of KSHV with roles in viral gene expression. However, more studies are still needed to characterize lncRNAs that are reported in large-scale studies, to identify their exact molecular functions.

LONG NONCODING RNAS AND EPSTEIN–BARR VIRUS

EBV is a ubiquitous virus that is usually asymptomatic but causes malignant diseases in some cases such as nasopharyngeal carcinoma and lymphoma. Study on EBV transcriptome conducted by O'Grady et al. (2013) revealed that almost all the EBV genome is bidirectionally transcribed into several ncRNAs during the lytic phase.

As described by Marquitz et al., the BamHI A rightward transcripts (BARTs) are alternatively spliced and give rise to large wide of viral polyadenylated RNAs in Epstein–Barr virus–related epithelial cancers. The BART introns are the origin of 44 miRNAs while the exons are spliced and polyadenylated to form nonprotein-coding RNAs with nuclear localization that regulate infection in a similar mechanism of lncRNAs (Marquitz et al., 2015). In addition, two other nuclear lncRNAs, oriPtLs and oriPtRs, are transcribed during EBV reactivation and contribute to the viral gene expression and evasion of the host immune response (Cao et al., 2015). LOC553103 is another lncRNA that is associated with EBV-related cancer as it is responsible for promoting cancer cell migration and invasion. This lncRNA is regulated by a miRNA EBV-miR-BART6-3p (He et al., 2016). In a study on lymphoblastoid cell line, lncRNA profiling showed different patterns of expression of H19, H19 antisense, 7SL, and p53 mRNA. These are also released in the exosomes. Huang et al. used RNA-Seq and identified five lncRNAs with specific expression patterns to EBV-associated gastric cancer. Among these, SNHG8 showed interactions with several EBV proteins such as LF3, BHLF1, BHRF1, and BNLF2a and interacted with genes and pathways targeted by EBV such as transcription and mRNA metabolism (Huang et al., 2016).

OTHER EXAMPLES

Studies on other viruses also showed specific expression profiles of different lncRNAs following emerging viral infections. For example, in enterovirus 71 (EV71) infection, Yin et al. identified 160 enhancer-like lncRNA and mRNA pairs that are differentially expressed and hypothesized the enhancing role of these molecules upon neighboring genes. In addition, they used functional enrichment analysis and demonstrated the role of lncRNA–mRNA partners in many processes such as alternative splicing and acetylation (Yin et al., 2013). RNA-Seq analysis further confirmed by qPCR identified lncRNA BISPR and mRNA BST pair in hepatitis E virus (HEV) infection, as differentially expressed after 24 and 72 hour postinfection with HEV and correlated its upregulation with HEV viral replication.

Altogether, these findings highlight the different roles of lncRNAs in regulating the viral life cycle epigenetically, especially through regulation of viral gene expression, replication, and latency, and, thus, increasing viral pathogenesis. This is mediated by the interference of lncRNAs with viral and host genes.

LONG NONCODING RNAS IN VIRAL–HOST INTERACTION

The mechanisms of host immune systems are in constant evolution along with viral mechanisms of infecting different cellular organisms. Understanding both strategies and their interactions is of great importance in order to develop effective therapeutic approaches against viral infectious diseases with minimal inflammatory damage.

IMMUNE RESPONSE EMPLOYS LONG NONCODING RNAS AGAINST VIRUSES

Despite the extensive studies on host antiviral strategies, the exact underlying mechanisms are not fully understood. Studies on the involvement of lncRNAs in viral–host interactions added an additional layer of complexity and provided new insights into understanding responsible molecular pathways (Peng et al., 2010; Imamura et al., 2014; Li et al., 2017). A summary of main mechanisms is provided in Table 6.2.

First, Peng et al. (2010) correlated the expression levels of lncRNAs to the host antiviral immune response and suggested that lncRNAs may regulate immunity through their association with chromatin modifying complexes and the modulation of neighboring genes expression. Further studies were conducted to better understand this regulation. For example, Ouyang et al. showed that LncRNA NRAV negatively regulates INF-stimulated genes (ISGs) through a mechanism that appears to involve histone modification. During IAV infection, infected host downregulates lncRNA NRAV to allow ISGs production and repress viral replication (Ouyang et al., 2014). In addition, Barriocanal et al. reported that two lncRNAs, lncISG15, and lncBST2/BISPR, were induced by INF production during IAV and hepatitis C virus (HCV) infections, along with other ISGs such as BST2, IRF1, and ISG15. They found that LncBST2 positively regulates BST2/Tetherin, which have influence on INF release. On the other hand, Kambara et al. (2014) found that lncRNA-CMPK2 was actively upregulated after INF treatment and negatively regulates transcription of ISGs and they proposed a mechanism of this repression that involves RNA–protein interactions with chromatin modifying complexes or transcription factors. The negative regulation of IFN is important to maintain their controlled production and prevent inflammatory damage to the host (Fig. 6.1).

In addition, NEAT1, as mentioned previously, induces paraspeckle bodies' formation. SFPQ, a component of paraspeckles, binds to IL-8 gene promoter and represses its transcription. Imamura et al. (2014) found that NEAT1 activates IL-8 expression through a mechanism involving paraspeckles, as this lncRNA interacts with SFPQ and relocates it back to the paraspeckle, away from IL-8 promoter, which activates the expression of this cytokine. Recent findings indicate that NEAT1 employs the same mechanism and remove SFPQ from gene promoter to activate RIG-I expression and thus producing INFs (Ma et al., 2017a,b). Morchikh et al. further studied the role of NEAT1 in the regulation of immune system. In fact, they indicated NEAT1, in complex with HEXIM1, which is required for the assembly of HDP-RNP subunits composed of HEXIM1, DNA-PK, and paraspeckles. Subsequently, the HDP−PNP complex interacts with cGAS and PQBP1 to mediate immune response to pathogens (Morchikh et al., 2017).

Atianand et al. demonstrated the repression effect of long intergenic noncoding RNA (lincRNA)-EPS on immunity-related GTPase. They suggest different possible mechanisms such as blocking transcription via RNA−RNA interactions, direct interaction with target DNA sequences or by targeting genes through a ribonucleoprotein complex with hnRNPL (Atianand et al., 2016). Likewise, Nishitsuji et al. indicated that hnRNPU binds and stabilizes lncRNA#32. This lncRNA further binds to activating transcription factor to induce ISGs transcription, and, thus, repressing viral replication (Nishitsuji et al., 2016).

Using microarray analysis on mouse brains, Li et al. (2017) identified 1007 mRNAs and 518 lncRNAs during infection with Japanese encephalitis virus. Furthermore, Gene Ontology and network analysis revealed that the pathways associated with lncRNAs and mRNAs upregulation were related to antiviral innate immunity, inflammatory response, apoptotic, acetylation, and nucleotide binding. Moreover, the coexpression network analysis of selected lncRNAs and mRNAs indicated their interaction with three mRNAs, Nod1, Tap2, and Col4a1, which are involved in activating JNK cascade and NFκB production (Li et al., 2017).

Collectively, these data show the important role of lncRNAs in mediating antiviral response of infected hosts.

LONG NONCODING RNAS MODULATE VIRAL EVASION OF IMMUNITY

In counterpart, viruses are continuously adapting their mechanism to evade antiviral immunity. Therefore viral pathogens hijack molecules in order to complete their life cycle and assure their survival. For example, Xiong et al. (2015) indicated that a lncRNA Lethe is activated by the transcriptional activator STAT3 in HCV-infected cells, represses IFN-1 produced by the antiviral immunity in order to promote HCV replication. Moreover, HCV appears to activate proviral lncRNAs within their host. HCV induces the upregulation of lncRNA EGOT, also known as CSR32. Depletion of EGOT reduced viral replication by producing ISGs. In fact, many CSRs are upregulated upon HCV infection and development of liver cancer, which demonstrate that HCV employs cellular lncRNA EGOT to repress immunity and promote cell proliferation (Carnero et al., 2016).

Interestingly, viral pathogens encode their own lncRNAs as well. Thus knowing that viral genome is limited in length, viral encoded noncoding transcripts must play important roles for maintaining their life cycle and producing new progeny. Moreover, RNAs, especially lncRNAs, are more able to evade host immunity compared to protein elements. Therefore viruses might preferably employ these noncoding molecules to invade their hosts (Ding et al., 2016). Consistent with this hypothesis, Moon et al. found that flaviviruses encoded a lncRNA, referred to as sfRNA. This viral encoded lncRNA represses antiviral immune response through its interaction with RNA interference machinery, used by the host to degrade viral RNAs. Using coimmunoprecipitation analysis, they found an interaction of sfRNA with the RNase Dicer and Ago2, which are important players of RNA interference.

LONG NONCODING RNAS AS NEW CANDIDATES FOR VIRAL BIOMARKER AND THERAPY

As stated above, transcriptome analysis indicated that different expression patterns of large lncRNAs sets correlate to viral infection compared to normal controls. This raises the possibility of using such molecules as biomarker and to developing therapeutic strategies for viral diseases. In this chapter, we present related studies in the case of emerging viruses, represented mostly by HCV that has been mostly studied in recent years.

Several studies focused on lncRNAs related to HCV especially in HCC. Zhang et al. found significant difference in expression of seven lncRNAs between preneoplastic lesions and HCC. In comparison to dysplasia, LINC01419 and AK021443 lncRNAs presented elevated and decreased expressions, respectively, at early stage HCC. Further validation with qRT-PCR confirmed the significant overexpression of LINC01419 in HCC with hepatitis B virus (HBV) and HCV etiology compared to normal liver tissues (Zhang et al., 2015). Later, changes in expression levels of eight lncRNAs were significantly related to HCC compared to adjacent normal tissues. Notably, the aberrant expression of lncRNAs was specific to the related hepatitis virus, as lncRNA

TABLE 6.2 Mechanisms of Long Noncoding RNAs (lncRNAs) in Antiviral Immune Response

Immunity pathway	Related lncRNA(s)	Mechanism	Author/year
Interleukins	Nonidentified	• Chromatin modification • Regulation of neighboring genes	Peng et al. (2010)
Activation of interleukins	NEAT1	Relocation of SFPQ to paraspeckles	Imamura et al. (2014)
Suppression of ISGs	NRAV	Histone modification	Ouyang et al. (2014)
Suppression of antiviral interference	sfRNA	Interaction with Dicer and Ago2	
Activation of ISGs	lncRNA#32	ATF2	
Repression of IRGs	LincRNA-EPS	• Chromatin remodeling • Interaction with heterogeneous nuclear ribonucleoprotein L	Atianand et al. (2016)
Activation of RIG-I	NEAT1	Relocation of SFPQ to paraspeckles	Ma et al. (2017a,b)
Inflammatory response	E52329 and N54010	Reducing JNK and MKK4 phosphorylation	Li et al. (2017)
DNA-mediated immunity (cGAS–STING–IRF3 pathway)	NEAT1	Formation of a complex with HEXIM1, DNA-PK, and paraspeckle	Morchikh et al. (2017)

ATF2, Activating transcription factor 2; *IRGs*, immunity-related GTPase; *ISGs*, interferon-stimulated genes; *LincRNA*, long intergenic noncoding RNA; *LncRNA*, Long noncoding RNA; *RIG-I*, retinoic acid-inducible gene I.

PCAT-29 was correlated to HBV, while aHIF and PAR5 were related to HCV and Y3 to hepatitis D virus–induced HCC (Zhang et al., 2016).

Interestingly, recent studies suggest potential use of lncRNAs as serum biomarkers for HCV-induced HCC. As reported by Kamel et al. in their study of lncRNAs expression in serum of 160 participants, expression patterns of lncRNA–UCA1 and lncRNA–WRAP53 were significantly different among sera of HCC, chronic HCV infection, and healthy controls, with a higher expression in advanced stage of HCC. They further found improved sensitivity when combining the studied lncRNAs expression with serum alpha-fetoprotein level (Kamel et al., 2016). Rheumatoid arthritis (RA) is also induced by HCV. Using qRT-PCR, elevated levels of lncRNA-AF085935 in sera of HCV-induced RA

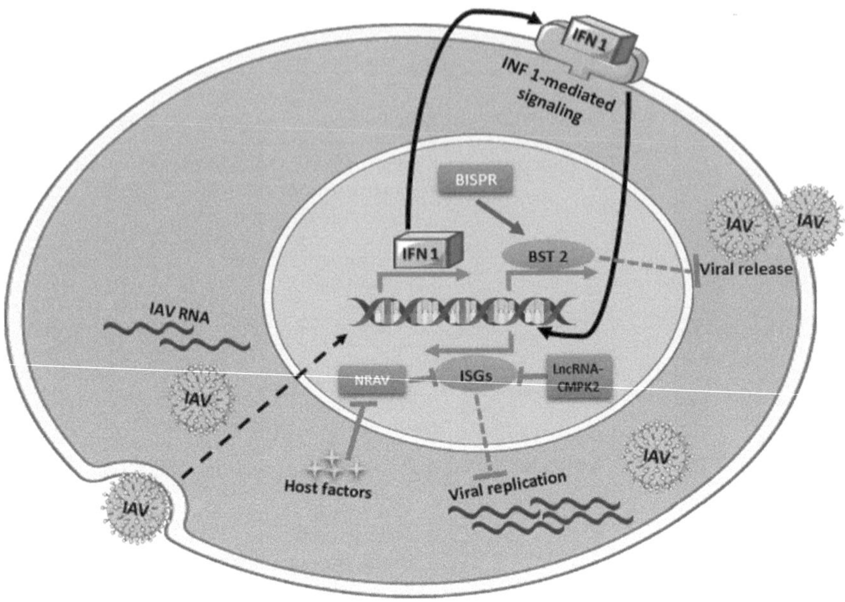

FIGURE 6.1 Following IAV infection, the infected cell downregulates NRAV, which is a negative regulator of ISGs, in order to allow ISGs activation and thus repressing viral replication. On the other hand, lncRNA-CMPK2 represses ISGs, which is important to maintain the balance of inflammatory response and prevent damage to the cell. *IAV*, Influenza A virus; *ISGs*, interferon-stimulated genes; *lncRNAs*, long noncoding RNAs.

patients were reported, suggesting lncRNA-AF085935 as candidate biomarker for early detection of HCV-related RA (Sabry et al., 2017). Another lncRNA, referred to as ncRNA-HEIH, is expressed in a different manner in sera and exosomes during HCV infection and is proposed as an indicator of HCC. In fact, reported expression levels of this lncRNA were higher in HCV-induced HCC patients compared to those in HCV-induced cirrhosis patients and less expression levels were observed in chronic hepatitis C patients (Zhang et al., 2017).

Recently, a study using microarray analysis to detect expression changes of coding and noncoding transcripts in HIV-1 and HIV-2-infected monocyte-derived macrophages revealed significant different expression profiles. Actually, chr2: 165509129-165519404 and lincRNA: chr12: 57761837-57762303 correlated better to HIV-1 infection and lincRNA: chr10:128586385-128592960, XLOC_001148 and lincRNA: chr5:87580664-87583451, related more to HIV-2 infection.

These data reveal significant importance of lncRNAs as new biomarker for detection, diagnosis, and prognostic of viral pathologies. However, these efforts have to extend to other viral pathologies, and more investigations are needed to further confirm their specificity and sensitivity.

CONCLUSION

Existing data on the lncRNAs associated to viral infections give more insight to cellular pathways involved in their regulation. In addition, large-scale transcriptomic studies identified thousands of lncRNAs that need to be studied and characterized to better understand their mechanism of action. This is providing a novel and promising research field that can lead to new molecular network and pathway discoveries, enabling a deeper understanding of these pathologies. Moreover, the tissue specificity and the difference of expression patterns of lncRNAs among different viral infections suggest promising roles as diagnosis biomarkers.

Acknowledgments

The Ministry of High Education of Morocco, the University Hassan II of Casablanca, and Faculty of Sciences and Techniques Mohammedia support this work, as well as the staff of the Laboratory of Virology, Microbiology, Quality and Biotechnologies/Ecotoxicology and Biodiversity for their technical assistance.

The authors would like to thank Khalid El Bairi for the writing advices and his assistance in editing this chapter.

CONFLICT OF INTEREST

The authors declare no conflict of interest.

References

Atianand, M.K., Hu, W., Satpathy, A.T., Shen, Y., Ricci, E.P., Alvarez-Dominguez, J.R., et al., 2016. A long noncoding RNA lincRNA-EPS acts as a transcriptional brake to restrain inflammation. Cell 165 (7), 1672–1685.

Borah, S., Darricarrère, N., Darnell, A., Myoung, J., Steitz, J.A., 2011. A viral nuclear noncoding RNA binds re-localized poly(A) binding protein and is required for late KSHV gene expression. PLoS Pathog. 7 (10), e1002300.

Borah, S., Nichols, L.A., Hassman, L.M., Kedes, D.H., Steitz, J.A., 2012. Tracking expression and subcellular localization of RNA and protein species using high-throughput single cell imaging flow cytometry. RNA 18 (8), 1573–1579.

Campbell, M., Kim, K.Y., Chang, P.C., Huerta, S., Shevchenko, B., Wang, D.H., et al., 2014. A lytic viral long noncoding RNA modulates the function of a latent protein. J. Virol. 88 (3), 1843–1848.

Cao, S., Moss, W., O'Grady, T., Concha, M., Strong, M.J., Wang, X., et al., 2015. New noncoding lytic transcripts derived from the Epstein-Barr virus latency origin of replication, oriP, are hyperedited, bind the paraspeckle protein, NONO/p54nrb, and support viral lytic transcription. J. Virol. 89 (14), 7120–7132.

Carnero, E., Barriocanal, M., Prior, C., Pablo Unfried, J., Segura, V., Guruceaga, E., et al., 2016. Long noncoding RNA EGOT negatively affects the antiviral response and favors HCV replication. EMBO Rep. 17 (7), 1013–1028.

Conrad, N.K., 2016. New insights into the expression and functions of the Kaposi's sarcoma-associated herpesvirus long noncoding PAN RNA. Virus Res. 212, 53–63.
DiStefano, J.K., 2018. The emerging role of long noncoding RNAs in human disease. Dis. Gene Ident. 1706, 91–110.
Ding, Y.Z., Zhang, Z.W., Liu, Y.L., Shi, C.X., Zhang, J., Zhang, Y.G., 2016. Relationship of long noncoding RNA and viruses. Genomics 107 (4), 150–154.
Fan, H., Zhang, Q., Zhao, X., Lv, P., Liu, M., Tang, H., 2017. Transcriptomic profiling of long non-coding RNAs in hepatitis B virus-related hepatocellular carcinoma. Oncotarget 8 (39), 65421–65434.
Hasegawa, Y., Nakagawa, S., 2015. Long noncoding RNA in epigenetic gene regulation. Long Noncoding RNAs 133–149.
He, B., Li, W., Wu, Y., Wei, F., Gong, Z., Bo, H., et al., 2016. Epstein-Barr virus-encoded miR-BART6-3p inhibits cancer cell metastasis and invasion by targeting long non-coding RNA LOC553103. Cell Death Dis. 7 (9), e2353.
Huang, T., Ji, Y., Hu, D., Chen, B., Zhang, H., Li, C., et al., 2016. SNHG8 is identified as a key regulator of Epstein–Barr virus (EBV)-associated gastric cancer by an integrative analysis of lncRNA and mRNA expression. Oncotarget 7 (49), 80990–81002.
Imam, H., Bano, A.S., Patel, P., Holla, P., Jameel, S., 2015. The lncRNA NRON modulates HIV-1 replication in a NFAT-dependent manner and is differentially regulated by early and late viral proteins. Sci. Rep. 5, 8639.
Imamura, K., Imamachi, N., Akizuki, G., Kumakura, M., Kawaguchi, A., Nagata, K., et al., 2014. Long noncoding RNA NEAT1-dependent SFPQ relocation from promoter region to paraspeckle mediates IL8 expression upon immune stimuli. Mol. Cell 53 (3), 393–406.
Josipovic, I., Pflüger, B., Fork, C., Vasconez, A.E., Oo, J.A., Hitzel, J., et al., 2018. Long non-coding RNA LISPR1 is required for S1P signaling and endothelial cell function. J. Mol. Cell. Cardiol. 116, 57–68.
Josset, L., Tchitchek, N., Gralinski, L.E., Ferris, M.T., Eisfeld, A.J., Green, R.R., et al., 2014. Annotation of long non-coding RNAs expressed in collaborative cross founder mice in response to respiratory virus infection reveals a new class of interferon-stimulated transcripts. RNA Biol. 11 (7), 875–890.
Kambara, H., Niazi, F., Kostadinova, L., Moonka, D.K., Siegel, C.T., Post, A.B., et al., 2014. Negative regulation of the interferon response by an interferon-induced long non-coding RNA. Nucleic Acids Res. 42 (16), 10668–10680.
Kamel, M.M., Matboli, M., Sallam, M., Montasser, I.F., Saad, A.S., El-Tawdi, A.H.F., 2016. Investigation of long noncoding RNAs expression profile as potential serum biomarkers in patients with hepatocellular carcinoma. Transl. Res. 168, 134–145.
Laurent, G.S., Wahlestedt, C., Kapranov, P., 2015. The landscape of long noncoding RNA classification. Trends Genet. 31 (5), 239–251.
Li, J., Chen, C., Ma, X., Geng, G., Liu, B., Zhang, Y., et al., 2016. Long noncoding RNA NRON contributes to HIV-1 latency by specifically inducing tat protein degradation. Nat. Commun. 7, 11730.
Li, Y., Zhang, H., Zhu, B., Ashraf, U., Chen, Z., Xu, Q., et al., 2017. Microarray analysis identifies the potential role of long non-coding RNA in regulating neuroinflammation during Japanese encephalitis virus infection. Front. Immunol. 8, 1237.
Liu, W., Ding, C., 2017. Roles of LncRNAs in viral infections. Front. Cell. Infect. Microbiol. 7, 205.
Ma, H., Han, P., Ye, W., Chen, H., Zheng, X., Cheng, L., et al., 2017a. The long noncoding RNA NEAT1 exerts antihantaviral effects by acting as positive feedback for RIG-I signaling. J. Virol. 91 (9), e02250-16.

Ma, Q., Li, L., Tang, Y., Fu, Q., Liu, S., Hu, S., et al., 2017b. Analyses of long non-coding RNAs and mRNA profiling through RNA sequencing of MDBK cells at different stages of bovine viral diarrhea virus infection. Res. Vet. Sci. 115, 508–516.

Marquitz, A.R., Mathur, A., Edwards, R.H., Raab-Traub, N., 2015. Host gene expression is regulated by two types of noncoding RNAs transcribed from the Epstein-Barr virus BamHI a rightward transcript region. J. Virol. 89 (22), 11256–11268.

Massimelli, M.J., Kang, J.G., Majerciak, V., Le, S.Y., Liewehr, D.J., Steinberg, S.M., et al., 2011. Stability of a long noncoding viral RNA depends on a 9-nt core element at the RNA 5′end to interact with viral ORF57 and cellular PABPC1. Int. J. Biol. Sci. 7 (8), 1145.

Massimelli, M.J., Majerciak, V., Kruhlak, M., Zheng, Z.M., 2013. Interplay between polyadenylate-binding protein 1 and Kaposi's sarcoma-associated herpesvirus ORF57 in accumulation of polyadenylated nuclear RNA, a viral long noncoding RNA. J. Virol. 87 (1), 243–256.

Morchikh, M., Cribier, A., Raffel, R., Amraoui, S., Cau, J., Severac, D., et al., 2017. HEXIM1 and NEAT1 long non-coding RNA form a multi-subunit complex that regulates DNA-mediated innate immune response. Mol. Cell 67 (3), 387–399.

Nguyen, Q., Carninci, P., 2016. Expression specificity of disease-associated lncRNAs: toward personalized medicine. Curr. Top. Microbiol. Immunol. 394, 237–258.

Nishitsuji, H., Ujino, S., Yoshio, S., Sugiyama, M., Mizokami, M., Kanto, T., et al., 2016. Long noncoding RNA# 32 contributes to antiviral responses by controlling interferon-stimulated gene expression. Proc. Natl. Acad. Sci. U.S.A. 113 (37), 10388–10393.

Ouyang, J., Zhu, X., Chen, Y., Wei, H., Chen, Q., Chi, X., et al., 2014. NRAV, a long non-coding RNA, modulates antiviral responses through suppression of interferon-stimulated gene transcription. Cell Host Microbe 16 (5), 616–626.

O'Grady, T., Cao, S., Strong, M.J., Concha, M., Wang, X., Splinter BonDurant, S., et al., 2013. Global bidirectional transcription of the Epstein–Barr virus genome during reactivation. J. Virol. 88 (3), 1604–1616.

Paliwal, D., Joshi, P., Panda, S.K., 2017. Hepatitis E virus (HEV) egress: role of BST2 (Tetherin) and interferon induced long non-coding RNA (lncRNA) BISPR. PLoS One 12 (11), e0187334.

Peng, X., Gralinski, L., Armour, C.D., Ferris, M.T., Thomas, M.J., Proll, S., et al., 2010. Unique signatures of long noncoding RNA expression in response to virus infection and altered innate immune signaling. MBio 1 (5), e00206–e00210.

Postler, T.S., Pantry, S.N., Desrosiers, R.C., Ghosh, S., 2017. Identification and characterization of a long non-coding RNA up regulated during HIV-1 infection. Virology 511, 30–39.

Saayman, S., Ackley, A., Turner, A.M.W., Famiglietti, M., Bosque, A., Clemson, M., et al., 2014. An HIV-encoded antisense long noncoding RNA epigenetically regulates viral transcription. Mol. Ther. 22 (6), 1164–1175.

Sabry, D., Elamir, A., Mahmoud, R.H., Abdelaziz, A.A., Fathy, W., 2017. Role of LncRNA-AF085935, IL-10 and IL-17 in rheumatoid arthritis patients with chronic hepatitis C. J. Clin. Med. Res. 9 (5), 416–425.

Takayama, K., Inoue, S., 2015. The role of androgen-regulated long noncoding RNAs in prostate cancer. Long Noncoding RNAs 191–210.

Tang, W.-G., Hu, B., Sun, H.-X., Sun, Q.-M., Sun, C., Fu, P.-Y., et al., 2017. Long non-coding RNA00364 represses hepatocellular carcinoma cell proliferation via modulating p-STAT3-IFIT2 signaling axis. Oncotarget 8 (60), 102006–102019.

Trypsteen, W., Mohammadi, P., Van Hecke, C., Mestdagh, P., Lefever, S., Saeys, Y., et al., 2016. Differential expression of lncRNAs during the HIV replication cycle: an underestimated layer in the HIV-host interplay. Sci. Rep. 6, 36111.

Wang, X.-J., He, C., Jiang, S.-C., Deng, S.-Q., Peng, H.-J., Wei, H.-X., 2017. The differential expression and possible function of long noncoding RNAs in liver cells infected by dengue virus. Am. J. Trop. Med. Hyg. 97 (6), 1904–1912.

Wang, Y., Chen, L., Chen, B., Li, X., Kang, J., Fan, K., et al., 2013. Mammalian ncRNA-disease repository: a global view of ncRNA-mediated disease network. Cell Death Dis. 4 (8), e765.

Winterling, C., Koch, M., Koeppel, M., Garcia-Alcalde, F., Karlas, A., Meyer, T.F., 2014. Evidence for a crucial role of a host non-coding RNA in influenza A virus replication. RNA Biol. 11 (1), 66–75.

Xiong, Y., Yuan, J., Zhang, C., Zhu, Y., Kuang, X., Lan, L., et al., 2015. The STAT3-regulated long non-coding RNA Lethe promote the HCV replication. Biomed. Pharmacother. 72, 165–171.

Yin, Z., Guan, D., Fan, Q., Su, J., Zheng, W., Ma, W., et al., 2013. LncRNA expression signatures in response to enterovirus 71 infection. Biochem. Biophys. Res. Commun. 430 (2), 629–633.

Zapata, J.C., Campilongo, F., Barclay, R.A., DeMarino, C., Iglesias-Ussel, M.D., Kashanchi, F., et al., 2017. The Human Immunodeficiency Virus 1 ASP RNA promotes viral latency by recruiting the polycomb repressor complex 2 and promoting nucleosome assembly. Virology 506, 34–44.

Zhang, C., Yang, X., Qi, Q., Gao, Y., Wei, Q., Han, S., 2017. lncRNA-HEIH in serum and exosomes as a potential biomarker in the HCV-related hepatocellular carcinoma. Cancer Biomark. 21, 1–9.

Zhang, H., Zhu, C., Zhao, Y., Li, M., Wu, L., Yang, X., et al., 2015. Long non-coding RNA expression profiles of hepatitis C virus-related dysplasia and hepatocellular carcinoma. Oncotarget 6 (41), 43770–43778.

Zhang, Q., Chen, C.Y., Yedavalli, V.S., Jeang, K.T., 2013. NEAT1 long noncoding RNA and paraspeckle bodies modulate HIV-1 posttranscriptional expression. MBio 4 (1), e00596-12.

Zhang, Q., Lai, M.-M., Lou, Y.-Y., Guo, B.-H., Wang, H.-Y., Zheng, X.-Q., 2016a. Transcriptome altered by latent human cytomegalovirus infection on THP-1 cells using RNA-seq. Gene 594 (1), 144–150.

Zhang, Q., Matsuura, K., Kleiner, D.E., Zamboni, F., Alter, H.J., Farci, P., 2016b. Analysis of long noncoding RNA expression in hepatocellular carcinoma of different viral etiology. J. Transl. Med. 14 (1), 328.

CHAPTER

7

Scientific Advances in the Diagnosis of Emerging and Reemerging Viral Human Pathogens

Rahma Ait Hammou[1], *Mustapha Benhassou*[1,2,3], *Hlima Bessi*[1] *and Moulay Mustapha Ennaji*[1]

[1]Laboratory of Virology, Microbiology, Quality, Biotechnologies/Eco-Toxicology and Biodiversity, Faculty of Sciences and Techniques, Mohammedia, University Hassan II of Casablanca, Casablanca, Morocco [2]Mohamed VI University of Health Sciences of Casablanca, Casablanca, Morocco [3]School of Medicine and Pharmacy of Casablanca, University Hassan II of Casablanca, Casablanca, Morocco

ABBREVIATIONS

CDC Centers for Disease Control and Prevention
DNA deoxyribonucleic acid
EBOV Ebola virus
EVD Ebola virus disease
FDA Food and Drug Administration
HBV hepatitis B virus
HCV hepatitis C virus
HIV human immunodeficiency virus
HSV herpes simplex virus

Emerging and Reemerging Viral Pathogens
DOI: https://doi.org/10.1016/B978-0-12-814966-9.00007-X

ICU	intensive care unit
LRN	Laboratory Response Network
miRNA	microRNAs
PCR	polymerase chain reaction
RNA	ribonuleic acid
RT-PCR	reverse transcription-polymerase chain reaction
TB	tuberculosis
VZV	varicella-zoster virus

CLINICAL APPLICATION OF MOLECULAR METHODS

The diagnosis of any infectious disease requires active communication between clinicians and clinical laboratory personnel, usually in the form of "consultations" from clinicians to the laboratory director "on call" or directly with the laboratory technologists. Arriving at a correct diagnosis in a timely fashion begins with the acquisition of an adequate clinical specimen (blood, cerebrospinal fluid, urine, respiratory secretions, bronchoalveolar lavage, feces, etc.) and its transportation to the laboratory in an appropriate container. Specimens for molecular diagnostic tests sometimes require different methods of transportation/ preservation from those of regular specimens so as to ensure that the techniques, used efficiently, will detect the suspected infectious agent. The classic diagnostic principles in clinical medicine still apply when using molecular methods of diagnosis. Patients and populations are evaluated clinically and epidemiologically, and case definitions are created so that test parameters, such as sensitivity, specificity, and predictive values, can be defined. Likewise, understanding the concept of screening versus confirmatory testing is also important. The three major steps performed in molecular assays are specimen processing, nucleic acid amplification or hybridization, and product detection (Nolte and Caliendo, 2007; Persing et al., 2004). Processing of the specimen is one of the most important steps for the successful detection of nucleic acids. Protocols vary, depending on the specimen received, and the specimen "matrix" plays an important role in nucleic acid extraction, owing to the presence of extraneous material in some matrices, that could potentially interfere with the molecular assays being performed. The most widely known example of such interference is the presence of *Taq* polymerase inhibitors in nucleic acid extracts, leading to false-negative results. Nucleic acid extraction protocols are usually cumbersome, and automation has been relatively challenging. However, several instruments have been designed for the clinical laboratory in which automated extraction protocols have been incorporated. Automation has also been incorporated for nucleic acid amplification methods, including conventional and real-time polymerase chain reaction (PCR), branched DNA, and isothermal technologies that obviate the usual thermal cycling and potentially speed up the time required for amplification. Automated platforms, however, have decreased the cycling time by changing more quickly the temperatures required for different PCR cycles. Real-time PCR systems are usually coupled with automated detection systems that also vary in complexity. A few instruments are also capable of complete automation, including extraction, nucleic acid amplification, and detection systems. Tables 7.1 and 7.2 present a summary of the techniques available for the detection of nucleic acids from infectious agents.

TABLE 7.1 Classification of Nucleic Acid–Based Molecular Testing

Type	Advantages	Disadvantages
NONAMPLIFICATION TECHNIQUES		
Nucleic acid	Simple to perform	Low analytic sensitivity
Peptide nucleic acid probes	Bind to target more avidly	Low analytic sensitivity
SIGNAL AMPLIFICATION		
Hybrid capture assays	Simple to perform	Insensitive
Branched-DNA assays	Quantitative or qualitative—no risk of amplicon contamination	Potentially less sensitive than target amplification assays
TARGET AMPLIFICATION		
PCR and variants	High sensitivity, specificity depends on primers and amplicon contamination	Risk of amplicon contamination. Need thermocyclers
Reverse transcriptase–PCR	RNA targets susceptible to ribonucleases Multiple detection of pathogens from single specimen	Requires thermostable DNA polymerase Primer design is critical
PCR multiplex	Powerful technology when coupled with liquid-based microarray beads (Luminex, Toronto, Ontario, Canada) Highly sensitive and specific	Potentially lower sensitivity owing to competition for reagents in the PCR mixture High risk of amplicon contamination if transfer occurs
Nested PCR	Target amplification and detection in the same tube	Technically complex

(Continued)

TABLE 7.1 (Continued)

Type	Advantages	Disadvantages
Real-time (kinetic) PCR	Multiple detection systems: SYBER Green, FRET probes, dual hybridization probes, molecular beacons, dark quencher probes Low risk of amplicon contamination Decreased time of analysis	Probe design is critical for success Limiting multiplexing capabilities
Transcription-based amplification methods	Isothermal, rapid kinetics, RNA product single stranded: no denaturation for detection	Poor performance for DNA targets
Nucleic acid sequence–based amplification	No amplicon contamination (labile RNA) Isothermal, rapid kinetics, RNA product single stranded: no denaturation for detection No amplicon contamination (labile RNA)	Multienzyme system (stability) Poor performance for DNA targets
Transcription-mediated amplification	Isothermal. Fast amplification time	Multienzyme system (stability) Nonspecific primer hybridization owing to low temperature used
Strand displacement amplification		
PROBE AMPLIFICATION		
Cleavase invader assays (Hologic Inc. Third Wave Technologies, Bedford, MA)	Detection of point mutations based on primer design	Less prone to amplicon contamination
Cycling probe assays[a]		
Ligase chain reaction[a]		

[a]*Not available in the United States.*
DNA, Deoxyribonucleic acid; *PCR*, polymerase chain reaction.

TABLE 7.2 Detection and Analysis Platforms for Amplification Tests

Agarose and polyacrylamide gels
Colorimetric microtiter plates
Conventional hybridization
Nucleic acid sequencing • Conventional • Pyrosequencing • CLIP (integrated amplification and sequencing)
Hybridization microarrays
Bead-based flow cytometric assays
Electrospray ionization and mass spectrometry

FIRST-TIME DIAGNOSIS

Fortunately, the biotechnology boom of the late 1990s and early 2000s fueled the development of highly automated nucleic acid–based testing methods, which had important implications for the identification of infectious pathogens in human specimens (Monecke and Ehricht, 2005). One of these technologies, commonly referred to as real-time PCR, has gained considerable popularity. This method combines nucleic acid amplification and fluorescent detection of the amplified product in the same closed system (Morse, 1995; Relman et al., 1990; Hjelle et al., 1994; Aragon et al., 2006). The promulgation of real-time PCR as an important testing platform in clinical microbiology was catapulted by US homeland security efforts to produce rapid, reliable testing methods for identifying potential agents of bioterrorism. The Laboratory Response Network (LRN), an integrated group of public health, armed forces, and private referral laboratories, was created by the Centers for Disease Control and Prevention (CDC) to serve as a reference laboratory network for identifying and confirming agents of bioterrorism. In a very short period, scientists at CDC successfully developed a number of real-time PCR assays for the detection of agents of bioterrorism, and these assays are now available at many of the LRN laboratories.

Therefore molecular tools have improved the initial diagnosis of emerging and reemerging infectious diseases especially for pathogens that are uncultivated or difficult to isolate in clinical laboratories (Nolte and Caliendo, 2007; Persing et al., 2004). In addition, conventional diagnostic tests for infectious agents are slow, sensitive, expensive, or unavailable. Therefore since 1990 the identification of several pathogens has been obtained with molecular techniques that have contributed primarily to the diagnosis of these infectious agents (Persing et al., 2004; Relman, 1998). Nucleic acid–based assays are now gold standards for several infectious agents, including hepatitis C virus (HCV), *enterovirus*, *Bordetella pertussis*, and *herpes simplex* virus (HSV) (in the context of herpes encephalitis, *Chlamydia trachomatis*, and others). Cost-effectiveness is undoubtedly an extremely important factor when introducing molecular diagnostics into the clinical laboratory to replace conventional techniques. The best example is that of common bacterial pathogens, some of which require, in addition to identification, sensitivity tests to guide clinicians in choosing the most effective antibiotic.

One of the most promising platforms in clinical microbiology laboratories is that of molecular diagnostics using multiplexing technology (Loeffelholz, 2004; Smithn et al., 1998). The ability of real-time PCR to amplify and detect the product to be amplified, using specific probes at the same time, allows for multiple amplifications with the same clinical sample. The design of primers is the most critical step in multiplex PCR so that efficiency and specific amplification are not compromised. Real-time multiplex coupling PCR with liquid-stained microarrays provides the multiplex detection of pathogens from clinical syndromes. FDA-approved diagnostic products are already available for viral respiratory pathogens.

We could imagine the use of multiplex systems to diagnose enteric pathogens, nosocomial infections, including identification of genes responsible for antibiotic resistance (according to the hospital's epidemiological data), pathogens of the central nervous system, hemorrhagic fever, and sepsis. We could also consider the inclusion of emerging and reemerging infectious agents in some of the syndrome-based diagnostic products to establish true incidence and prevalence of these highly neglected infectious diseases.

Furthermore, numerous reports have described the utility of this user-friendly technology for the rapid (same-day) and accurate detection of many emerging (new) and reemerging pathogens as well as pathogens commonly encountered in medical practice. A search for all articles published in the *Journal of Clinical Microbiology* from 2000 through 2003, which evaluated real-time PCR as a test method for pathogen detection and/or identification of genes or mutations associated with antimicrobial resistance in pathogens, revealed in 109 articles. Among these articles, 84 described assays with the LightCycler instrument (Roche Diagnostics Corporation, Indianapolis, IN); 21 described assays with the ABI PRISM 7000, 7700, or 7900H instrument (Applied Biosystems, Foster City, CA); 2 described assays with the SmartCycler instrument (Cepheid, Sunnyvale, CA); and 2 described assays with the iCycler instrument (Bio-Rad Laboratories, Hercules, CA). The availability of nucleic acid–based technology, such as real-time PCR, along with conventional staining and culture methods and immunoassays, can provide laboratories of many sizes with a comprehensive and responsible approach for the detection of both commonly encountered and emerging or reemerging pathogens.

EXAMPLES OF MOLECULAR DETECTION OF VIRAL EMERGING AND REEMERGING DISEASES

Variola Virus

The clinical presentations of concerned patients infected with common viruses that cause cutaneous vesicular lesions [HSV, varicella-zoster virus (VZV), enterovirus, or disseminated vaccinia virus following smallpox vaccination] might mimic those of patients with smallpox. Another complicating feature is that some recipients of the smallpox vaccine may develop erythema multiforme, which can also present as vesicular lesions. It has been our experience that by using real-time PCR assays, one can rapidly discriminate among these possibilities. Analyte specific reagents (ASRs) or kits for the detection of HSV or VZV with the LightCycler instrument are available from at least two vendors [RealArt HSV 1/2 kit and RealArt VZV PCR kit (Artus); LightCycler Herpes Simplex Virus 1/2 and LightCycler VZV ORF29 (Roche Diagnostics Corporation)]. Kits are also available for testing for VZV (RealArt VZV PCR kit; Artus) with the ABI PRISM 7000, 7700, and 7900H instruments. Researchers have used the assays with the Roche LightCycler instrument to routinely detect HSV and VZV and have developed an in-house real-time PCR assay for poxviruses, including variola virus that uses the LightCycler instrument. These assays have been invaluable for providing a rapid result, especially for military personnel who have developed cutaneous vesicular lesions as a complication of receiving the smallpox vaccine and who have been on assignment in areas of the world at significant risk for bioterrorism events. Importantly, the home-brewed real-time assay that they have developed can discriminate among several poxviruses and was useful in the identification of viremia in a recent case of monkey poxvirus disease in a patient from the Upper Midwest. An ASR for the detection of variola virus with the LightCycler instrument is also available from Artus (RealArt Orthopox PCR kit).

West Nile Virus

West Nile virus, a RNA virus of the family Flaviviridae, has a predilection for the central nervous system and can be associated with significant morbidity and mortality. The first human cases of West Nile virus infection occurred in the northeastern United States in the summer of 1999; since then the disease has progressed relentlessly from east to west across the continental United States. Yet, no effective therapy has been defined.

Traditionally, during the summer and early fall in the United States, viral central nervous system disease is most frequently caused by enterovirus. In most regions of the United States, West Nile virus infection must now also be considered during this time of the year. HSV can cause encephalitis at any time of the year, and antiviral therapy is available and effective. Therefore ruling out HSV infection should be a priority, especially when encephalitis is encountered. Real-time PCR has replaced viral culture as the gold standard for the rapid and accurate detection of HSV in cerebrospinal fluid. As mentioned previously, ASRs or kits for the detection of HSV are available from Artus and Roche. Artus also has a kit that can be used to test for enterovirus (RealArt Enterovirus RT-PCR kit) with the LightCycler instrument.

Limited studies have shown that the PCR detection of West Nile virus in cerebrospinal fluid is less sensitive than immunoassay for immunoglobulin M antibodies. Currently, only a few referral and public health laboratories have the capability to perform immunoassays. At least two companies offer ASRs or kits for real-time PCR [RealArt WNV RT-PCR kit (Artus); LightCycler WNV Detection Kit (Roche Applied Science)] with the LightCycler platform. If effective antiviral therapy becomes available, the rapid on-site diagnosis of West Nile virus disease in areas of endemicity may be desirable.

SARS CoV

One important lesson learned from the 2002–03 winter outbreaks of severe acute respiratory syndrome (SARS) was that the early identification and quarantine of individuals with suspected cases of SARS were essential for controlling the disease, especially in institutional settings. This effective approach toward the control of a communicable infectious disease adds credence to the concept that similar measures can be effective for controlling and preventing nosocomial VRE and methicillin-resistant *Staphylococcus aureus* (MRSA) outbreaks. No laboratory tests were available for the detection of SARS coronavirus (SARS CoV) during much of the outbreak, as the etiological agent was not confirmed until early March 2003. Eventually, real-time PCR tests were developed and were available commercially from at least two manufacturers for use with several real-time PCR testing platforms [RealArt HPA-Coronavirus RT-PCR Kits (Artus) for use with the LightCyler instrument, the ABI PRISM 7000, 7700, and 7900H instruments, and the Rotor-Gene instrument; and LightCycler SARS CoV (Roche Diagnostics Corporation) for use with the LightCycler instrument]. During the outbreak, it was important to rule out treatable influenza virus types A or B infections, whose clinical presentations can mimic those of SARS CoV.

Rapid antigen tests for the detection of influenza virus (both types A and B) are relatively easy to perform and may be useful in the local setting for the detection of cases of influenza; however, these tests lack sensitivity. As infections due to both influenza virus types A and B are now treatable, rapid on-site diagnostic capabilities are important. Recently, a real-time PCR assay that uses the LightCycler platform was demonstrated to have much greater sensitivity than antigen detection (100% and 44%, respectively) for the detection of influenza virus type A infections.

Following the 2002–03 SARS outbreak, many LRN member laboratories developed the capability to detect SARS CoV. Should another outbreak occur, this public health laboratory network should facilitate the laboratory diagnosis of cases, especially when testing at the local level is not available.

DISEASE PROGNOSIS

The prognosis of infectious diseases has also been significantly affected by the availability of molecular techniques. Viral load in several chronic infections, such as human immunodeficiency virus (HIV), human papillomavirus, hepatitis C, BK virus, and cytomegalovirus, plays an important role in predicting better or worse outcomes and the need to start specific antiviral therapy to improve survival (Humar et al., 1999; Mellors et al., 1996; Reidn et al., 1987; Walsh et al., 1997). Viral and bacterial quantification by molecular methods could also be used in the future to differentiate a mild subclinical infection from a disease caused by the agent under certain conditions.

DIAGNOSIS BY MICROARRAYS

A microarray is a "collection of microscopic features," usually DNA, which can be probed with target molecules to produce quantitative or qualitative data (Miller and Tang, 2009). Indeed, the diagnosis of infectious diseases is most often qualitative (presence or absence of the pathogen), although the quantitative data become more numerous under certain conditions, such as the HIV/immunodeficiency syndrome acquired and hepatitis C.

The microarray platforms can be classified as follows: printed, synthesized in situ, high-density bead networks, electronic suspension, and liquid bead (Loeffelholz, 2004; Miller and Tang, 2009). All microarray platforms have the ability to multiplex, a characteristic that becomes indispensable in the diagnosis of infectious diseases.

For all approaches the probe is the DNA sequence bound to the solid surface of the array, and the target is the nucleic acid to be detected. The probes are synthesized and immobilized on microscopic spots.

It has been reported that of all platforms, suspension beads probably offer the most practical technology to clinical microbiology labs (Loeffelholz, 2004; Miller and Tang, 2009). Universal heel sets are available, and user-defined applications are easily implemented because of their flexibility. Therefore the use of wide-range or multiplex PCR followed by microarrays provides an excellent platform for the rapid and efficient identification of bacterial, viral, or fungal pathogens (Mikhailovich et al., 2008).

For the diagnosis of bacterial infections, microarrays are also incorporated in clinical laboratories for the rapid detection and characterization of MRSA, determination of antimicrobial drug resistance in several pathogens, such as *Enterococcus* and *Mycobacterium tuberculosis*, and the diagnosis of sepsis (Mancini et al., 2010; Aragon et al., 2006; Monecke and Ehricht, 2005). Several difficult problems remain before the microarrays can be widely used in the clinical setting, including pre- and postanalytical variables, such as the type of clinical specimens (matrices), extraction techniques, labeling, and quality controls (Miller and Tang, 2009). The introduction into microbiology laboratories of additional diagnostic tests based on molecular technology will depend heavily on automation. Many molecular tests remain difficult because of complex and lengthy problems of processing and interpretation of samples.

MOLECULAR DIAGNOSIS AND LATEST GENERATION SURVEILLANCE SYSTEMS

Biosurveillance has become an important health priority because of the increased risk of emerging and reemerging pathologies and their effects on human populations (Sintchenko and Gallego, 2009). The potential consequences of large-scale epidemics or epidemics can be economically, socially, and environmentally profound, as evidenced by the global spread of coronavirus associated with severe acute respiratory syndrome. Traditional biosurveillance relies heavily on disease reporting by clinical laboratories to local, state, and federal health-care organizations (Sintchenko and Gallego, 2009).

Most clinically diagnosed cases are confirmed using standard microbiological methods, such as serologic markers and cultures, followed by biochemical identification, if necessary, or DNA-based typing methods, once a pure isolate is obtained. As a result, these methods cannot be described as slow and insensitive. Traditional biosurveillance has improved through electronic reporting to some extent, with an increase in the proportion of diagnosed cases reported in all cases occurring in the community (Panackal et al., 2002; Effler et al., 1999).

On the other hand, other factors include variability in case definitions (clinical and laboratory criteria), screening practices, contact tracking methods, and the quality of performance of the diagnostic tests used.

The second surveillance system is said to be syndromic and is based on the collection of vast amount of data from the various services, including emergency departments, intensive care units (ICUs), admission and hospitalization systems, and clinical laboratories (Sintchenko and Gallego, 2009; Bravata et al., 2004; Lewis et al., 2002). Indeed, the purpose of this system is to play the role of an alarm signal when the models obtained over time move away from the "normal" state. The normal state is determined by the collected data and by creating "baselines" for a given area or community (Wang et al., 2005; Berger et al., 2006).

As expected, the specificity of these systems is rather low but can be improved by combining syndromic surveillance systems and laboratory surveillance (Sintchenko and Gallego, 2009; Buehler et al., 2004; Weber and Pitrak, 2003).

The latest generation of surveillance is based on the genomics of surveillance systems and is essentially based on the development of rapid molecular tests applied to the diagnosis of infectious diseases. A technique based on molecular subtyping has increased sensitivity and specificity in the investigation of epidemics at all levels (hospital, local, state, and national) (Hedberg and Besser, 2006; Monecke et al., 2007; Mellmann et al., 2006). Similarly, modern typing methods, based on direct data, such as sequencing and powerful software, to analyze gene clusters and genetic evolution, have also made it possible to detect possible epidemics by new pathogens or reemerging agents (Sintchenko and Gallego, 2009). Typical examples include surveillance systems for influenza A viruses and severe acute respiratory syndrome. Powerful molecular techniques cannot be used in isolation. Modern biomonitoring systems are important elements, including traditional epidemiological tools and newly developed "computer-based" systems in which complete integration of microbial profiling is merged with epidemiological surveillance and spatial surveillance (depending on the use of systems geographical information). Such a "network" or "global laboratory" would provide an exquisitely comprehensive view of potential biotherapies (Urwin and Maiden, 2003; Casadevall and Relman, 2010; Reis et al., 2007; O'Connor et al., 2003; Layne and Beugelsdijk, 1998; Heymann and Rodier, 2001).

NANOTECHNOLOGY DIAGNOSIS

Nanotechnology is a broad term that encompasses several disciplines and techniques, some of which currently have or will have a major impact on health care. For the purpose of this review, nanobiotechnology and its applications in molecular diagnostics are discussed. Nanotechnology is the study of the control and manipulation of matter at the atomic and molecular scales (Johnson et al., 2008), the materials used are 100 nm or less in at least one of the dimensions of the material. On the other hand, molecular diagnosis is an essential element in the development of personalized medicine, thus presenting punctual performance of diagnostic procedures. This report focuses on the application of several technologies in the clinical laboratory setting. The

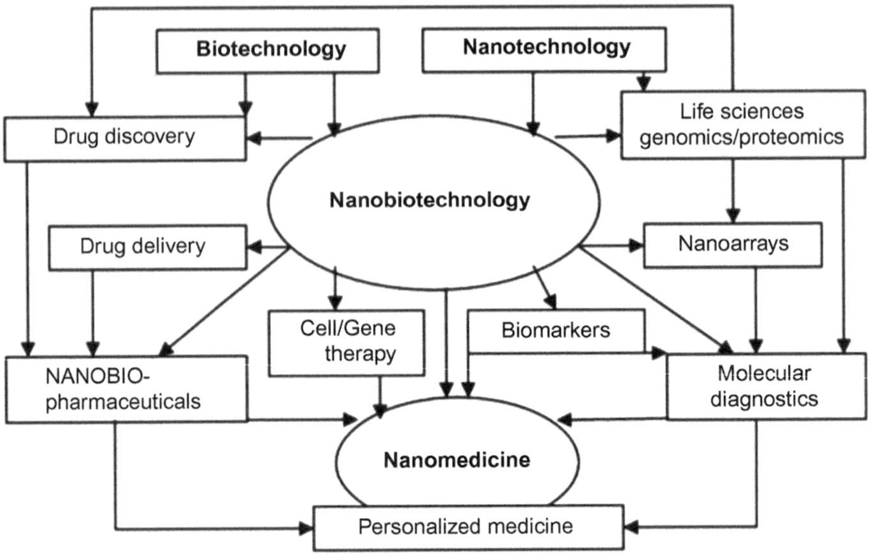

FIGURE 7.1 **The interdependence of various technologies that contribute to clinical nanodiagnosis.**

different interrelationships between nanotechnology and molecular diagnostics and their role in nanomedicine and personalized medicine are shown in Fig. 7.1.

Array Based on Nanotechnology

These are nanofluidic networks for the isolation and analysis of biomolecules, such as nucleic acids or proteins. The networks use nanotubes for the isolation of molecules and the detection of molecules trapped with electrode-based systems. The volumes required for such tests are very small, compared to those required with conventional instruments, and turnaround times are significantly reduced (Jain, 2007).

Whenever a single molecule of DNA moves in the nanotube, the electric current changes abruptly. The current returns to its reference value when the DNA molecule exits the nanotube. Nanofluidic technology is expected to have wide applications in systems biology, personalized medicine, pathogen detection, drug development, and clinical research.

Nanoparticles

Gold nanoparticles, nanocrystals (also called quantum dots) and magnetic nanoparticles are the main examples in this category. Gold nanoparticles range from 0.8 to 250 nm and can be functionalized with a variety of biomolecules including antibodies, nucleic acids, peptides, proteins, and carbohydrates. Thus gold nanoparticles have better optical absorption and scattering properties because of a unique property known as plasma resonance. Like nanocrystals (see the next section), their optical properties can be modified by changing size, shape, and composition.

On the other hand the surface of nanocrystals can also be functionalized with nucleic acids, antibodies, proteins, and peptides, allowing adaptation to several diagnostic platforms (Bruchez et al., 1998; Hotz, 2005; Michalet et al., 2005). The magnetic nanoparticles are made of iron and incorporated into copolymer beads. The surface charge of the beads can be manipulated with the polymer coating to enhance nonspecific protein adsorption to the surface of the beads, increasing the specificity of the assay (Johnson et al., 2008; Jain, 2007).

Nanobiosensors

Nanobiosensors are nanosensors that are allowed for the detection of chemical or biological materials. These materials are extremely sensitive (Jain, 2003). Prototype sensors have demonstrated the detection of nucleic acids, proteins, and ions.

These sensors can operate in liquid or gaseous phase, which allows them to be used for different downstream applications. These sensors are inexpensive in their manufacture and are portable. Thus they can serve even as bases for the implementation of implantable devices for detection and monitoring.

Cantilever Biosensors (Cantilever)

This technology provides an alternative approach to PCR and complements current DNA and protein microarray methods. Using this method, it is not necessary to label or copy the target molecules. The advantages of cantilevers are that they provide fast, unlabeled recognition of specific DNA sequences for single-nucleotide polymorphisms, oncogenes, and genotyping. Nanocantilevers could be crucial in the design of a new class of ultrasonic sensors to detect viruses, bacteria, and other pathogens (Gupta et al., 2006). Finally, a real-time cantilever biosensor can provide continuous monitoring of clinical parameters in personalized medicine.

Viral Nanobiosensors

The virus particles are biological nanoparticles. Giving examples of HSV and adenovirus that can be used to trigger the assembly of nanomagnetic beads as nanobiosensors for clinically relevant viruses

(Perez et al., 2003). Thus this technique offers the possibility of detecting up to five virus particles in a 10 mL serum sample. This system has the advantage of having an increased sensitivity compared to the enzyme-linked immunosorbent assay (ELISA)-based methods as well as an improvement over the PCR-based detection because it is less expensive and faster and has fewer artifacts.

MICRORNAS AND EMERGING VIRAL INFECTIONS

History of MicroRNA

MicroRNAs (MiRNAs) represent a recently uncovered class of small noncoding RNAs from 20 to 22 nucleotides that function as posttranscriptional regulators of gene expression. These miRNAs are coded by separate genes that are localized in the nonprotein coding part of the genome. MicroRNAs predict to regulate about 30%–60% of human genes (Griffiths-Jones et al., 2008; Malumbres, 2013).

These small noncoding RNAs bind their target messenger RNAs (mRNAs) in the untranslated region (UTR) and coding sequence (CDS) regions and act as negative regulators. This action affects various biological processes including cell growth, differentiation, signal transduction, metabolism, and development (Inui et al., 2010).

Dysregulation of miRNA expression has been described in several diseases, and their role appears to be pivotal in driving tumorigenesis (Iorio and Croce, 2012). MiRNA dysregulation is considered to be an early event in tumorigenesis (Cortez et al., 2011).

It was discovered in 1993 in the *Caenorhabditis elegans* nematode of a DNA fragment containing a small sequence that encodes a 22-nucleotide RNA, Lin4 that regulates the transition from larval stage L1 to stage L2.

This regulation is achieved by repressing the expression of the proteins LIN14 and LIN28 by this miRNA by binding to the 3′ noncoding regions (3′-UTR) of the corresponding mRNAs.

In 2000 another miRNA Let7 was identified, which regulates the transition from larval stage L4 to adult stage by the same mechanism of action. Since then, several discoveries have been made and have demonstrated the conservation of these miRNAs in several species. Studies have revealed that more than 500 miRNAs can potentially target nearly 30% of mRNAs in humans (Friedman et al., 2009).

Structure and Biogenesis

In the nucleus the gene of miRNA is transcribed into a pri-miRNA by the action of a RNA polymerase type II or III (Borchert et al., 2006;

Yoontae et al., 2003). Pri-miRNA will then be cleaved by the nuclear microprocessor complex formed by the association of the enzyme Drosha and the DGCR8 (DiGeorge critical region 8) protein into an intermediate precursor called a pre-miRNA (Han et al., 2004).

Once the complex is attached to primer-miRNA, the DGCR8 protein will allow to define the cleavage distance that generally measures 11 base pairs from this junction, while the two Drosha domains will cleave the 3′ and 5′ ends of the pri-miRNA. The resulting pre-miRNA is a loop stem having two unmatched nucleotides at the 3′ end and a phosphate at the 5′ end, this asymmetry is specific for type III RNAses. The pre-miRNA will migrate to the cytoplasm by the exportin 5 (Huang et al., 2011; Garza, 2011; Sotillo and Thomas-Tikhonenko, 2011).

In the cytoplasm the pre-miRNA undergoes another stage of maturation where the RLC (RISC loading complex) is composed by RNaseDICER acytoplasmic endonuclease that performed the second stage of processing, which includes a leading strand or miR and a passenger strand or miR*.

For miRNAs with a high degree of complementarity throughout the loop stem, an additional cleavage at the middle of the strand is effected by the triteness activity of the Ago2 protein before that cleaved by the enzyme dicer in order to generate an ac-pre-miRNA (Ago2-cleaved precursor miRNA) (Eulalio et al., 2008).

Then, the resulting miRNA will form a RISC complex, one of the two strands called the "passenger strand" is degraded, whereas the strand having the least stable 5′ end called the "guide strand" is retained—it is the mature miRNA. The evidences have shown that the strand with more unstable 50 (weaker base pairing) has lower chance of degradation (Sotillo and Thomas-Tikhonenko, 2011) (Fig. 7.2).

MicroRNAs in Emerging Disease

Although new established methods such as reverse transcription quantitative PCR (RT-qPCR) have been used to detect viral infections, there is still a lack of robust biomarkers for early diagnosis and prognosis of the infectious disease. Increasing evidence indicates that cell free miRNAs are present in body fluids including blood and saliva. They are produced endogenously in response to the molecular change in cells, and therefore they can be used as diagnostic reporters for various diseases such as cancer and viral infections (Zhu et al., 2014). MiRNA is a class of small (18–24 nucleotides in length) noncoding RNAs, which involves in gene regulation and plays important roles in cell proliferation, differentiation, apoptosis, and tumorigenesis (Kaladhar, 2015). From the previous reports, miRNAs have distinct expression profiles in virus-infected cells in comparison to their healthy counterparts

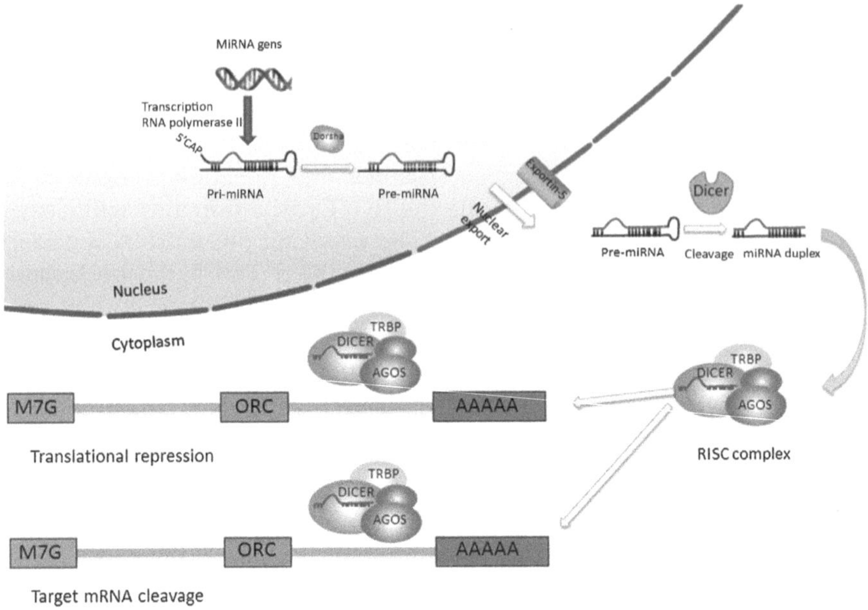

FIGURE 7.2 An overview of the miRNA biogenesis and its functional mechanism. RNA polymerase II produces a 500–3000 nucleotide transcript, called pri-miRNA, that is mature to pre-miRNA hairpin by a DROSHA (_60–100 nucleotides). This double-stranded hairpin structure is exported from the nucleus into cytoplasm by exportin 5 (RAN GTPase). Lastly, the pre-miRNA is processed by DICER1 and produces sense and antisense strands, approximately 20 nucleotides in length, the effective strand called antisense and known as mature miRNA and short-lived complementary sequence called passenger strand (miR*). The antisense-stranded miRNA is combined into RISC, which then targets it to the target 30 untranslated region mRNA sequence to facilitate repression and cleavage. *AA*, Poly A tail; *m7G*, 7-methylguanosine cap; *miRNAs*, microRNA; *ORF*, open reading frame; *RNA*, ribonucleic acid.

(Tambyah et al., 2013; Zhu et al., 2014). Moreover, the expression levels of many miRNAs in virus-infected samples and normal controls exhibit fold change difference (Song et al., 2010).

This studies show that in addition to regulating multiple processes, miRNAs themselves may be independent effectors of innate immunity by directly targeting viral transcripts. In vitro studies show miRNA target influenza; vesicular stomatitis virus, human T-cell leukemia virus 1; human papillomavirus; and enterovirus 71, and they inhibit viral replication (Fig. 7.3), which facilitates the clearance or potentiating viral latency of the pathogen (Bai and Nicot, 2015; Hen et al., 2014; Zheng et al., 2013). MiRNAs were also essential to show malaria transcripts translocation into the parasite (150). Indeed, due to their targeting of viral transcripts, miRNAs have the potential to partially dictate the cell tropism of a virus, the resistance of resting T-cells to human T-cell leukemia virus appears to be due to their expression of mir-28-3p

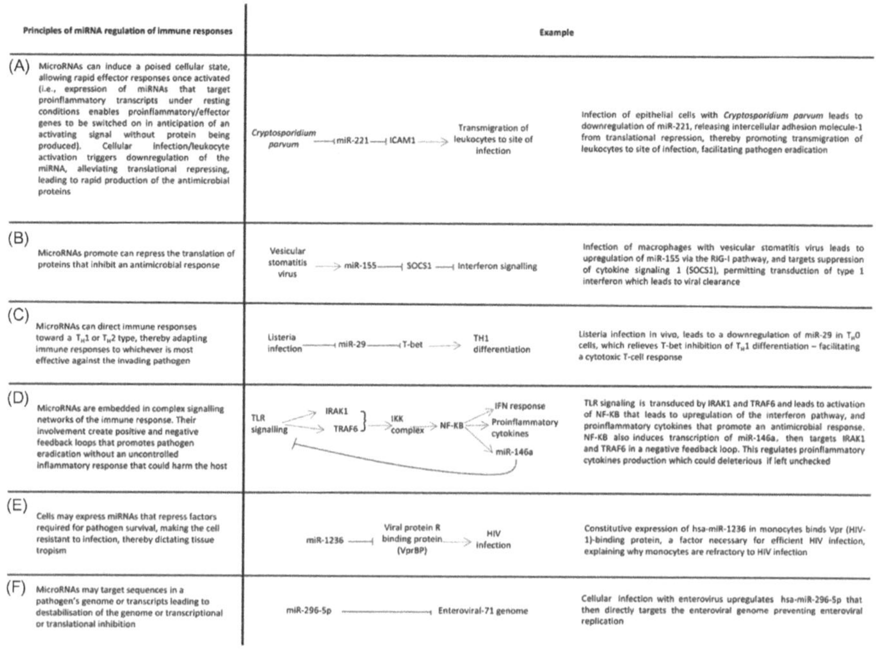

FIGURE 7.3 Key ways miRNAs mediate immune responses to pathogens. (A–F) A variety of ways in which miRNAs regulate immune responses. *miRNAs*, microRNAs.

(Bai and Nicot, 2015). Although a few scientists agree on the direct interaction between miRNAs and viral transcripts, Bogerd et al. (2014) argue that cellular miRNAs do not target viruses as global downregulation of host cell miRNAs (via DICER knockout) does not lead to the enhancement of 11 viruses in human embryonic kidney cell line 293T. However, Bogerd et al.'s model is problematic as viruses may be dependent on cell mechanisms that are controlled by miRNAs, and the usual host cell of the viruses in their study is not human embryonic kidney cells. Contrary to Bogerd et al.'s study, there is evidence that direct targeting of viral genome/transcripts occurs in vivo as several groups have successfully attenuated viral vaccines by incorporating human miRNA seed sites in viral genome (see The Clinical Applications of miRNAs: Improving Vaccines) (Barnes et al., 2008). The relative importance of miRNA direct targeting of viruses in innate immunity remains to be seen however as in vivo and in vitro studies show viral mutation of miRNA seed sites in viral genomes means viruses can quickly evolve to avoid being targeted by miRNAs (Zheng et al., 2013; Heiss et al., 2012).

There is another category of miRNAs, which are secreted from cells and called extracellular miRNAs (ex-miRNAs, circulating miRNAs).

This ex-miRNAs can be isolated from most biological fluids (de Candia et al., 2013; Irmak et al., 2012; Weber et al., 2010).

Often, we find these ex-miRNAs in extracellular vesicles (exosomes, microvesicles, and apoptotic bodies), and through their association with Argonaute protein (a component of the RISC complex—see Fig. 7.2) and high-density lipoprotein (de Candia et al., 2013; Arroyo et al., 2011; Vickers et al., 2011; Zernecke et al., 2009). Their biological function is debated since their secretion may be activated as intercellular communicators of gene regulation, or as cellular waste disposal method, or passively secreted as a by-product of cell death (Turchinovich and Cho, 2014). Although these three theories can be valid, there is increasing evidence that ex-miRNAs are functional, can be passed between leukocytes in vitro and in vivo, and play a role in disease (Alexander et al., 2015; Bell and Taylor, 2017; Lehmann et al., 2012; Mittelbrunn et al., 2011). Regardless of their functions, one clinical application of ex-miRNA is the use of these as biomarkers of infectious disease.

Furthermore, these miRNAs have shown the potential to be used as a biomarker for the prognosis and therapy of infectious diseases. miRNAs have also shown a significantly altered expression during infection. The altered expression of miRNA level in an infected human can be identified by the use of advanced diagnostic tools. In addition to their availability in numerous body fluids, miRNAs are highly stable in these fluids. These features make miRNAs, in single or in a combination (Peng et al., 2016), ideally suited as biomarkers for disease diagnosis.

Presently, only few standardized procedures are available for the isolation and characterization of specific miRNA. Experimental research and its observation have shown that small interfering RNA, premature miRNAs, and transfer RNA may interfere with specific miRNA during the process of isolation and characterization. Therefore this interference leads to the false-positive result, which should be taken care of during diagnosis. The necessity of the large amount of RNA input for the Northern blot technique can generate difficulties in the quantification of the miRNA. RT-qPCR, microarray profiling, and next-generation sequencing have been found to be useful for the identification of novel miRNA. An experienced researcher having a good knowledge of molecular biology as well as bioinformatics should do the characterization of miRNA.

Applications of MicroRNAs in the Treatment of Infectious Diseases

The immunomodulatory functions of miRNAs represent a promising application of miRNAs in the target of promoting antimicrobial pathways during infection and controlling dysregulated inflammatory responses during sepsis. In the lungs of mice infected with nontypeable *Haemophilus influenza*, Hock et al. noted that the physiological downregulation of miR-328-3p promotes phagocytosis by neutrophils and macrophages and bacterial killing and found that it boosts this downregulation by intratracheally administering an antagomiR of miR-328-3p enhanced bacterial killing when they made challenge with nontypeable *H. influenza*. In another case, Alexander et al. observed that mice inflammatory response to endotoxin in vivo was ameliorated and enhanced with administration of exosomes containing miR-146a and miR-155, respectively, which prompts the authors to conclude that such treatments could be useful adjuncts in managing sepsis (in the case of miR-146a) or vaccination (in the case of miR-155). Moreover, another work by Wang et al. (2015) supports the idea that miRNAs can be used in diseases therapy, especially sepsis, when they administrate mesenchymal stem cell exosomes containing miR-223, which confer cardiac protection in septic mice.

Overall, using miRNA-based therapies to leverage immune response may prove useful adjuncts to standard antimicrobial therapies, for example, in multidrug-resistant Gram-negative infections, or chronic viral infections such as hepatitis C. Nevertheless, there are significant challenges in implementing miRNA-based antimicrobial therapeutics, which include devising methods of administration, and drug design that will protect miRNA mimics/antagomiRs from circulating RNAse enzymes. Delivery systems have to ensure targeted efficient delivery of miRNAs to the site of infection, because, as noted earlier, a miRNA may appear in many cell types, serving very different functions making off-target effects a real possibility, limiting efficacy, and safety (Chen et al., 2015).

A detailed analysis of the outcomes of phase 1 trials of two miRNA-based cancer treatments will provide more important data on the feasibility of miRNA-based therapeutics generally (Beg et al., 2017; Reid et al., 2013). As noted in the earlier RG-101 trial, viral mutation and resistance is an issue that will need tackling (Van der Ree et al., 2017).

Applications of MicroRNAs as Biomarkers of Infectious Disease

The class of ex-miRNAs is ideal biomarker candidates due to their possible isolation from biological fluids (Boon and Vickers, 2013). Molecular methods such as RT-PCR are already used routinely in the clinical setting to quickly identify infections (e.g., respiratory infections in babies with bronchiolitis) and could be used to quantify ex-miRNAs in patient samples.

A search of the literature identified 57 studies assessing ex-miRNAs in infectious diseases through whole micronome profiling and candidate miRNA approaches. Thus the huge majority of these works are based on serum and plasma, but ex-miRNAs in cerebrospinal fluid (CSF), saliva, and sputum have also been implicated. Until now, most of the studies have focused on HCV, hepatitis B virus, HIV, TB, and sepsis with the aim to improve diagnosis and prognosticate infection outcome (e.g., death in sepsis, liver cirrhosis in hepatitis) or treatment response.

These infection studies are based on comparison of ex-miRNA profile of patients and healthy controls. Then, many studies identify that expressions of ex-miRNA are highly predictive of infection. In the study by Zhang et al. (2013), they found a differentiated expression of miR-378, miR-483-5p, miR-22, miR-29c, miR-101, and miR-320b between pulmonary TB and healthy controls with a high sensitivity and specificity going up to 95% and 91.8%, respectively. The limitation of healthy controls used for comparison is that differentially expressed ex-miRNAs may represent a nonspecific marker of infection, and this leads to limited clinical translatability of these studies given most people undergoing tests are symptomatic of some disease process. Although, to solve this problem, a handful of studies have chosen more pragmatic comparator groups and promisingly suggest ex-miRNA signatures can differentiate particular infectious disease from other differential diagnoses.

Furthermore, a promising application of ex-miRNA biomarker work may be to differentiate viral from bacterial infection, identify or prognosticate sepsis, and in monitoring of response to antimicrobial treatment. There are substantial interstudy discrepancies in miRNAs identified as potential biomarkers. This may be due to heterogeneity in study design (e.g., data normalization methods) and confounders such as hospital differences in defining sepsis and ICU admission criteria. Differences in the lengths of illness between patients create noise in the data; longitudinal studies that measure serial miRNA levels would provide temporal information on miRNA expression in sepsis and may resolve some conflicting findings.

However, there are challenges in using miRNAs as biomarkers of infectious disease, and this is underlined by a lack of interstudy cross validation of many results. Conflicting study results may arise from heterogeneity in study design including differences in populations and control groups, methods of miRNA extraction and the circulating fraction under investigation (serum, plasma, microvesicles, or exosomes), micronome expression profiling platforms (next-generation sequencing, probe-based hybridization microarrays, or RT-PCR arrays) and the dearth of miRNAs assessed, the limited statistical power of many studies at the profiling stage, data normalization methods, whether *P*-values were adjusted to take account of multiple testing issues (usually not done), and whether confirmatory cohorts were used to validate results. Given there is good evidence that miRNA contained in exosomes is functionally secreted as intercellular mediators of gene regulation, it is tempting to speculate that biomarker studies which profile miRNAs in exosomes rather than ex-miRNA in total plasma/serum (which will include a background of miRNA present from dead cells) could be more sensitive or specific biomarkers; comparisons of different extraction methods within the same biomarker study could help resolve this possibility.

Examples of MicroRNAs in Diagnosis of Emerging Diseases

Case of H7N9

In the previous studies, it was reported that miRNAs play an important role in influenza virus–host interaction. In a study by Fang Peng et al., the miRNA expression profiles in the sera of H7N9-infected patients and healthy controls were analyzed using miRNA microarray. Among the 94 miRNAs that were significantly differentially expressed in H7N9 serum samples when compared with that of healthy controls, 53 miRNAs were upregulated and 41 downregulated. Five serum miRNA candidates (hsa-miR-197-5p, hsa-miR-320a, hsa-miR-320d, hsa-miR-320e, and hsa-miR-765) were further verified by RT-qPCR. Receiver operating characteristic curve analysis was performed to evaluate the potential use of these miRNAs for the H7N9 infection diagnosis from the serum samples.

In this study, miRNA microarray assays revealed differential expression of 94 miRNAs in H7N9 patients' serum samples when compared to that in healthy controls. They identified five miRNAs that can be used for the diagnostic biomarkers for the early detection of the H7N9 infection, and this miRNA signature will advance our understanding of the molecular mechanisms involved in the influenza H7N9 infectious disease.

Ebola Virus

Rapid and accurate diagnosis of highly transmissible, lethal illnesses such as Ebola virus (EBOV) disease (EVD) is critical for restricting pathogen spread and for applying appropriate therapeutic strategies. As demonstrated by the recent EVD outbreak in Western Africa, early detection and confirmation of suspected cases are essential to halting disease spread (Blackley et al., 2014).

Current diagnostics rely on identifying EBOV in blood samples by targeting viral antigens using enzyme immunoassays or by amplifying specific viral sequences through quantitative reverse transcriptase PCR (RT-PCR) (Trombley et al., 2010).

In the study by Janice Duy et al., the authors analyzed the expression of 752 circulating miRNA sequences in archived plasma from rhesus macaques exposed to EBOV infected either through intramuscular injection or through aerosol inhalation. We identified miRNAs that showed significant changes in abundance during lethal EBOV infection for each group. They found 15 miRNAs correlated with viral titer in both rhesus macaque as well as human samples. As a proof of concept for a host miRNA–driven diagnostic, they identified eight miRNA classifiers predictive of acute infection with high accuracy in both nonhuman primates and humans, and this classifier identified half of the presymptomatic macaque hosts.

Keeping these caveats aside, this work shows that miRNAs are potential diagnostic candidates via a proof of concept acute EVD classifier while also establishing the potential basis for presymptomatic or asymptomatic diagnosis of the disease.

CONCLUSION

Clinical microbiology laboratories at the local level have an increasing responsibility to provide rapid and accurate diagnostic services for emerging (new) and reemerging infectious diseases, especially those diseases for which significant mortality or morbidity may occur as the result of a delay in diagnosis. Rapid, accurate diagnosis of emerging and reemerging infectious diseases may also be critical at the local level to ensure optimal infection control. Detection of these pathogens has often required esoteric procedures such as conventional PCR, which could be performed only at referral laboratories or, recently, at public health laboratories.

Recent technical advances in molecular diagnostics have resulted in the development of user-friendly automated testing platforms, such as real-time PCR. These novel-testing methods can be used to detect emerging and reemerging pathogens as well as common pathogens and have the potential for broadscale use in smaller laboratories in close proximity to the delivery of care.

While writing this review, a large outbreak of influenza virus type A (H3N2) was peaking in the United States, and new influenza virus type A strains (H5N1, H9N2) have been associated with both avian and human influenza in regions of the Far East. The apparent significant morbidity and mortality associated with these new influenza virus strains emphasize the need for rapid, accurate laboratory diagnostic capabilities at the local level. As is the case for SARS, agents of bioterrorism, and the other pathogens, rapid diagnostic methods, such as real-time PCR, and microarray will likely play a major role in the early and sensitive detection of emerging and reemerging infectious diseases encountered in the future.

Otherwise, a class of small RNAs implicated in the diagnosis of these diseases is miRNAs and is considered an essential mediator of host response to pathogens. Since several microbes have evolved to exploit their pleiotropic characteristics, identification of key genes and pathways in terms of activation, enhancement, repression, or silent, which are essential to facilitate the immune response, is based on the elucidation of the roles of miRNAs in host response to infectious disease. The complex regulatory network within which miRNAs are embedded makes unpicking the roles of miRNAs tough but not impossible. Integrating large miRNA and mRNA datasets using advanced statistical techniques (in a "systems biology" approach) will facilitate the unpicking of these complex networks.

Overall, miRNAs have multiple targets, and therefore any vaccines or treatments that harness miRNAs may produce off-target effects compromising safety; however, there are challenges that must be overcome. With the objective to improve the cross-study reproducibility of the findings, especially in the context of ex-miRNA biomarkers identification, universal endogenous controls are needed, and a more standardized approach to biomarker studies may also help. Initiatives devoted to harnessing the diagnostic and therapeutic potential of extracellular RNAs such as The National Institute for Health Extracellular Communication Consortium can facilitate this.

As the literature and experimental studies on miRNAs are developing, the potential for new miRNA therapeutics, diagnostics/prognostics, and vaccines becomes tangibly closer. Translating the insights of miRNA studies into improving the lives of patients is the critical next step.

Acknowledgments

All the authors are thankful to the contributors of the Team of Virology, Oncology, and Medical Biotechnologies of Laboratory of Virology, Microbiology, Quality, and Biotechnologies/ETB and also to the Fondation Lalla Salma de lutte contre le cancer.

References

Alexander, M., Hu, R., Runtsch, M.C., Kagele, D.A., Mosbruger, T.L., Tolmachova, T., et al., 2015. Exosome-delivered microRNAs modulate the inflammatory response to endotoxin. Nat. Commun. 6, 7321.

Aragon, L.M., Navarro, F., Heiser, V., Garrigo, M., Espanol, M., Coll, P., 2006. Rapid detection of specific gene mutations associated with isoniazid or rifampicin resistance in

Mycobacterium tuberculosis clinical isolates using non-fluorescent low-density DNA microarrays. J. Antimicrob. Chemother. 57 (5), 825–831.

Arroyo, J.D., Chevillet, J.R., Kroh, E.M., Ruf, I.K., Pritchard, C.C., Gibson, D.F., et al., 2011. Argonaute2 complexes carry a population of circulating microRNAs independent of vesicles in human plasma. Proc. Natl. Acad. Sci. U.S.A. 108, 5003–5008.

Bai, X.T., Nicot, C., 2015. miR-28-3p is a cellular restriction factor that inhibits human T cell leukemia virus, type 1 (HTLV-1) replication and virus infection. J. Biol. Chem 290, 5381–5390.

Barnes, D., Kunitomi, M., Vignuzzi, M., Saksela, K., Andino, R., 2008. Harnessing endogenous miRNAs to control virus tissue tropism as a strategy for developing attenuated virus vaccines. Cell Host Microbe 4, 239–248.

Beg, M.S., Brenner, A.J., Sachdev, J., Borad, M., Kang, Y.K., Stoudemire, J., et al., 2017. Phase I study of MRX34, a liposomal miR-34a mimic, administered twice weekly in patients with advanced solid tumors. Invest. New Drugs 35, 180–188.

Bell, E., Taylor, M.A., 2017. Functional roles for exosomal microRNAs in the tumour microenvironment. Comput. Struct. Biotechnol. J. 15, 8–13.

Berger, M., Shiau, R., Weintraub, J.M., 2006. Review of syndromic surveillance: implications for waterborne disease detection. J. Epidemiol. Community. Health 60 (6), 543–550.

Blackley, D.J., Lindblade, K.A., Kateh, F., Broyles, L.N., Westercamp, M., Neatherlin, J.C., et al., 2015. Rapid intervention to reduce Ebola transmission in a remote village—Gbarpolu County, Liberia, 2014. Morb. Mortal. Wkly. Rep. 64 (7), 175.

Bogerd, H.P., Skalsky, R.L., Kennedy, E.M., Furuse, Y., Whisnant, A.W., Flores, O., et al., 2014. Replication of many human viruses is refractory to inhibition by endogenous cellular microRNAs. J. Virol. 88, 8065–8076.

Boon, R.A., Vickers, K.C., 2013. Intercellular transport of microRNAs. Arterioscler. Thromb. Vasc. Biol. 33, 186–192.

Borchert, G.M., Lanier, W., Davidson, B.L., 2006. RNA polymerase III transcribes human microRNAs. Nat. Struct. Mol. Biol. 13 (12), 1097–1101.

Bravata, D.M., McDonald, K.M., Smith, W.M., et al., 2004. Systematic review: surveillance systems for early detection of bioterrorism-related diseases. Ann. Intern. Med. 140 (11), 910–922.

Bruchez Jr., M., Moronne, M., Gin, P., Weiss, S., Alivisatos, A.P., 1998. Semiconductor nanocrystals as fluorescent biological labels. Science 281 (5385), 2013–2016.

Buehler, J.W., Hopkins, R.S., Overhage, J.M., Sosin, D.M., Tong, V., 2004. Framework for evaluating public health surveillance systems for early detection of outbreaks: recommendations from the CDC Working Group. Morb. Mortal. Wkly Rep 53 (RR05), 1–11.

Casadevall, A., Relman, D.A., 2010. Microbial threat lists: obstacles in the quest for biosecurity? Nat. Rev. Microbiol 8 (2), 149–154.

Chen, Y., Gao, D.Y., Huang, L., 2015. In vivo delivery of miRNAs for cancer therapy: challenges and strategies. Adv. Drug Deliv. Rev. 81, 128–141.

Cortez, M.A., Bueso-Ramos, C., Ferdin, J., Lopez-Berestein, G., Sood, A.K., Calin, G.A., 2011. MicroRNAs in body fluids—the mix of hormones and biomarkers. Nat. Rev. Mol. Cell Biol. 8 (8), 467–477.

de Candia, P., Torri, A., Gorletta, T., Fedeli, M., Bulgheroni, E., Cheroni, C., et al., 2013. Intracellular modulation, extracellular disposal and serum increase of miR-150 mark lymphocyte activation. PLoS One 8, e75348.

Effler, P., Ching-Lee, M., Bogard, A., Ieong, M.C., Nekomoto, T., Jernigan, D., 1999. Statewide system of electronic notifiable disease reporting from clinical laboratories: comparing automated reporting with conventional methods. JAMA 282 (19), 1845–1850.

Eulalio, A., Huntzinger, E., Izaurralde, E., 2008. Getting to the root of miRNA-mediated gene silencing. Cell 132 (1), 9–14.

Friedman, R.C., et al., 2009. Most mammalian mRNAs are conserved targets of microRNAs. Genome Res. 19 (1), 92–105.

Garza, E.V., 2011. MicroRNA: biogenesis, functions and objectives identification. Rev. Hematol 12 (1).

Griffiths-Jones, S., Saini, H.K., van Dongen, S., Enright, A.J., 2008. miRBase: tools for microRNA genomics. Nucleic Acids Res. 36, D154–D158 (Database issue).

Gupta, A.K., Nair, P.R., Akin, D., Ladisch, M.R., Broyles, S., Alam, M.A., et al., 2006. Anomalous resonance in a nanomechanical biosensor. Proc. Natl. Acad. Sci. U.S.A 103, 13362–13367.

Han, J., et al., 2004. The Drosha-DGCR8 complex in primary microRNA processing. Genes Dev 18 (24), 3016–3027.

Hedberg, C.W., Besser, J.M., 2006. Commentary: cluster evaluation, PulseNet, and public health practice. Foodborne Pathog. Dis. 3 (1), 32–35.

Heiss, B.L., Maximova, O.A., Thach, D.C., Speicher, J.M., Pletnev, A.G., 2012. MicroRNA targeting of neurotropic flavivirus: effective control of virus escape and reversion to neurovirulent phenotype. J. Virol. 86, 5647–5659.

Hen, J., Ji, Y., Li, A., Zhang, Q., Song, W., Li, Y., et al., 2014. miR-122 directly inhibits human papillomavirus E6 gene and enhances interferon signaling through blocking suppressor of cytokine signaling 1 in SiHa cells. PLoS One 9, e108410.

Heymann, D.L., Rodier, G.R., 2001. Hot spots in a wired world: WHO surveillance of emerging and re-emerging infectious diseases. Lancet Infect. Dis. 1 (5), 345–353.

Hjelle, B., Jenison, S., Torrez-Martinez, N., et al., 1994. A novel hantavirus associated with an outbreak of fatal respiratory disease in the southwestern United States: evolutionary relationships to known hantaviruses. J. Virol. 68 (2), 592–596.

Hotz, C.Z., 2005. Applications of quantum dots in biology: an overview. Methods Mol. Biol. 303, 1–17.

Huang, Y., Shen, X.J., Zou, Q., Wang, S.P., Tang, S.M., Zhang, G.Z., 2011. Biological functions of microRNAs: a review. J. Physiol. Biochem. 67 (1), 129–139.

Humar, A., Gregson, D., Caliendo, A.M., et al., 1999. Clinical utility of quantitative cytomegalovirus viral load determination for predicting cytomegalovirus disease in liver transplant recipients. Transplantation 68 (9), 1305–1311.

Inui, M., Martello, G., Piccolo, S., 2010. MicroRNA control of signal transduction. Nat. Rev. Mol. Cell Biol. 11, 252–263.

Iorio, M.V., Croce, C.M., 2012. MicroRNA involvement in human cancer. Carcinogenesis 33, 1126–1133.

Irmak, M.K., Erdem, U., Kubar, A., 2012. Antiviral activity of salivary microRNAs for ophthalmic herpes zoster. Theor. Biol. Med. Modell 9, 21.

Jain, K.K., 2003. Current status of molecular biosensors. Med. Device Technol. 14, 10–15.

Jain, K.K., 2007. Applications of nanobiotechnology in clinical diagnostics. Clin. Chemother 53 (11), 2002–2009.

Johnson, C.J., Zhukovsky, N., Cass, A.E., Nagy, J.M., 2008. Proteomics, nanotechnology and molecular diagnostics. Proteomics 8 (4), 715–730.

Kaladhar, B.R., 2015. MicroRNA (miRNA) in cancer. Cancer Cell Int. 15, 38.

Layne, S.P., Beugelsdijk, T.J., 1998. Laboratory firepower for infectious disease research. Nat. Biotechnol. 16, 825–829.

Lehmann, S.M., Krüger, C., Park, B., Derkow, K., Rosenberger, K., Baumgart, J., et al., 2012. An unconventional role for miRNA: let-7 activates toll-like receptor 7 and causes neurodegeneration. Nat. Neurosci. 15, 827–835.

Lewis, M.D., Pavlin, J.A., Mansfield, J.L., et al., 2002. Disease outbreak detection system using syndromic data in the greater Washington DC area. Am. J. Prev. Med. 23 (3), 180–186.

Loeffelholz, M.J., 2004. Solid and liquid phase array technologies. In: Persing, D., Tenover, F., Versalovic, J., Tang, Y.W., Unger, E.R., Relman, D.A., White, T.J. (Eds.), Molecular Microbiology: Diagnostic Principles and Practice. ASM Press, Washington, DC, pp. 281–284.

Malumbres, M., 2013. MiRNAs and cancer: an epigenetics view. Mol. Aspects Med. 34 (4), 863–874.

Mancini, N., Carletti, S., Ghidoli, N., Cichero, P., Burioni, R., Clementi, M., 2010. The era of molecular and other non-culture-based methods in diagnosis of sepsis. Clin. Microbiol. Rev 23 (1), 235–251.

Mellmann, A., Friedrich, A.W., Rosenkotter, N., et al., 2006. Automated DNA sequence-based early warning system for the detection of methicillin-resistant *Staphylococcus aureus* outbreaks. PLoS. Med. 3 (3), e33.

Mellors, J.W., Rinaldo, C.R., Gupta, P., White, R.M., Todd, J.A., Kingsley, L.A., 1996. Prognosis in HIV-1 infection predicted by the quantity of virus in plasma. Science 272 (5265), 1167–1170.

Michalet, X., Pinaud, F.F., Bentolila, L.A., et al., 2005. Quantum dots for live cells, in vivo imaging, and diagnostics. Science 307 (5709), 538–544.

Mikhailovich, V., Gryadunov, D., Kolchinsky, A., Makarov, A.A., Zasedatelev, A., 2008. DNA microarrays in the clinic: infectious diseases. Bioessays 30 (7), 673–682.

Miller, M.B., Tang, Y.W., 2009. Basic concepts of microarrays and potential applications in clinical microbiology. Clin. Microbiol. Rev. 22 (4), 611–633.

Mittelbrunn, M., Gutiérrez-Vázquez, C., Villarroya-Beltri, C., González, S., Sánchez-Cabo, F., González, M.Á., et al., 2011. Unidirectional transfer of microRNA-loaded exosomes from T cells to antigen-presenting cells. Nat. Commun. 2, 282.

Monecke, S., Ehricht, R., 2005. Rapid genotyping of methicillin-resistant *Staphylococcus aureus* (MRSA) isolates using miniaturised oligonucleotide arrays. Clin. Microbiol. Infect 11 (10), 825–833.

Monecke, S., Berger-Bachi, B., Coombs, G., et al., 2007. Comparative genomics and DNA array-based genotyping of pandemic *Staphylococcus aureus* strains encoding Panton-Valentine leukocidin. Clin. Microbiol. Infect. 13 (3), 236–249.

Morse, S.S., 1995. Factors in the emergence of infectious diseases. Emerg. Infect. Dis 1 (1), 7–15.

Nolte, F.S., Caliendo, A.M., 2007. Molecular detection and identification of microorganisms. In: Murray, P.R., Baron, E.J., Jorgensen, J.H., Landry, M.L., Pfaller, M.A. (Eds.), Manual of Clinical Microbiology, vol. 1. ASM Press, Washington, DC, pp. 218–244.

O'Connor, M.J. Buckeridge, D.L. Choy, M. Crubezy, M. Pincus, Z. Musen, M.A., 2003. BioSTORM: a system for automated surveillance of diverse data sources. In: AMIA Annual Symposium Proceedings, p. 1071.

Panackal, A.A., M'Ikanatha, N.M., Tsui, F.C., et al., 2002. Automatic electronic laboratory-based reporting of notifiable infectious diseases at a large health system. Emerg. Infect. Dis. 8 (7), 685–691.

Peng, F., He, J., Loo, J.F.C., Yao, J., Shi, L., Liu, C., et al., 2016. Identification of microRNAs in throat swab as the biomarkers for diagnosis of influenza. Int. J. Med. Sci. 13 (1), 77–84.

Perez, J.M., Simeone, F.J., Saeki, Y., Josephson, L., Weissleder, R., 2003. Viral-induced self-assembly of magnetic nanoparticles allows the detection of viral particles in biological media. J. Am. Chem. Soc. 125, 10192–10193.

Persing, D.H., Tenover, F.C., Versalovic, J., Tang, Y.W., Unger, E., Relman, D.A., 2004. Molecular Microbiology: Principles and Practice. ASM Press, Washington, DC.

Reid, G., Pel, M.E., Kirschner, M.B., Cheng, Y.Y., Mugridge, N., Weiss, J., et al., 2013. Restoring expression of miR-16: a novel approach to therapy for malignant pleural mesothelioma. Ann. Oncol. 24, 3128–3135.

Reidn, R., Greenberg, M., Jenson, A.B., et al., 1987. Sexually transmitted papillomaviral infections, I: the anatomic distribution and pathologic grade of neoplastic lesions associated with different viral types. Am. J. Obstet. Gynecol 156 (1), 212–222.

Reis, B.Y., Kirby, C., Hadden, L.E., et al., 2007. AEGIS: a robust and scalable real-time public health surveillance system. J. Am. Med. Inform. Assoc. 14 (5), 581–588.

Relman, D.A., 1998. Detection and identification of previously unrecognized microbial pathogens. Emerg. Infect. Dis. J 4 (3), 382–389.

Relman, D.A., Loutit, J.S., Schmidt, T.M., Falkow, S., Tompkins, L.S., 1990. The agent of bacillary angiomatosis: an approach to the identification of uncultured pathogens. N. Engl. J. Med 323 (23), 1573–1580.

Sintchenko, V., Gallego, B., 2009. Laboratory-guided detection of disease outbreaks: three generations of surveillance systems. Arch. Pathol. Lab. Med. 133 (6), 916–925.

Smithn, P.L., WalkerPeach, C.R., Fulton, R.J., DuBois, D.B., 1998. A rapid, sensitive, multiplexed assay for detection of viral nucleic acids using the FlowMetrix system. Clin. Chem. 44 (9), 2054–2056.

Song, L., Liu, H., Gao, S., et al., 2010. Cellular microRNAs inhibit replication of the H1N1 influenza A virus in infected cells. J. Virol. 84 (17), 8849–8860.

Sotillo, E., Thomas-Tikhonenko, A., 2011. Shielding the messenger (RNA): microRNA-based anticancer therapies. Pharmacol. Ther. 131 (1), 18–32.

Tambyah, P.A., Sepramaniam, S., Swaminathan, P., et al., 2013. microRNAs in circulation are altered in response to influenza A virus infection in humans. PLoS One 8 (10), e76811.

Trombley, A.R., Wachter, L., Garrison, J., Buckley-Beason, V.A., Jahrling, J., Hensley, L.E., et al., 2010. Comprehensive panel of real-time TaqMan polymerase chain reaction assays for detection and absolute quantification of filoviruses, arenaviruses, and New World hantaviruses. Am. J. Trop. Med. Hyg. 82 (5), 954–960.

Turchinovich, A., Cho, W.C., 2014. The origin, function and diagnostic potential of extracellular microRNA in human body fluids. Front. Genet. 5, 30.

Urwin, R., Maiden, M.C., 2003. Multi-locus sequence typing: a tool for global epidemiology. Trends Microbiol. 11 (10), 479–487.

Van der Ree, M.H., de Vree, J.M., Stelma, F., Willemse, S., van der Valk, M., Rietdijk, S., et al., 2017. Safety, tolerability, and antiviral effect of RG-101 in patients with chronic hepatitis C: a phase 1B, double-blind, randomised controlled trial. Lancet 389 (10070), 709–717.

Vickers, K.C., Palmisano, B.T., Shoucri, B.M., Shamburek, R.D., Remaley, A.T., 2011. MicroRNAs are transported in plasma and delivered to recipient cells by high-density lipoproteins. Nat. Cell Biol. 13, 423–433.

Walsh, E.E., McConnochie, K.M., Long, C.E., Hall, C.B., 1997. Severity of respiratory syncytial virus infection is related to virus strain. J. Infect. Dis 175 (4), 814–820.

Wang, L., Ramoni, M.F., Mandl, K.D., Sebastiani, P., 2005. Factors affecting automated syndromic surveillance. Artif. Intell. Med. 34 (3), 269–278.

Wang, X., Gu, H., Qin, D., Yang, L., Huang, W., Essandoh, K., et al., 2015. Exosomal miR-223 contributes to mesenchymal stem cell-elicited cardioprotection in polymicrobial sepsis. Sci. Rep 5, 13721.

Weber, S.G., Pitrak, D., 2003. Accuracy of a local surveillance system for early detection of emerging infectious disease. J. Am. Med. Assoc. 290 (5), 596–598.

Weber, J.A., Baxter, D.H., Zhang, S., Huang, D.Y., Huang, K.H., Lee, M.J., et al., 2010. The microRNA spectrum in 12 body fluids. Clin. Chemother 56, 1733–1741.

Yoontae, L., et al., 2003. The nuclear RNase III Drosha initiates microRNA processing. Nature 425 (6956), 415–419.

Zernecke, A., Bidzhekov, K., Noels, H., Shagdarsuren, E., Gan, L., Denecke, B., et al., 2009. Delivery of microRNA-126 by apoptotic bodies induces CXCL12-dependent vascular protection. Sci. Signal. 2, ra81.

Zhang, X., Guo, J., Fan, S., Li, Y., Wei, L., Yang, X., et al., 2013. Screening and identification of six serum microRNAs as novel potential combination biomarkers for pulmonary tuberculosis diagnosis. PLoS One 8, e81076.

Zheng, Z., Ke, X., Wang, M., He, S., Li, Q., Zheng, C., et al., 2013. Human microRNA hsa-miR-296-5p suppresses enterovirus 71 replication by targeting the viral genome. J. Virol. 87, 5645–5656.

Zhu, Z., Qi, Y., Ge, A., Zhu, Y., Xu, K., Ji, H., et al., 2014. Comprehensive characterization of serum microRNA profile in response to the emerging avian influenza A (H7N9) virus infection in humans. Viruses 6, 1525–1539.

CHAPTER

8

Introduction to Computational and Bioinformatics Tools in Virology

Pramodkumar Pyarelal Gupta[1], *Yassine Kasmi*[2,3] *and Črtomir Podlipnik*[4]

[1]School of Biotechnology and Bioinformatics, D Y Patil Deemed to be University, Navi Mumbai, India [2]Faculty of Science and Techniques Mohammedia, University of Hassan II Casablanca, Casablanca, Morocco [3]Moroccan Foundation for Advanced Science, Innovation and Research, Rabat, Morocco [4]Faculty of Chemistry and Chemical Technology, University of Ljubljana, Ljubljana, Slovenia

ABBREVIATIONS

DDEs Delay differential equations
MD Molecular dynamics
ML Maximum likelihood
NJ Neighbor-joining
ODEs Ordinary differential equations
SNP Single nucleotide polymorphisms

INTRODUCTION

Bioinformatics, chemoinformatics, and computational biology are emerging fields that have become more and more familiar in scientific research studies. These specialties are nowadays present in the everyday work of researchers in several scientific fields. The main aims of the

Emerging and Reemerging Viral Pathogens
DOI: https://doi.org/10.1016/B978-0-12-814966-9.00008-1

mentioned scientific fields are to reduce the time and the costs of the scientific research. The application of bioinformatics and computational tools allows us to understand better the living organism's mechanisms applying algorithms based on the knowledge of biology, physics, chemistry, informatics, and mathematics.

In this chapter, we intend to present a resume of an introductory course on bioinformatics, chemoinformatics, and computational biology, while the other chapters from this book will discuss individual cases of bioinformatics, chemoinformatics, and computational biology.

THE FIRST QUESTION ASKED, WHAT ARE BIOINFORMATICS, CHEMOINFORMATICS, AND COMPUTATIONAL BIOLOGY?

Several definitions are developed and published by researchers and founders of these sciences, despite this, several new researchers find confusion in the distinction between the different constituents of biological modeling. Among the definition of bioinformatics, we find the following quotes:

> Bioinformatics: Research, development, or application of computational tools for the use of biological, medical, behavioural or health including those to acquire, store, organise, archive, analyse, or visualise such data (Huerta et al., 2000).
>
> Bioinformatics is a new multidisciplinary field that comes out from the combination of other sciences and fields like biology, computer science, statistics, chemistry, mathematics and even more (Chowdhary et al., 2016).
>
> Bioinformatics is introduced since the mid-1980 to descript the using of informatics technologies in life sciences. Later, bioinformatics came to be somewhat prosaically defined as the use of computers to retrieve, process, analysis, and simulate biological information. An even narrower definition was the application of information technology to the management of biological data (Ramsden, 2015).

In this educational document, we define bioinformatics as the use of computer tools to understand and answer to one or more scientific questions that attract the interest of researchers. These questions are specifically related to genomics (alignment, assembly, genes function), proteomics [protein identification, prediction of proteins' three-dimensional (3D) structures, posttranslational modifications], and drug design. The important field of bioinformatics is structural proteomics, the main task of which is the analysis of proteins' structures at large scale. Structural information of proteins is obtained using different experimental techniques such as X-ray crystallography and nuclear magnetic resonance (NMR) spectroscopy with a combination of computational and informatics tools. The structural analysis of proteins is, for example, essential for understanding how the drug binds to the protein, how the proteins interact with each other. Such analysis also helps us to interpret the gene regulatory network and role of the proteins in signaling pathways.

Computational biology is an interdisciplinary field that uses mathematical (computational) modeling to treat large collections of biological data (genetic sequences, cell populations, protein samples). The analysis of biological data is fundamental for the better understanding of complicated biological systems, such as the human body, which consists of several subunits and several systems that have diverse, complex, and stochastic regulations modes. The typical tasks of computational biologists are the development and application of data analytics, theoretical methods, mathematical modeling, and computational technical simulation to the study of biological, behavioral, and social systems (Huerta et al., 2000).

The central mathematical apparatus of mathematical biology is the theory of differential equations and mathematical statistics. Mathematical biology uses methods of applied mathematics, including mathematical modeling of biological processes and phenomena. The problems that we deal with methods of mathematical biology are usually very complicated, and therefore we solve them using computers. Another theoretical subdivision of mathematical biology is mathematical physics of biological objects, which studies the effect of physical laws on the biological level of the organization of matter and energy.

Chemoinformatics is a branch of theoretical chemistry that combines the field of chemistry with methods of computer science and informational techniques, applied to a range of problems in the field of chemistry. The main tasks of chemoinformatics are related to design, creation, organization, management, retrieval, analysis, dissemination, and visualization of chemical information with the goal to solve chemical problems. Unlike quantum chemistry considering molecules as an ensemble of electrons and nuclei, or molecular mechanics [molecular dynamics (MD) simulations] based on classical molecular model ("atoms" and "bonds"), chemoinformatics represents molecules as objects in a chemical space defined by either molecular graphs or some numerical parameters called molecular descriptors. Chemoinformatics uses available experimental data to establish quantitative relationships linking the chemical structure of molecules with their biological activities or physicochemical properties. These models, in turn, could efficiently be employed in virtual screening for computer-aided design of new molecules possessing desirable properties (Baskin and Varnek, 2009).

BIOMATHEMATICS AND COMPUTATIONAL BIOLOGY

The importance of mathematics in the modeling of biological phenomena is well known. Today, the quasitotality of theses in biology uses the statistics to analyze, interpret, and validate the results and the conclusions of the lab experiments. The biomathematics sector is not emergent, but their application in the last 100 years created a scientific revolution for theoreticians and biologists. In this section, we will outline different aspects and uses of biomathematics and computational biology.

Mathematics and Virology

The applications of mathematics in virology are vital for the better understanding of the mechanisms and actions of viral infections, assembly of viruses, the genetic analysis of viruses, the epidemiology, and other essential topics in virology.

Several chapters in this book describe the extensive use of mathematics to solve the mentioned problems of virology. For example, Allali and Ennaji in Chapter 17, Human Immunodeficiency Virus as Emergent Viral Infection with the Presence of the Immune Adaptive Response: Viral Dynamics, present the implication of the differential equation on HIV and its transmission. Their mathematical models are describing the HIV with the presence of the adaptive immune response. They describe the basic viral dynamics model with a system of three nonlinear differential equations representing the interaction between uninfected CD4 + T cells, the infected ones, and the free HIV viruses. Another example is Chapter 18, Mathematical Modeling in Virology, by Hattaf and Yousfi, where the authors present the application of ordinary differential equations (ODEs) and delay differential equations (DDEs) in virology and epidemiology. The primary focus of their research is the modeling and analysis the epidemiology of infectious diseases by using ODEs, DDEs, partial differential equations, and differential equations in their contracts and their control strategies.

Computational genomics is another method that requires the application of various statistical and mathematical algorithms to study genomes of cells and organisms. The massive amount of genomic data requires enormous storage capability and high-performance computing.

Another field related to use mathematical algorithms for modeling biological systems is molecular modeling. This field of research involves theoretical and computer-based methods of modeling and simulating the behavior of a wide range of molecular systems (from single molecule consists of few atoms to several million particles system of the whole virus) (Freddolino et al., 2006).

System biology, which aims to simulate networks of large-scale biological interactions (for an entire cell or even an entire organism), often uses differential equations. The prediction of protein structures and structural genomics makes attempts to systematically calculate accurate 3D models of protein structures that have not yet been experimentally obtained.

Subsections of calculating biochemistry and biophysics, usually using modeling and structural simulation techniques, such as MD or Boltzmann sampling technique (in turn based on the Monte Carlo method), shed light on the kinetics and thermodynamics of protein functions.

DATABASE AND INFORMATICS

In this section, we will introduce biological databases and bioinformatics tools used to collect and explore data related to virology. Biological databases are collections of relevant life sciences information, extracted from scientific experiments, published literature, and computational analysis. They contain information from different biosciences including proteomics, genomics, metabolomics, and phylogenomics. The information contained in biological databases includes functions, structure, and location (both cellular and chromosomal) of gene, clinical effects of mutations, and similarities of sequences and biological structures.

To understand the biological databases the knowledge about computer relational databases and information retrieval is essential. The design of biological databases, their development, and their long-term management is one of the most critical tasks of bioinformatics. The data includes gene sequences, textual descriptions, ontological attributes and classifications, annotations, and tabular data. These are often described as semistructured data and can be represented as arrays, delimited vital documents, and XML structures. They are cross-common between databases using access numbers (protein sequences of unique identifiers or DNA) as one of the foreign key references.

Description of Databases

Biological databases have become a valuable tool to help scientists understand and explain some biological phenomena of biomolecular structure and interaction, to complement the metabolism of organisms, and to understand the evolution of species. This knowledge facilitates the fight against the disease, helps in the development of drugs, and the discovery of fundamental relationships between species in the history of life.

Biological knowledge is distributed among several general and specialized databases. This fact sometimes makes it difficult to ensure consistency of information. Biological databases are cross-referenced with other databases with the access number as a way of linking their knowledge-related references.

An essential resource for finding biological databases is an annual issue of the journal Nucleic Acids Research (NAR).

The Most Common Types of Biological Data

Data is the primary objects that are stored in the database, therefore in this section, we will describe the most common types of biological data. The biological data could be classified as follows:

- Experimental data—data from the results of laboratory experiments, digital images, for example, the results of the observation.
- Phylogenetic data—data on the evolutionary relationship between different groups of organisms. Information is obtained through molecular sequencing (multiple sequence alignment) data and morphological data.
- Metabolic data—data on the metabolic pathways of organisms (enzymatic reaction in living organisms, etc.).
- Raw data—data that has never been subjected to manipulation or transformation.
- Sequence—DNA sequence data or protein sequences, multiple sequence alignment.
- Structure data—the 3D structure of proteins, DNA, RNA, or small molecules.
- Expressed sequence tags—these are short DNA sequences (about 300–1000 bp), which are derived from cDNA. Represent genes expressed in tissues from which it is derived from cDNA (transcription). These data allow us to base comparisons between different organisms, differentiating gene families, information on genes expressed in specific tissues, or depending on the response to external influences are used for the identification of gene transcripts, and help with gene discovery and sequence determination, etc.
- Single nucleotide polymorphisms (SNPs) and other variants—the individual nucleotide gaps in a DNA sequence. SNPs are most often found genetic variation in the human population. Each SNP represents a difference of a single building block of DNA. Most SNPs are unlikely to have a direct effect on human health but may alter the effect of the drug or the body's response to stress exposure to chemical or biological contaminants in the environment. Also, the SNP can be used to map laws governing inheritance in families with a high incidence of certain diseases, for example, diabetes, cardiovascular diseases, and cancer.

Access to Database

Most biological databases can be accessed via the website; biological data is arranged so that users can easily view online and download them in different formats. Biological data is stored in many formats (e.g., text, sequence data, and protein structure). Each type of format is usually found in some database, for example,

- Text formats are available on PubMed and OMIM.
- Sequence data is available on GenBank DNA and UniProt, Swiss-Prot for Proteins.
- Protein structures are available on APB, SCOP, and CATH.

ALIGNMENT AND PHYLOGENY

Pairwise alignment of sequences is used to identify regions of similarity, which may indicate functional, structural, and evolutionary relationships between two biological sequences (proteins or nucleic acids). Sequence alignment in bioinformatics is a way of representing and comparing two or more sequences or strings of DNA, RNA, or primary protein structures to highlight their areas of similarity, which could indicate functional or evolutionary relationships between the genes or proteins consulted. The aligned sequences are written with letters that represent the amino acids (or nucleotides) in rows of a matrix in which, where appropriate, spaces are inserted into the areas with identical or similar aligned structures.

If two sequences are in one part of the alignment of a common ancestor, the disparities can be interpreted as point mutations (substitutions), gaps that indels (insertion or deletion mutations) introduced in one or two lineages in the time since diverged. In the alignment of protein sequences the degree of similarity between the amino acids occupying a specific position in the sequence can be interpreted as an approximate measure of conservation in a particular region or pattern sequence, between lineages. The absence of substitutions, or the presence of highly conserved substitutions (replacement of amino acids which side chains have similar chemical properties), in a particular region of the sequence indicates that the material is of structural and functional importance. Although the nucleotide bases of DNA and RNA are more similar to each other than to amino acids, the associated conservation bases could indicate similar functional or structural roles. Sequence alignment can be used with any of the biological similarities, and the identification of a series of letters and words in human language or analysis of the sequence of financial data.

Very short or very similar sequences can be manually aligned. Many interesting problems of bioinformatics require alignment of numerous long and highly variable sequences, which cannot be aligned by humans. Human knowledge is mainly used in the construction of alignment algorithms that produce high quality, and the adjustment from time to time the final result to represent the models that are difficult to introduce into the algorithms (especially in the case of nucleotide sequences). Computer approaches to sequence alignment are divided into two categories: global alignment and local alignment. Computing a global alignment is a form of overall optimization that forces the alignment to occupy the total length of all introduced sequences (problem sequences). In comparison, local alignments identify similar regions in more extended sequences that are typically strongly divergent from each other. Often local alignments are preferred, but they may be more difficult to calculate because the challenge of identifying regions of greater similarity is added. Variety of computer algorithms for the problem of sequence alignment methods are applied as slow, but such optimization as dynamic programming, and heuristic or probabilistic methods adequate, but not exhaustive designed to evolve search in databases.

Local and Global Alignments

Image of a local and global alignment showing a tendency to make global alignment gaps if the sequences are not very similar.

Global alignments, trying to align each residue in each sequence, are most useful when the initial problem sequences are similar and of the same size (not to say that global alignments cannot end gaps). A general global alignment strategy is the Needleman–Wunsch algorithm based on dynamic programming. Local alignments are more useful for different sequences in which it is suspected that there are similar regions or patterns of sequences in a broader context very similar. The Smith–Waterman algorithm is a general local alignment method based on dynamic programming. With sufficiently similar sequences, there is no difference between global and local alignments.

Hybrids, known as semiglobal methods or "glocal" methods, try to find the best possible alignment that includes the beginning and the end of the other sequence. It can be particularly useful when the "upstream" part of a sequence overlaps the "downstream" part of the other. In this case, neither global nor local alignment is ideal: a global alignment tries to force the alignment to extend beyond the overlap area, while the local alignment does not entirely cover the overlap area.

Some of the most often used tools for the alignment of sequences are collected in Table 8.1.

The example of a sequence from UniProt database is shown as follows (Figs. 8.1–8.4):

TABLE 8.1 Alignment Tool

Sr no.	Tool name	Alignment type	Web URL
1	BLAST Nucleotide-nucleotide BLAST (blastn) Nucleotide 6-frame translation-protein (blastx) Nucleotide 6-frame translation-nucleotide 6-frame translation (tblastx) Protein-nucleotide 6-frame translation (tblastn) Protein–protein BLAST (blastp) PSI-BLAST Quick BLASTP (Accelerated protein–protein BLAST) New PHI-BLAST DELTA-BLAST Mega Blast	Pairwise	https://blast.ncbi.nlm.nih.gov/Blast.cgi
2	FASTA FASTX FASTY	Pairwise	https://www.ebi.ac.uk/Tools/sss/fasta/
3	Clustal omega	MSA	https://www.ebi.ac.uk/Tools/msa/clustalo/
4	COFEE tool	MSA	http://tcoffee.crg.cat/

DELTA, Domain Enhanced Lookup Time Accelerated BLAST; *PHI-BLAST*, Pattern Hit Initiated BLAST; *PSI-BLAST*, Position-Specific Iterative BLAST.

```
Sequence: UniProtKB - Q05127 (VP35_EBOZM)
>sp|Q05127|VP35_EBOZM Polymerase cofactor VP35 OS = Zaire ebolavirus
(strain Mayinga-76) GN = VP35 PE = 1 SV = 1
MTTRTKGRGHTAATTQNDRMPGPELSGWISEQLMTGRIPVSDIFCDIENNPGLCYASQMQ
QTKPNPKTRNSQTQTDPICNHSFEEVVQTLASLATVVQQQTIASESLEQRITSLENGLKP
VYDMAKTISSLNRVCAEMVAKYDLLVMTTGRATATAAATEAYWAEHGQPPPGPSLYEESA
IRGKIESRDETVPQSVREAFNNLNSTTSLTEENFGKPDISAKDLRNIMYDHLPGFGTAFH
QLVQVICKLGKDSNSLDIIHAEFQASLAEGDSPQCALIQITKRVPIFQDAAPPVIHIRSR
GDIPRACQKSLRPVPPSPKIDRGWVCVFQLQDGKTLGLKI
```

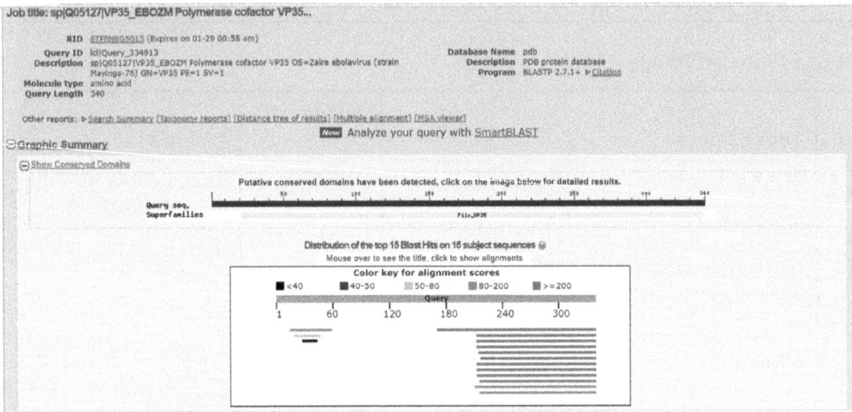

FIGURE 8.1 **BLASTp output for sequence viral sequence.**

Download ⌄ GenPept Graphics

Chain A, Crystal Structure Of Reston Ebolavirus Vp35 Rna Binding Domain

Sequence ID: 3KS4_A **Length:** 184 **Number of Matches:** 1

▷ See 5 more title(s)

Range 1: 15 to 184 GenPept Graphics ▼ Next Match ▲ Previous Match

Score	Expect	Method	Identities	Positives	Gaps
302 bits(774)	1e-103	Compositional matrix adjust.	137/170(81%)	157/170(92%)	0/170(0%)

```
Query  171   PGPSLYEESAIRGKIESRDETVPQSVREAFNNLNSTTSLTEENFGKPDISAKDLRNIMYD  230
             PGP+LYEE+A++GKI+   +   VP +V+EA+ NL+ST++LTEENFGKP ISAKDL+ IMYD
Sbjct  15    PGPALYEENALKGKIDDPNSYVPDAVQEAYKNLDSTSTLTEENFGKPYISAKDLKEIMYD  74

Query  231   HLPGFGTAFHQLVQVICKLGKDSNSLDIIHAEFQASLAEGDSPQCALIQITKRVPIFQDA  290
             HLPGFGTAFHQLVQVICK+GKD+N LD IHAEFQASLA+GDSPQCALIQITKRVPIFQD
Sbjct  75    HLPGFGTAFHQLVQVICKIGKDNNLLDTIHAEFQASLADGDSPQCALIQITKRVPIFQDV  134

Query  291   APPVIHIRSRGDIPRACQKSLRPVPPSPKIDRGWVCVFQLQDGKTLGLKI  340
              PP+IHIRSRGDIPRACQKSLRP PPSPKIDRGWVC+F++QDGKTLGLKI
Sbjct  135   PPPIIHIRSRGDIPRACQKSLRPAPPSPKIDRGWVCLFKMQDGKTLGLKI  184
```

Download ⌄ GenPept Graphics

Chain A, Structure Of The Ebola Vp35 Interferon Inhibitory Domain

Sequence ID: 3FKE_A **Length:** 129 **Number of Matches:** 1

▷ See 25 more title(s)

Range 1: 2 to 129 GenPept Graphics ▼ Next Match ▲ Previous Match

Score	Expect	Method	Identities	Positives	Gaps
266 bits(681)	2e-90	Compositional matrix adjust.	126/128(98%)	127/128(99%)	0/128(0%)

```
Query  213   NFGKPDISAKDLRNIMYDHLPGFGTAFHQLVQVICKLGKDSNSLDIIHAEFQASLAEGDS  272
             + GKPDISAKDLRNIMYDHLPGFGTAFHQLVQVICKLGKDSNSLDIIHAEFQASLAEGDS
Sbjct  2     HMGKPDISAKDLRNIMYDHLPGFGTAFHQLVQVICKLGKDSNSLDIIHAEFQASLAEGDS  61

Query  273   PQCALIQITKRVPIFQDAAPPVIHIRSRGDIPRACQKSLRPVPPSPKIDRGWVCVFQLQD  332
             PQCALIQITKRVPIFQDAAPPVIHIRSRGDIPRACQKSLRPVPPSPKIDRGWVCVFQLQD
Sbjct  62    PQCALIQITKRVPIFQDAAPPVIHIRSRGDIPRACQKSLRPVPPSPKIDRGWVCVFQLQD  121

Query  333   GKTLGLKI  340
             GKTLGLKI
Sbjct  122   GKTLGLKI  129
```

FIGURE 8.2 **Sequence alignment between the query sequence and output sequence in BLASTp.**

```
CLUSTAL O(1.2.4) multiple sequence alignment


sp|P35259|VP35_MABVM      -------------------MWDSSYMQQVSEGLMTGKVPIDQVFGANPL--EKLYKRRKP
sp|Q05127|VP35_EBOZM      MTTRTKGRGHTAATTQNDRMPGPELSGWISEQLMTGRIPVSDIFCDIENNPGLCYASQMQ
sp|Q8JPY0|VP35_EBORR      -----------MYNNKLKVCSGPETTGWISEQLMTGKIPVTDIFIDIDNKPDQMEVRLKP
                                               . .    :** ****::*: ::*

sp|P35259|VP35_MABVM      KGTVGLQCSPCLMSKATSTDDIIWDQLI--VKRTLADLLIPINRQISDIQSTLSEVTTRV
sp|Q05127|VP35_EBOZM      QTKPNP--KTRN---SQTQTDPICNHSFEEVVQTLASLATVVQQQTIAS----ESLEQRI
sp|Q8JPY0|VP35_EBORR      SSRSST--RTCT---SSSQTEVNYVPLLKKVEDTLTMLVNATSRQNAAI----EALENRL
                          .  .           : : :      :  *  **: *    .:*          . :  *:

sp|P35259|VP35_MABVM      HEIERQLHEITPVLKMGRTLEAISKGMSEMLAKYDHLVISTGRTTAPAAAFDAYLNEHGV
sp|Q05127|VP35_EBOZM      ---TSLENGLKPVYDMAKTISSLNRVCAEMVAKYDLLVMTTGRATATAAATEAYWAEHGQ
sp|Q8JPY0|VP35_EBORR      ---STLESSLKPIQDMGKVISSLNRSCAEMVAKYDLLVMTTGRATSTAAAVDAYWKEHKQ
                                    :.*: .*.:.:.::.:  :**:**** **::***:*: *** :**  **

sp|P35259|VP35_MABVM      PPPQPAIFKDLGVAQQACSKGTMVKNATTDAADKMSKVLELSEETFSKPNLSAKDLALLL
sp|Q05127|VP35_EBOZM      PPPGPSLYEESAIRGKIESRDETVPQSVREAFNNLNSTTSLTEENFGKPDISAKDLRNIM
sp|Q8JPY0|VP35_EBORR      PPPGPALYEENALKGKIDDPNSYVPDAVQEAYKNLDSTSTLTEENFGKPYISAKDLKEIM
                          *** *::::: .:  :  . .  * ::. :* .::...  *:**.*.** :*****  ::

sp|P35259|VP35_MABVM      FTHLPGNNTPFHILAQVLSKIAYKSGKSGAFLDAFHQILSEGENAQAALTRLSRTFDAFL
sp|Q05127|VP35_EBOZM      YDHLPGFGTAFHQLVQVICKLGKDSNSLDIIHAEFQASLAEGDSPQCALIQITKRVPIFQ
sp|Q8JPY0|VP35_EBORR      YDHLPGFGTAFHQLVQVICKIGKDNNLLDTIHAEFQASLADGDSPQCALIQITKRVPIFQ
                          : **** .* ** *.**:.*:. ...  . :    *:  *::*:. *.** :::: .  *

sp|P35259|VP35_MABVM      GVVPPVIRVKNFQTVPRPSQKSLRAVPPNPTIDKGWVCVYSSEQGETRALKI
sp|Q05127|VP35_EBOZM      DAAPPVIHIRSRGDIPRACQKSLRPVPPSPKIDRGWVCVFQLQDGKTLGLKI
sp|Q8JPY0|VP35_EBORR      DVPPPIIHIRSRGDIPRACQKSLRPAPPSPKIDRGWVCLFKMQDGKTLGLKI
                          .. **:*:::.    :** .***** .**.*.**:****::. ::*:* .***
```

FIGURE 8.3 **Multiple sequence alignments of VP35 protein from three different strain of Ebola virus, that is, Marburg, Reston, and Zaire.**

Phylogenetic Tree

This is a Neighbour-joining tree without distance corrections.

Download Phylogenetic Tree Data

Branch length: ◉ Cladogram ○ Real

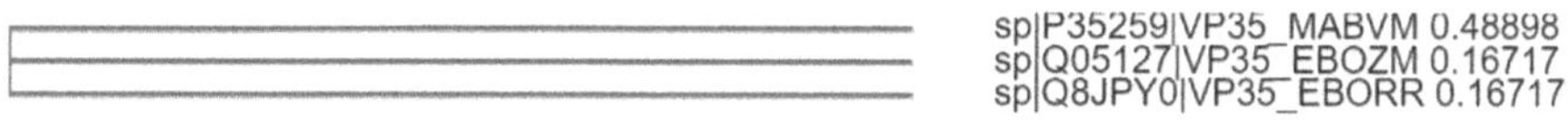

Tree Data

```
(
sp|P35259|VP35_MABVM:0.48898,
sp|Q05127|VP35_EBOZM:0.16717,
sp|Q8JPY0|VP35_EBORR:0.16717);
```

FIGURE 8.4 **Tree diagram from multiple sequence alignment of VP35 protein from three different strain of Ebola virus, that is, Marburg, Reston, and Zaire.**

A wide range of topics is covered in the sequence analysis such as the following:

1. The pairwise and multiple comparisons of the sequence to find the similarity and identity.
2. Identification of intrinsic features on the sequence such as an exon, intron, and regulatory regions.
3. Identification of reading frames, gene, posttranslational modification, and active and catalytic site.
4. Identification of SNP and point mutations.
5. Modeling and identification of genetic diversity
6. The sequence to 3D structure modeling, by selecting appropriate templates from the desired output of pairwise alignment.

INTRODUCTION TO PHYLOGENETIC ANALYSIS

Phylogenetics is the study of evolutionary relationships. Phylogenetic analysis is the way to infer or estimate these relationships. The evolutionary history deduced from the phylogenetic analysis is usually portrayed as branching, tree-like diagrams that represent an estimated pedigree of inherited relationships among molecules (gene trees), organisms, or both. Phylogenetics is sometimes called cladistic because the word "clade," a set of descendants of a common ancestor, is derived from the Greek word for the branch. However, cladistics is a particular method of formulating hypotheses about evolutionary relationships. The fundamental principle behind cladistics is that members of a group or clade share a mutual evolutionary history and are more connected to each other than members of another group. A given group is recognized by sharing unique characteristics that were not in distant ancestors. These characteristics shared derivatives, may be all that can be observed and described of two organisms that have developed a two-sequence spine that has developed a mutation to a particular pair of a basic gene.

Typically, cladistic analysis is performed by comparing several characteristics or characters at once, either several phenotypic characters or several base pairs or amino acids in a sequence.

There are three underlying cladistic hypotheses:

- Every group of organisms is linked by the descendants of a common ancestor (fundamental principle of the theory of evolution).
- There is a model of bifurcation of cladogenesis. This hypothesis is controversial.
- Change in characteristics occurs in lineages over time. This is a necessary condition for working a cladistic.

A phylogenetic tree most often represents relationships resulting from the cladistic analysis. Several types of data can be used to build phylogenetic trees:

- Traditionally, phylogenetic trees have been constructed from morphological features (e.g., bill shapes, feathers, number of legs). Today, we mainly use molecular data such as DNA sequences and protein sequences.
- Discrete characters: Each character has a finite number of states. For example, discrete characters include the number of legs of an organism, or a column in an alignment of DNA sequences. In the latter case the number of states for the character of the column is four (A, C, T, and G).
- Numerical comparative data: This data encodes the distances between objects and is usually based on sequence data. For example, we could hypothetically say, distance (man, mouse) = 500 and distance (man, chimpanzee) = 100.

Definitions and Terminology

In this section the essential terms and definitions used in the phylogenetic analysis are collected.

- *Maximum parsimony*: The tree chosen to minimize the number of changes needed to explain the data.
- *Maximum plausibility*: As part of a sequence evolution model, the tree that gives the highest probability of the data is found.
- *Analogic*: A characteristic that appears close in two taxa that come from two different ancestors.
- *Ancestor*: Any organism, population, or species from which other organisms, the population, or species have descended by reproduction.
- *Apomorphic*: Specialized (=derived) characters of an organism.
- *Base group*: The earlier divergent group in a clade, for example, to hypothesize that sponges are basic animals is to suggest that the lineage(s) leading to sponges diverged from the lineage that gave rise to all other animals.
- *Biological classification*: The orderly arrangement of organisms in the hierarchical system that perfectly reflects evolutionary history.
- *State of Character*: Characters are usually described regarding their states, for example, "hair presents" versus "hair absents," where "hair" is the character, and "present" and "absent" are its states.
- *Clade*: A monophyletic taxon, a group of organisms, which includes the last common ancestor of all its members and descendants of this

last common ancestor. The word clade originates from the ancient Greek word "Klados," which means branch or twig.

- *Cladogenesis*: The development of a new clade, the fractionation of a single line into two distinct lines, speciation.
- *Cladogram*: A diagram, resulting from a cladistic analysis, representing a possible branching sequence of lineages leading to taxa under study. The branch points in a cladogram are called nodes. All taxa occur at the ends of the cladogram.
- *Convergence*: Similarities that have arisen independently in two or more organizations that are not closely related and contrasted with homology. As several genomes are sequenced, we become more interested in learning about the evolution of the protein or gene (for example, the phylogenetic investigation of genes, rather than organismic phylogeny). This can facilitate our understanding of the function of proteins and genes. Protein studies and gene evolution involve the comparison of homologous sequences that have common origins but may or may not have a common activity. Sequences that share an arbitrary threshold level of similarity determined by corresponding base alignment are called counterparts. They are inherited from a common ancestor that had a similar structure, although the ancestor's structure may be complicated to determine because descent modified it.

Homologs are most often either orthologs, paralogs, or xenologs.

- *Orthologs* are homologs produced by speciation. They represent genes derived from a common ancestor, which have diverged due to the divergence of the organisms with which they are associated. They tend to have a similar function.
- *Paralogs* are homologs produced by gene duplication. They represent genes derived from a common ancestral gene, which are duplicated in an organism and then diverged. They tend to have different functions.
- *Xenologs* are homologs resulting from horizontal gene transfer between the two organisms. The question of whether a gene of interest has recently been transferred to the current host by horizontal gene transfer is often tricky. Sometimes the %(guanine-cytosine content (GC)) content, if it differs from the gene in the actual host, can be an average to conclude that the external origin is almost unavoidable, but often it is unclear whether a gene has horizontal origins. The function of xenologists can vary depending on how the gene moves horizontally; however, in general, the function tends to be similar.

The construction methods of the phylogenetic tree assume some evolutionary models. For a set of data, these models may be violated because of events such as the transfer of genetic material between organisms. Thus when interpreting a given analysis, one must always consider the model used and its assumptions and entertain other possible explanations for the observed results. For example, consider the figure tree given later. A survey of organismal relationships in the tree suggests that eukaryotic one is more related to bacteria than other eukaryotes. The various construction methods of the phylogenetic tree are listed as follows:

- *Neighbor-joining (NJ) methods* apply the general data of clustering techniques to sequence analysis using genetic distance as a measure of cluster management. The simple NJ method produces trees, but it does not assume a constant rate of evolution (i.e., a molecular clock) in all lineages. Its parent, unweighted group pair with arithmetic mean produces rooted trees and requires constant-rate support, that is, it assumes an ultrametric tree in which distances from the root to each end of the branch are equal.
- *The maximum likelihood (ML) method* uses standard statistical techniques to derive probability distributions from assigning probabilities of possible phylogenetic trees. The method requires a surrogate model to evaluate the probability of particular mutations; roughly speaking, a tree that requires more mutations at the interior nodes to account for the observed phylogeny is rated as having a lower probability. This method is mainly similar to the maximum parsimony method but allows maximum statistical likelihood more flexibility to allow for varying rates of evolution across lineages and sites. In fact, the method requires that evolution at different locations and along different lineages must be statistically independent. ML is, therefore, well suited to the analysis of related sequences, but, because officially, one must search for all possible combinations of topology length and tree branch, it is expensive regarding performing on several sequences. The "pruning" algorithm, a variety of dynamic programming, is often used to reduce the search space to compute the probability of subtrees efficiently. The method calculates the probability for each site in a "linear" way, starting with a node which only descendants are the leaves (i.e., the tips of the tree) and working backward toward the "nested" sets node.

So for the question of NJ or ML, NJ is faster regarding computation but rather weak compared to ML. ML uses the more complex evolution model regarding mathematical relations and calculates method by performing global alignments, so "closer to real life." The mathematics behind ML/Bayesian methods is known to be stronger than NJ to reconstruct sequence stories, use in the first step of the NJ method to define a useful dataset or to test new things. Thanks to ML, we will have a robust phylogeny for editing, it will be tough to publish in a document pure phylogeny as a simple NJ tree, ... or only as a pedagogical figure.

However, the trees produced by the method rest only if the substitution model is irreversible, which is not generally true for biological systems. The search for the ML tree also includes a branch length optimization component that is difficult to improve algorithmically; general global optimization tools such as the Newton–Raphson method are often used. The search for tree topologies defined by probability has not been shown to be NP-complete but remains extremely difficult because branch-and-bound search is not yet sufficient for trees represented this way.

BAYESIAN AND MAXIMUM LIKELIHOOD OF PHYLOGENY

Different new difficult problems in pattern matching appear in different scientific fields, including bioinformatics and image analysis. In a class of shape analysis problems, it is assumed that points in two or more configurations are tagged, and these configurations must be associated with filtering a particular transformation. Usually, the transformation is a rigorous transformation or similarity. Several new problems occur when configuration points are not tagged and labeling is ambiguous, and some points do not appear in each configuration (Green and Mardia, 2006).

The Bayesian approach uses a likelihood function to create a quantity called posterior probability of trees using an evolution model, based on some prior probabilities, producing the most likely phylogenetic tree for the data. It has become popular, and it has helped answer these questions because of advances in computer speeds and the integration of Monte Carlo algorithms, Poisson process, Von Mises–Fisher distribution, and Markov chains (Green and Mardia, 2006).

The main advantages of the Bayesian approach compared to maximum probability methods and the maximum economy are the efficiency of the calculation, the ability to work with complex models of evolution, as well as the fact that, unlike the methods of indication of the best for a given criterion tree. It allows us to choose several phylogenetic trees with the highest posterior probability (Holder and Lewis, 2003).

PHYLODYNAMICS

The phylodynamics is defined according to as rapidly evolving pathogens are unique in that their ecological and evolutionary dynamics occur on the same timescale and can therefore potentially interact (Pybus and Rambaut, 2009). It is also an interaction between genetic diversity (phylogenetics and population genetics), virus phenotype (molecular virology and immunology), and epidemic dynamics (mathematical and spatial epidemiology). In the studies of phylodynamics, we often depend on the Bayesian statistical tools.

GENE PREDICTION METHODS

The gene identification and prediction are one of the primary tasks for researchers working with both novel and well-studied partial or whole length genomes (Besemer and Borodovsky, 2005). The fundamental basis of molecular evolution analysis and phylogenetic relations is dependent upon the relationship between genes and their conserveness (Klasberg et al., 2016). Accurate prediction of genes is one of the essential and fundamental steps in all metagenomic projects (Liu et al., 2013).

In in silico biology, gene finding or prediction is a method to identify the region in the genomic DNA, which encodes a gene, includes protein-coding genes or RNA genes. Gene prediction also includes functional elements such as regulatory regions. Gene prediction is closely associated with the target search or locating a specific binding site within the genome of an organism. Numerous aspects of gene prediction are based on existing knowledge of biochemical processes in the cell such as transcription, translation, regulation, and protein–protein interaction network (Redding and Greene, 2013; Sokolov et al., 2005).

Two broad approaches are mainly used for prediction of genes from metagenomic DNA fragments; the first one is the evidence-based method that significantly depends upon the homology searches, the methodology implements a comparison against known protein database by homology search packages such as BLAST. In homology-based method, only known genes were mined out as a predicted genes from the database against the query input sequence, and novel genes might be overlooked due to lack of knowledge or genes, which are not reported to the databases (Hoff et al., 2008; Hyatt et al., 2012; Kelley, Liu et al., 2012; Liu et al., 2013; Noguchi et al., 2006; Noguchi et al., 2008; Rho et al., 2010; Zhu et al., 2010).

However, the second approach is ab initio–based algorithm that gives higher sensitivity and optimum output. In recent years ab initio methods have specially designed for metagenomics fragments (Liu et al., 2013; Zhu et al., 2010).

The recent classification of genes based on phylogenetic analysis is listed as follows:

- *Orphan* (or taxonomically restricted) *genes* are classified based on a given phylogeny. A gene that is only found in a single species or a branch, but not outside, is orphan in that specific branch (Klasberg et al., 2016).
- *Novel genes* are classified by their age. Genes that have emerged inside a defined time frame are novel genes. The time frame is not fixed and needs to be defined for each study. All novel genes are orphan in a specific clade, but, depending on the time frame, not all orphan genes are classified as a novel (Klasberg et al., 2016).
- *De novo genes* are defined based on their mechanism of emergence, out of previously noncoding DNA. This gene might occur via the acquisition of transcriptional regulation, consecutive point mutations, or genomic rearrangements (Klasberg et al., 2016).

INTRODUCTION TO MOLECULAR MODELING

Molecular modeling encompasses computer-assisted chemical modeling techniques to mimic or model a given biological or chemical behavior. The design of new molecules, target structure and their modeling, is a branch of molecular modeling.

These techniques allow in addition to the spatial representation of the most simple of the most complex molecules, the calculation of their physicochemical properties, and they are essential in medicinal chemistry, the goal of which is to optimize the structures of new active substances (homology modeling). In this field, molecular modeling is increasingly being supplemented by combinatorial chemistry, virtual screening, QSAR, CoMFA, CoMSIA, simulation of biological molecules, and their nature by using force fields such as AMBER, CHARMM, OPLS.

Mechanistic (semiempirical) and statistical methods (MD, the Monte Carlo method) are more suitable for large molecules or interactions between a large number of molecules; however, chemical quantum calculation methods are much more accurate, due to the higher computational complexity than small and medium size first choice molecules. More and more processing power of modern computers is benefiting molecular modeling. Molecular modeling is a tool for researchers concerned about the structure and reactivity of molecules. Knowledge of the structure of molecular edifices makes it possible to understand what is achieved in a physical, chemical, or biological transformation. It can also make it possible to envisage such transformations. Both understanding and forecasting are greatly facilitated when visualizing structures. A molecule is correctly described by its geometry and its thermodynamic properties. Visualization must account for all of these characteristics. The essential question is to represent a molecule on the screen as close as possible to "reality." The use of computers has made it possible to develop a robust tool: molecular modeling.

PROTEIN STRUCTURE MODELING AND PREDICTION

Protein structure prediction is the implication of two-dimensional and 3D structure of a protein from its amino acid sequence. The prediction solely depends on its configuration of amino acid, that is, a primary structure that later undergoes in folding process and transforms to secondary and finally into a tertiary structure (DM, 2004; Huang and Brutlag, 2001; Pirovano and Heringa, 2010).

Protein structure prediction is one of the most significant goals of bioinformatics and theoretical biology. Numerous methods are implemented in the protein structure prediction such as homology modeling or comparative modeling and ab initio or de novo modeling. The assessment is carried out in every 2 years by Critical Assessment of Techniques for Protein Structure Prediction (McGuffin et al., 2018).

The protein secondary structure prediction aims to predict the local secondary structure such as alpha helices, beta sheets/strands or turns based only on the knowledge of their amino acid sequence. The most excellent method gives an accuracy of 80% in prediction, the secondary structure prediction depends on the refinement of sequence alignment, structural motifs, fold recognition, pattern matching, and ab initio methods (Huang and Brutlag, 2001; Pirovano and Heringa, 2010).

In the case of tertiary protein structure prediction, which implies a massive role of bioinformatics in structural biology, target modeling, and drug discovery in viroinformatics, numerous methodologies applied are homology/comparative modeling, threading techniques, de novo and ab initio based (Zhang, 2008).

As the protein structural folds are most evolutionary conserved than the amino acid sequences, a protein sequence target can be modeled with a distantly related template also. The significant difficulty in the homology modeling is the identification of appropriate template with a minimum 30%–40% of sequence similarity, whereas the homology modeling gives us the most optimum structure when the target and template have a high level of homology (Zhang and Skolnick, 2005).

Threading techniques: In case of missing residues or no optimum templates from the similarity searches, in threading technique the target sequence is scanned against the database of known or solved protein structure and thus relents a possible 3D structure (Bowie et al., 1991).

In ab initio or de novo modeling the 3D modeling of protein structure is built from scratch, the modeling procedure takes in account the physiochemical parameters rather than the previously solved structures (Lee et al., 2009).

MOLECULAR DYNAMICS AND FORCE FIELDS

In 1977 McCammon et al. first reported the application of MD simulation in the field of protein chemistry. The team has studied the dynamics of a folded globular protein (bovine pancreatic trypsin inhibitor) with an empirical potential energy function.

The spontaneous fluctuation of proteins in solution indicates that such fluctuation plays a vital role in the proteins' specific function (Ishima and Torchia, 2000; Ode et al., 2012) such as catalytic reaction, opening and closing of active site, and binding affinity of enzymes interaction with the other small molecule, cofactor, and biomolecules (Nicholson et al., 1995; Lu et al., 1998; Eisenmesser et al., 2005; Henzler-Wildman et al., 2007; Thorpe and Brooks, 2007 Abbondanzieri et al., 2008). Numerous experimental and computational methods are available to characterize and analyze the protein dynamics (Fig. 8.5), but it is too complicated to define the motions of protein and its dynamic phases at an atomic scale (Ode et al., 2012).

MD is the computational method to deal with either making a virtual framework or mimicking the natural condition out of the body (Dror et al., 2010; Henzler-Wildman et al., 2007). The techniques make

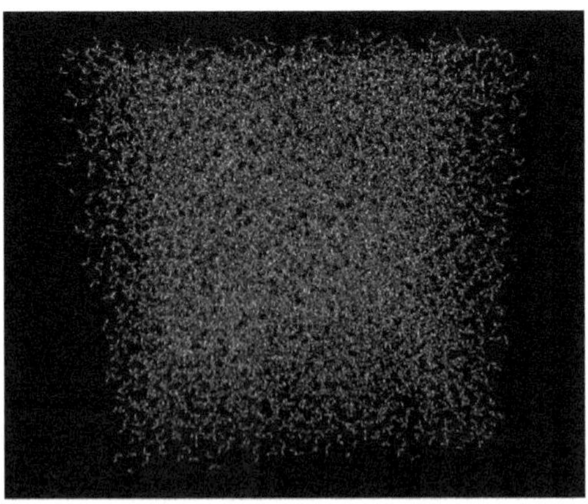

FIGURE 8.5 **Viral protein simulated in explicit water.**

possible for us calculate the rigid and flexible movements of atoms in a defined molecular system, such as protein in water (Adcock and McCammon, 2006; Karplus and Petsko, 1990).

Currently, MD simulation permits us to inspect the structural dynamics of proteins on a timescale of femtoseconds, picoseconds, nanoseconds to microseconds and in future mostly milliseconds (Dror et al., 2010; Henzler-Wildman et al., 2007). MD simulations help us in defining and also in refining the experimentally determined 3D structures of proteins, DNA, related biological macromolecules, and small chemical entities. It is also beneficial in modeling and constructing of poor or undescribed 3D structures of proteins in combination with homology modeling techniques (Baker and Sali, 2001; Martí-Renom et al., 2000; Sánchez et al., 2000).

Finally, the structural dynamics of the protein 3D structure can be addressed by the MD simulation over the time course from nanoseconds to microseconds. The trajectories and the snapshots of the protein dynamics give us the unique pattern and orientation of protein 3D structure in the specific interval of the period and condition concerning time, solvent system, and environmental condition.

MOLECULAR DOCKING

Molecular docking is a process that involves placing one molecule into the other with its appropriate conformation and configuration to interact with each other. In the field of molecular modeling, molecular docking refers to the study of how two molecule structures fit together. In 1890 Emil Fischer proposed the "Lock and Key" model that explained how the substrate fits into the active site/catalytic site/cavity of the macromolecule. Later in 1958 "Induced Fit Theory" was put forward by Daniel Koshland, where he explained how both ligand and target mutually adapt to each other through small conformation change (Molecular Docking, n.d.).

The most important docking systems are protein–ligand, protein–protein, and protein–nucleic acid. The outcome of the molecular docking is based on the scoring function and related physiochemical interactions between the target and ligand components, the interactions such as the following:

1. Electrostatic forces—dipole–dipole, charge–dipole, and charge–charge
2. Electrodynamics forces—Van der Waals interaction
3. Steric forces—caused by entropy
4. Solvent-related forces—hydrogen bond and hydrophobic interactions (Chaudhary and Mishra, 2016)

The scoring function furnishes a mode to the positioning of ligands proportional to some other. Ideally, the score should correspond directly to the binding affinity of the ligand for the protein, so that the best scoring ligands are the best binders. Scoring functions can be empirical, knowledge based, or molecular mechanics based. Scoring is compiled three different expressions applicable to docking and drug design: (1) generated configurations ranking by the docking search, (2) ranking different ligands against the protein (virtual screening), and (3) one or more ligands ranking against different proteins by their binding affinity (selectivity and specificity) (Abagyan et al., 1994; Friesner et al., 2004; Jones et al., 1997; Venkatachalam et al., 2003) (Table 8.2).

TABLE 8.2 List of Molecular Docking Online and Offline Tool

Sr. no.	Program	License	Sr no.	Program	License
1	1-Click Docking	Basic free version	16	FRED	Free for academic use
2	*AutoDock*	Freeware	17	GEMDOCK	Freeware
3	AutoDockVina	Open source	18	Glide	Commercial
4	BetaDock	Freeware	19	GOLD	Commercial
5	Blaster	Freeware	20	HADDOCK	Freeware
6	DOCK	Freeware for academic use	21	ICM-Dock	Commercial
7	DockingServer	Commercial	22	LigandFit	Commercial
8	*Docking Study with HyperChem*	Commercial	23	*MOE*	Commercial
9	DockVision	Commercial	24	MolDock	Academic
10	eHiTS	Commercial	25	MS-DOCK	Academic
11	FlexX	Commercial	26	ParDOCK	Freeware
12	FlexAID	Open source	27	PatchDock	Freeware
13	FlexPepDock	Freeware	28	Surflex-Dock	Commercial
14	FLIPDock	Free for academic use	29	SwissDock	Free webservice for academic use
15	FLOG	Academic	30	Vlife science docking engine	Commercial

MOE, Molecular Operating Environment.

References

Abagyan, R., Totrov, M., Kuznetsov, D., 1994. ICM—A new method for protein modelling and design: applications to docking and structure prediction from the distorted native conformation. J. Comput. Chem. 15 (5), 488–506. Available from: https://doi.org/10.1002/jcc.540150503.

Abbondanzieri, E.A., Bokinsky, G., Rausch, J.W., Zhang, J.X., Le Grice, S.F.J., Zhuang, X., 2008. Dynamic binding orientations direct activity of HIV reverse transcriptase. Nature 453 (7192), 184–189. Available from: https://doi.org/10.1038/nature06941.

Adcock, S.A., McCammon, J.A., 2006. Molecular dynamics: a survey of methods for simulating the activity of proteins. Chem. Rev. 106 (5), 1589–1615. Available from: https://doi.org/10.1021/cr040426m.

Baker, D., Sali, A., 2001. Protein structure prediction and structural genomics. Science (New York, NY) 294 (5540), 93–96. Available from: https://doi.org/10.1126/science.1065659.

Baskin, I., Varnek, A., 2009. Fragment descriptors in SAR/QSAR/QSPR studies, molecular similarity analysis and in virtual screening. ChemInform 40 (20), 1–49.

Besemer, J., Borodovsky, M., 2005. GeneMark: web software for gene finding in prokaryotes, eukaryotes and viruses. Nucleic Acids Res. 33 (Web Server), W451–W454. Available from: https://doi.org/10.1093/nar/gki487.

Bowie, J.U., Lüthy, R., Eisenberg, D., 1991. A method to identify protein sequences that fold into a known three-dimensional structure. Science (New York, NY) 253 (5016), 164–170. Retrieved from: <http://www.ncbi.nlm.nih.gov/pubmed/1853201>.

Chaudhary, K.K., Mishra, Nidhi, 2016. A review on molecular docking: novel tool for drug discovery. JSM Chem. 4 (3), 1029. Retrieved from: <https://dasher.wustl.edu/chem430/reading/jsmchem-4-1029-16.pdf>.

Chowdhary, M., Rani, A., Parkash, J., Mohd, S., Dev, D., 2016. Bioinformatics: an overview for cancer research. J. Drug Deliv. Ther. 6 (4), 69–72. Retrieved from: <http://jddtonline.info/index.php/jddt/article/view/1290>.

David, M., 2004. Bioinformatics: Sequence and Genome Analysis, second ed. Cold Spring Harbor Laboratory Press, Retrieved from: <https://www.amazon.com/Bioinformatics-Sequence-Analysis-David-Mount/dp/0879697121>.

Dror, R.O., Jensen, M.Ø., Borhani, D.W., Shaw, D.E., 2010. Exploring atomic resolution physiology on a femtosecond to millisecond timescale using molecular dynamics simulations. J. Gen. Physiol. 135 (6), 555–562. Available from: https://doi.org/10.1085/jgp.200910373.

Eisenmesser, E.Z., Millet, O., Labeikovsky, W., Korzhnev, D.M., Wolf-Watz, M., Bosco, D. A., et al., 2005. Intrinsic dynamics of an enzyme underlies catalysis. Nature 438 (7064), 117–121. Available from: https://doi.org/10.1038/nature04105.

Freddolino, P.L., Arkhipov, A.S., Larson, S.B., McPherson, A., Schulten, K., 2006. Molecular dynamics simulations of the complete satellite tobacco mosaic virus. Structure 14 (3), 437–449.

Friesner, R.A., Banks, J.L., Murphy, R.B., Halgren, T.A., Klicic, J.J., Mainz, D.T., et al., 2004. Glide: a new approach for rapid, accurate docking and scoring. 1. Method and assessment of docking accuracy. J. Med. Chem. 47 (7), 1739–1749. Available from: https://doi.org/10.1021/jm0306430.

Green, P.J., Mardia, K.V., 2006. Bayesian alignment using hierarchical models, with applications in protein bioinformatics. Biometrika 93 (2), 235–254. Available from: https://doi.org/doi.org/10.1093/biomet/93.2.235.

Henzler-Wildman, K.A., Lei, M., Thai, V., Kerns, S.J., Karplus, M., Kern, D., 2007. A hierarchy of timescales in protein dynamics is linked to enzyme catalysis. Nature 450 (7171), 913–916. Available from: https://doi.org/10.1038/nature06407.

Hoff, K.J., Tech, M., Lingner, T., Daniel, R., Morgenstern, B., Meinicke, P., 2008. Gene prediction in metagenomic fragments: a large scale machine learning approach. BMC Bioinf. 9, 217. Available from: https://doi.org/10.1186/1471-2105-9-217.

Holder, M., Lewis, P.O., 2003. Phylogeny estimation: traditional and Bayesian approaches. Nat. Rev. Genet. 4 (4), 275–284. Available from: https://doi.org/10.1038/nrg1044.

Huang, J.Y., Brutlag, D.L., 2001. The EMOTIF database. Nucleic Acids Res. 29 (1), 202–204. Retrieved from: <http://www.ncbi.nlm.nih.gov/pubmed/11125091>.

Huerta, M., Downing, G., Haseltine, F., Seto, B., Liu, Y., 2000. NIH Working Definition of Bioinformatics and Computational Biology. US National Institute of Health by BISTIC Definition Committee and released on July 17, 2000.

Hyatt, D., LoCascio, P.F., Hauser, L.J., Uberbacher, E.C., 2012. Gene and translation initiation site prediction in metagenomic sequences. Bioinformatics (Oxford, England) 28 (17), 2223–2230. Available from: https://doi.org/10.1093/bioinformatics/bts429.

Ishima, R., Torchia, D.A., 2000. Protein dynamics from NMR. Nat. Struct. Mol. Biol. 7 (9), 740–743. Available from: https://doi.org/10.1038/78963.

Jones, G., Willett, P., Glen, R.C., Leach, A.R., Taylor, R., 1997. Development and validation of a genetic algorithm for flexible docking. J. Mol. Biol. 267 (3), 727–748. Available from: https://doi.org/10.1006/jmbi.1996.0897.

Karplus, M., Petsko, G.A., 1990. Molecular dynamics simulations in biology. Nature 347 (6294), 631–639. Available from: https://doi.org/10.1038/347631a0.

Kelley, D.R., Liu, B., Delcher, A.L., Pop, M., Salzberg, S.L., 2012. Gene prediction with Glimmer for metagenomic sequences augmented by classification and clustering. Nucleic Acids Res. 40 (1), e9. Available from: https://doi.org/10.1093/nar/gkr1067.

Klasberg, S., Bitard-Feildel, T., Mallet, L., 2016. Computational identification of novel genes: current and future perspectives. Bioinf. Biol. Insights 10. Available from: https://doi.org/10.4137/BBI.S39950.

Lee, J., Wu, S., Zhang, Y., 2009. Ab initio protein structure prediction. Protein Structure to Function With Bioinformatics. Springer, pp. 3–25. Available from: https://doi.org/doi.org/10.1007/978-1-4020-9058-5_1.

Liu, Y., Guo, J., Hu, G., Zhu, H., 2013. Gene prediction in metagenomic fragments based on the SVM algorithm. BMC Bioinf. 14 (Suppl. 5), S12. Available from: https://doi.org/10.1186/1471-2105-14-S5-S12.

Lu, H.P., Xun, L., Xie, X.S., 1998. Single-molecule enzymatic dynamics. Science (New York, NY) 282 (5395), 1877–1882. Retrieved from: <http://www.ncbi.nlm.nih.gov/pubmed/9836635>.

Martí-Renom, M.A., Stuart, A.C., Fiser, A., Sánchez, R., Melo, F., Sali, A., 2000. Comparative protein structure modeling of genes and genomes. Annu. Rev. Biophys. Biomol. Struct. 29, 291–325. Available from: https://doi.org/10.1146/annurev.biophys.29.1.291.

McCammon, J.A., Gelin, B.R., Karplus, M., 1977. Dynamics of folded proteins. Nature 267 (5612), 585–590. Retrieved from: <http://www.ncbi.nlm.nih.gov/pubmed/301613>.

McGuffin, L.J., Shuid, A.N., Kempster, R., Maghrabi, A.H.A., Nealon, J.O., Salehe, B.R., et al., 2018. Accurate template-based modeling in CASP12 using the IntFOLD4-TS, ModFOLD6, and ReFOLD methods. Proteins: Struct. Funct. Bioinf. 86, 335–344. Available from: https://doi.org/10.1002/prot.25360.

Molecular Docking. (n.d.). <http://ncbr.muni.cz/~martinp/C3210/StructBioinf9.pdf> (retrieved 04.03.18).

Nicholson, L.K., Yamazaki, T., Torchia, D.A., Grzesiek, S., Bax, A., Stahl, S.J., et al., 1995. Flexibility and function in HIV-1 protease. Nat. Struct. Biol. 2 (4), 274–280. Retrieved from: <http://www.ncbi.nlm.nih.gov/pubmed/7796263>.

Noguchi, H., Park, J., Takagi, T., 2006. MetaGene: prokaryotic gene finding from environmental genome shotgun sequences. Nucleic Acids Res. 34 (19), 5623–5630. Available from: https://doi.org/10.1093/nar/gkl723.

Noguchi, H., Taniguchi, T., Itoh, T., 2008. MetaGeneAnnotator: detecting species-specific patterns of ribosomal binding site for precise gene prediction in anonymous prokaryotic and phage genomes. DNA Res. 15 (6), 387–396. Available from: https://doi.org/10.1093/dnares/dsn027.

Ode, H., Nakashima, M., Kitamura, S., Sugiura, W., Sato, H., 2012. Molecular dynamics simulation in virus research. Front. Microbiol. 3. Available from: https://doi.org/10.3389/fmicb.2012.00258.

Pirovano, W., Heringa, J., 2010. Protein secondary structure prediction. Methods Mol. Biol. (Clifton, NJ) 609, 327–348. Available from: https://doi.org/10.1007/978-1-60327-241-4_19.

Pybus, O.G., Rambaut, A., 2009. Evolutionary analysis of the dynamics of viral infectious disease. Nat. Rev. Genet. 10 (8), 540–550. Available from: https://doi.org/10.1038/nrg2583.

Ramsden, J., 2015. Bioinformatics, vol. 21. Springer London, London. Available from: https://doi.org/10.1007/978-1-4471-6702-0.

Redding, S., Greene, E.C., 2013. How do proteins locate specific targets in DNA? Chem. Phys. Lett. 570, 1–11. Available from: https://doi.org/10.1016/j.cplett.2013.03.035.

Rho, M., Tang, H., Ye, Y., 2010. FragGeneScan: predicting genes in short and error-prone reads. Nucleic Acids Res. 38 (20), e191. Available from: https://doi.org/10.1093/nar/gkq747.

Sánchez, R., Pieper, U., Melo, F., Eswar, N., Martí-Renom, M.A., Madhusudhan, M.S., et al., 2000. Protein structure modeling for structural genomics. Nat. Struct. Biol. 7, Suppl, 986–90. Available from: https://doi.org/10.1038/80776.

Sokolov, I.M., Metzler, R., Pant, K., Williams, M.C., 2005. Target search of N sliding proteins on a DNA. Biophys. J. 89 (2), 895–902. Available from: https://doi.org/10.1529/biophysj.104.057612.

Thorpe, I.F., Brooks, C.L., 2007. Molecular evolution of affinity and flexibility in the immune system. Proc. Natl. Acad. Sci. U.S.A. 104 (21), 8821–8826. Available from: https://doi.org/10.1073/pnas.0610064104.

Venkatachalam, C.M., Jiang, X., Oldfield, T., Waldman, M., 2003. LigandFit: a novel method for the shape-directed rapid docking of ligands to protein active sites. J. Mol. Graphics Modell. 21 (4), 289–307. Retrieved from: <http://www.ncbi.nlm.nih.gov/pubmed/12479928>.

Zhang, Y., 2008. Progress and challenges in protein structure prediction. Curr. Opin. Struct. Biol. 18 (3), 342–348. Available from: https://doi.org/10.1016/j.sbi.2008.02.004.

Zhang, Y., Skolnick, J., 2005. The protein structure prediction problem could be solved using the current PDB library. Proc. Natl. Acad. Sci. U.S.A. 102 (4), 1029–1034. Available from: https://doi.org/10.1073/pnas.0407152101.

Zhu, W., Lomsadze, A., Borodovsky, M., 2010. Ab initio gene identification in metagenomic sequences. Nucleic Acids Res. 38 (12), e132. Available from: https://doi.org/10.1093/nar/gkq275.

CHAPTER

9

Designing Antiviral Substances Targeting the Ebola Virus Viral Protein 24

Federico Dapiaggi[1], *Stefano Pieraccini*[1,2], *Donatella Potenza*[1], *Francesca Vasile*[1] *and Črtomir Podlipnik*[3]

[1]Department of Chemistry, University of Milano, Milano, Italy
[2]Institute of Molecular Science and Technologies, CNR, Milano, Italy
[3]Faculty of Chemistry and Chemical Technology, University of Ljubljana, Ljubljana, Slovenia

ABBREVIATIONS

CAS	computational alanine scanning
EBOV	Ebolavirus
EVD	Ebola viral disease
KPNA	karyopherin α
MD	molecular dynamics
MM–PBSA	molecular mechanics–Poisson Boltzmann surface area
PPIs	protein–protein interactions
VP24	viral protein 24

INTRODUCTION

Ebolavirus (EBOV), a member of the Filoviridae family, is a negative-sense ssRNA virus that has a filamentous appearance (Fig. 9.1B). EBOVs have five species named after the places of outbreaks: Zaire EBOV, Bundibugyo EBOV, Reston EBOV, Sudan EBOV, and Taï Forest EBOV. Among them, the Zaire strain is the most lethal. The outbreak of the EBOV, which lasted between December 2013 and April 2016, was

Emerging and Reemerging Viral Pathogens
DOI: https://doi.org/10.1016/B978-0-12-814966-9.00009-3

unprecedented in scale, being more extensive than all previous outbreaks combined. The outbreak generated more than 28,000 reported cases and at least 11,000 deaths in the population of Liberia, Guinea, and Sierra Leone and to a lesser extent in few other countries (Coltart et al., 2017). The virus is the causative agent of a zoonosis disease, which manifests as a life-threatening hemorrhagic fever in humans. The disease begins with the rapid development of fever and gastrointestinal issues; it continues with severe clotting disorders and finishes with multiorgan failure. Transmission of EBOV is mainly via physical contact with patients infected with the virus or its corpses. There is some evidence suggesting that it also spreads via droplets. We can limit the transmission by a proper combination of early diagnosis, contact tracing, patients' quarantine, infection control, and safe burial (Okware, 2016; Pandey et al., 2014). The mortality rate of the disease is very high, between 30% and 90%, with generally higher rates for Zaire EBOV than for the Bundibugyo and Sudan EBOV species (Bower et al., 2016). Surprisingly, the case–fatality ratios in the recent epidemic of Ebola viral disease (EVD) in West Africa tended to fall to around 40%. The better supportive care, earlier case detection and admission, could be the reason for lower fatality rate and the wane of this outbreak (Baseler et al., 2017). The disease is also associated with immunosuppression that is essential for its development. The suppression of interferon (IFN)-dependent antiviral response plays critical roles in the pathogenesis of Ebola viral infection. Fig. 9.1A is an electron micrograph of the

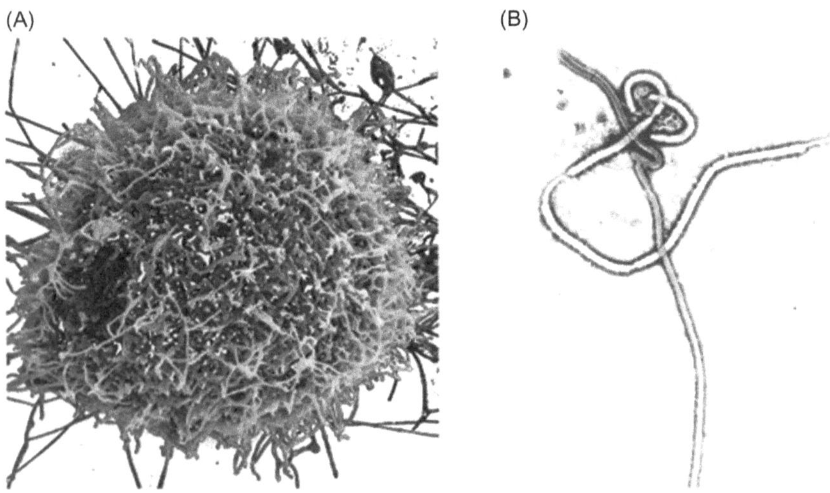

FIGURE 9.1 (A) An electron micrograph scan of Ebolavirus (filaments) particles, which are budding from an infected a cell (sphere in the center). (B) Filamentous structure of Ebolavirus.

viral attack to the mammalian cell (sphere in the center), particularly the budding off from the infected host cell.

EBOLA VIRUS—STRUCTURE AND FUNCTION

The RNA genome of EBOV's has seven genes, and it has the following structure: 3′-leader → NP → VP35 → VP40 → GP/sGp → VP30 → VP24 → L → 5′-trailer. These seven genes are coding at least 10 different proteins: nucleoprotein (NP); viral proteins (VP24, VP30, VP35, and VP40); glycoprotein (GP); soluble GP (sGP); small sGP; Δ-protein; and polymerase (L) (Yu et al., 2017). The EBOV's RNA that exists in negative-sense form cannot directly produce the virus proteins. The negative-sense RNA must be converted to its complementary copy by RNA polymerase (L) before translation. Like all other viruses, *EBOV* cannot replicate on its own, so it hijacks the host cell's machinery for its replication. In Fig. 9.2 the detailed cross section of the Ebola virion together with its building blocks (proteins) is shown.

The EBOV is composed of three layers from outer to inner: the viral envelope, the matrix space, and the nucleocapsid. The viral envelope is the outer layer of the EBOV; it consists of the lipid membrane hijacked

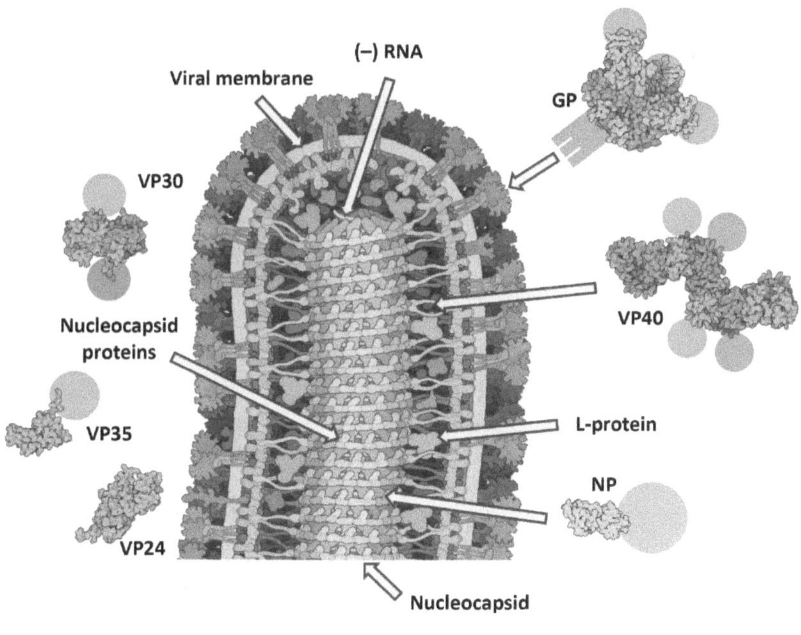

FIGURE 9.2 **Cross section of Ebola virus together with its building blocks (proteins)** (Goodsell et al., 2015).

from an infected cell. The numerous copies of the GP, distributed throughout whole viral membrane surface, are responsible for recognition and attaching to the host cell. The middle layer is a viral matrix consists of two proteins, VP40 and VP24. VP40, known as the major matrix protein, may rearrange into different forms, it can exist as a linear hexamer (matrix filament), octameric ring (viral transcription), and dimer (trafficking, matrix assembly, budding). As the VP24 is the main subject of this chapter, we will describe its role and action in the next section. The genetic material of the EBOV is encapsidated by NP and other viral proteins (VP35, VP30, VP24, and L protein) to form a helical nucleocapsid. Nucleocapsid also serves as a scaffold for virus assembly and as a template for genome transcription and replication (Yu et al., 2017).

Equipped with structural information about EBOV and knowing as much as possible about its action, one may propose different strategies for curing EBOV viral disease. The first possible strategy is to prevent adhesion of the virus to host cell; this can be done with monoclonal antibodies or with inhibitors of host-cell receptors (Misasi et al., 2016). Another promising strategy is to block viral escape from endosome, for example, with the inhibition of NPC1, or with disabling GP_1-GP_2 cleavage by protease inhibitors (Nyakatura et al., 2015). In addition, the inhibition of VP24 and VP35 may block the attack of the virus to the human immune system (Xu et al., 2014; Zhang et al., 2012). Then, one may target essential genes of the virus with small molecules such as TMPyP4 (Wang et al., 2016). The reader could find additional information related to the development of small molecule inhibitors against EBOV infection in the numerous reviews (Basu et al., 2015; Janeba, 2015; Litterman et al., 2015; Picazo and Giordanetto, 2015; Pleško and Podlipnik, 2016; Schafer et al., 2017).

STRUCTURE AND FUNCTION OF THE VIRAL PROTEIN 24

VP24 is the product of the sixth gene (Sanchez et al., 1993), and it is thought to be a minor matrix protein, in contrast to VP40, the most abundant EBOV protein. A virus matrix contains VP24 in small amounts. VP24 has typical characteristics of viral matrix proteins, due to its hydrophobic characteristics, the protein tends to associate with lipid bilayers, and it can oligomerize. For example, after singular expression in mammalian cells, it can form tetramers. High lipophilicity of the VP24 is essential for its function in virion assembly and budding (Han et al., 2003). VP24 also plays a vital role in intracellular assembly of the nucleocapsid. The electron microscopy experiments demonstrate that VP24 binds to NP and along with VP35 facilitate the assembly of the nucleocapsid and genome packaging (Banadyga et al., 2017; Noda et al., 2007). VP24 also acts as an essential regulatory factor in the process of the viral replication (Mateo et al., 2011; Watanabe et al., 2007).

Garcia-Dorival et al. used label-free quantitative proteomics to identify cellular proteins that had a high probability to interact with VP24. They confirmed several known interactions of VP24 with host cell proteins, such as interactions with karyopherins, which are responsible for nuclear import. They also discovered some new interactions, including one with the protein ATP1A1. This protein is essential for osmoregulation and cell signaling. The researchers have also shown that disruption of the ATP1A1 activity in EBOV-infected cells by small molecules could lead to a decrease in the virus progeny (Garcia-Dorival et al., 2014).

The cellular immune response to viral infection initiates with the binding of IFN to its receptors on the surface of the host cell. This binding leads to the activation of Janus tyrosine kinases (JAK1 and TYK2), which phosphorylate STAT1 protein. The phosphorylated STAT1 forms a stable dimer that interacts with karyopherin α5 (KPNA5). The KPNA5:STAT1:STAT1 complex then translocates from cytosol to the nucleus where STAT1 transcription factor activates expression of several hundred IFN-stimulated genes (Fig. 9.3). EBOV does not hinder the IFN-induced dimerization of STAT1. However, VP24 antagonizes KPNA5 and thus disables the formation of the complex between KPNA5 and STAT1 dimer, which is necessary for the translocation of STAT1 dimer to the nucleus. The measured affinity between KPNA5 and VP24 is in the low nanomolar range ($K_d = 1-10$ nM) in contrast to the STAT1 dimer that binds to KPNA5 with a micromolar dissociation constant (Xu et al., 2014). To sum up the EBOV protein, VP24, selectively inhibits nuclear import of STAT1 and as such blocks IFN-induced antiviral responses; this action of VP24 is shown schematically in Fig. 9.3 (Pleško and Podlipnik, 2016).

ENERGETICS OF VIRAL PROTEIN 24: KARYOPHERIN α5 COMPLEX

In this section, we will analyze the energetics of the interaction of VP24 with its binding partner KPNA5; this interaction is vital for the VP24 antagonistic function of the IFN-induced immune response.

We have used FoldX software (Schymkowitz et al., 2005), as a module of YASARA structure (Krieger and Vriend, 2014; Van Durme et al., 2011) for energy analysis of the interactions in the complex between viral VP24 and human KPNA5 (PDB-ID: 4U2X). As shown in Fig. 9.4, the VP24 is appearing as a trimer, and each VP24 monomer (chains A, B, C) interacts with the corresponding fragment of KPNA5 (chains D, E, F).

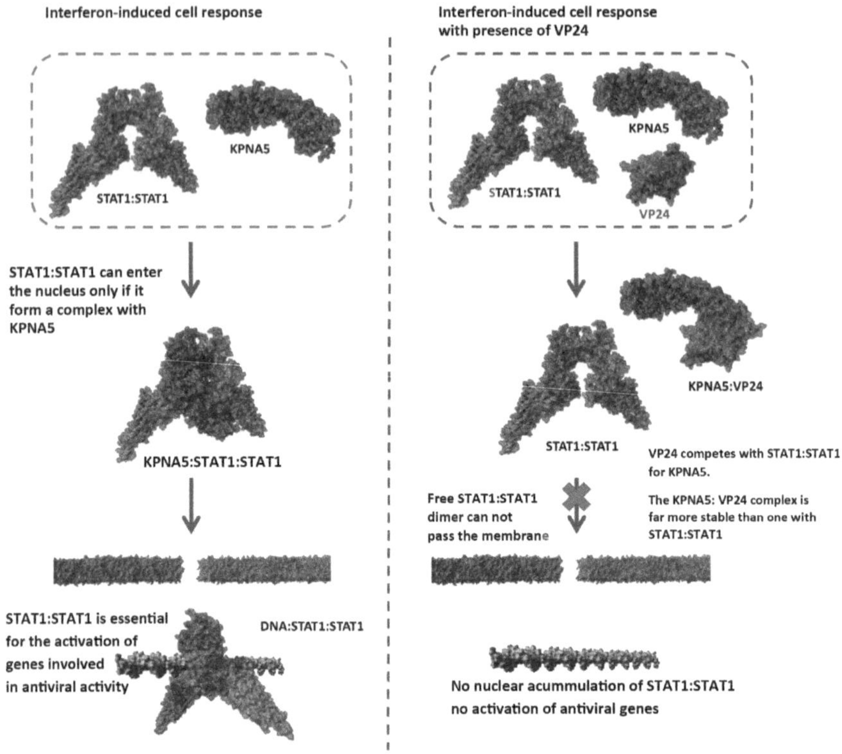

FIGURE 9.3 **The role of VP24 in the interferon-induced antiviral response** (Pleško and Podlipnik, 2016). *VP24,* Viral protein 24.

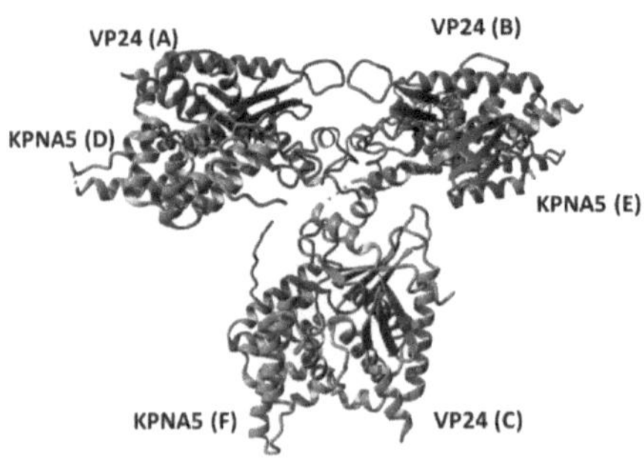

Chain	ΔG(kcal/mol)	ΔG(kcal/mol)
VP24 (A)	−46.1	−13.8
VP24 (B)	−47.6	−20.4
VP24 (C)	−33.1	−3.8
KPNA5 (D)	−32.3	−32.3
KPNA5 (E)	−27.2	−27.2
KPNA5 (F)	−29.3	−29.3

FIGURE 9.4 **The complex between VP24 and KPNA5 (PDB-ID: 4U2X).** *KPNA5,* Karyopherin α5; *VP24,* viral protein 24.

Structure of the complex between VP24 and KPNA5 along with FoldX's energy analysis is shown in Fig. 9.4.

We may observe that the binding energy of VP24 with its partner protein KPNA5 is around −30 kcal/mol. It is also evident that chains A and B of VP24 are more tightly bound to each other than any of these two chains of VP24 to chain C of VP24.

As the interaction between VP24 and KPNA5 plays an essential role in IFN antagonism, disruption of the interaction with binding a substance in KPNA5-binding region of VP24 could be a promising approach.

In the first instance, we will introduce the complete procedure for developing small peptides that successfully hinder interaction between VP24 and KPNA5 (Dapiaggi et al., 2017). In addition, we will review different attempts to find a small organic molecule that can successfully block the aggressive action of the virus. The majority of studies presented in this chapter are related to in silico screening of compounds that are present as an extract in different herbs regularly used in traditional medicine. Another option for accelerating the development of anti-Ebola treatment is repurposing already approved drugs (Sweiti et al., 2017).

DESIGN OF PEPTIDES INTERFERING VIRAL PROTEIN 24–KARYOPHERIN α INTERACTION

As previously explained, targeting the protein–protein complex VP24–KPNA could be an excellent therapeutic strategy to reduce Ebola virulence. Because of the peculiar characteristics of protein contact interfaces, which are often large and flat, lacking specific grooves and pockets suitable for the binding of small, drug-like molecules, targeting protein–protein interactions (PPIs) is a somewhat tricky task (Blundell et al., 2000; Lo Conte et al., 1999). Moreover, only a few natural compounds targeting PPIs have been discovered, so researchers in most cases lack molecular templates to drive their initial efforts. Finally, unlike enzymes, every PPI has to be considered a unique case, and the binding site does not appear to be preserved even among very similar interfaces (Chene, 2006). Despite all the drawbacks of PPIs as drug target, in recent years many encouraging results have been obtained, showing that exploiting PPIs as therapeutic targets is a viable, albeit challenging option (Raj et al., 2013; Scott et al., 2016), and several PPIs targeting molecules, both of peptidic (Fasan et al., 2004; Zinzalla and Thurston, 2009) and nonpeptidic nature (Arkin et al., 2003; Berg, 2003; Wells and McClendon, 2007), have been described.

Experimental and theoretical studies have shown that only a small fraction of the amino acids making up a protein–protein interface (around 10%) significantly contribute to the binding energy. These residues are referred to as "hot spots" (Keskin et al., 2008; Moreira et al., 2007b). Single mutations of such residues relevantly reduce the affinity for the partner protein, without changing the overall structure and integrity of the protein itself. Hot spots can be identified experimentally using alanine scanning mutagenesis. This approach consists in mutating sequentially residues located at the protein–protein interface into alanine and measuring the difference in binding free energies ($\Delta\Delta G$) between the wild-type and mutated complexes. This experimental procedure is, however, generally slow and labor intensive, since it requires DNA engineering, protein expression, and subsequent purification. For these reasons the alanine scanning mutagenesis is applied only to the preselected regions of the proteins, and blind alanine scanning for explorative purposes is unfeasible. In recent years, with the fast growth of computer science, computational alanine scanning (CAS) (Massova and Kollman, 1999; Zoete and Michielin, 2007) has become a valid alternative to the experimental one. This technique consists in calculating the binding Gibbs free energy between the subunits making up the wild-type complex (ΔG_{wt}) and the binding free energy upon mutation of each of the interfacial amino acids X into an alanine ($\Delta G_{Mut,X}$). Finally, for each residue X the $\Delta\Delta G$ can be obtained as

$$\Delta\Delta G = \Delta G_{Mut,X} - \Delta G_{wt}$$

If, the result of the mutation, $\Delta\Delta G$ is greater than a fixed threshold, generally between 2 and 4 kcal/mol, the mutated residue is defined as a hot spot.

There are, in principle, different ways to calculate a free energy of binding (Chipot and Pohorille, 2007). Many of them, ranging from score function–based methods (Trott and Olson, 2010) to more complex approaches relying on umbrella sampling (Lemkul and Bevan, 2010) and the Jarzynski equation (Cuendet and Michielin, 2008), have been applied to protein–protein complexes. Anyway, the molecular mechanics–Poisson Boltzmann surface area (MM–PBSA) approach (Hou et al., 2011; Massova and Kollman, 2000), which combines the molecular mechanics and the PBSA model in order to compute $\Delta G_{binding}$ of two species, has shown to be particularly efficient and accurate for computing binding free energy between proteins (Berhanu et al., 2013; Chen et al., 2016). This method is based on the generation of different structures of the complex through a molecular dynamics (MD) simulation and on the evaluation of the solvation contribution to binding free energy using an implicit solvent-based solvation model. For every snapshot extracted from the trajectory, the binding free energy is computed. Consider the binding of a ligand (L) and a receptor (R) forming a complex (C). In principle the $\Delta G_{binding}$ could be calculated in a single passage, using the explicit representation of the solvent. However, the significant contribution to this calculated energy would arise from the interactions between the solvent molecules, with the resulting fluctuation of the energy value virtually greater than the $\Delta G_{binding}$ itself. To bypass this problem, we can determine the $\Delta G_{binding}$ exploiting the thermodynamic cycle showed in Fig. 9.5, using the implicit solvent.

Different protocols exist to perform CAS. They differ both in the number of independent simulations they require to perform the calculation (Huo et al., 2002) and in the details of the simulation conditions. If we consider the number of MD simulations, we can distinguish the following two approaches:

Single-trajectory CAS uses only the MD simulation of the wild-type complex. The structures of the complex and its constituents are then extracted from this trajectory and postprocessed to obtain the mutated complexes. Monomers are obtained from the same simulation discarding the partner protein from the complex structure.

Double-trajectory CAS uses two simulations: one for the wild-type, and one for the mutated complex. Monomers are obtained as in the former case.

Single-trajectory CAS relies on the assumption that the considered point mutation, while modifying the binding energy of the complex, does not affect the protein three-dimensional (3-D) structure. This condition is often satisfied, and it has been shown that single-trajectory CAS generally gives results that are also affected by a lower statistical uncertainty (Massova and Kollman, 1999). Sometimes, point mutation can induce significant modifications in the 3-D structure of a complex, mainly when involving large residues. To keep this effect into account, double-trajectory approach can be used. This can lead to more accurate results, with a largely increased computational cost (one extra MD simulation is required for each point mutation) and a larger statistical error due to insufficient sampling. Moreover, using a single trajectory allows avoiding the explicit calculation of the entropic terms of the $\Delta\Delta G$

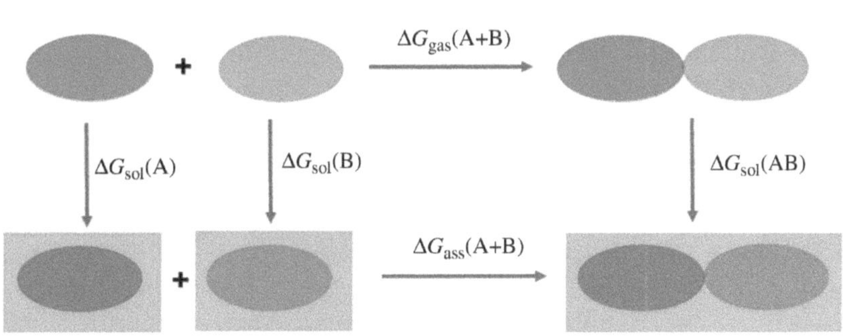

FIGURE 9.5 Thermodynamic cycle employed to calculate the binding free energy with the MM–PBSA approach. *MM–PBSA*, Molecular mechanics–Poisson Boltzmann surface area.

because both the wild-type and the mutated structure are obtained from the same trajectory, thus sampling the same phase space and leading to the mutual cancellation of this term.

Concerning the simulation conditions in MD, different approaches have been employed for alanine scanning computation. The MD simulations can be carried out either in an explicit or in an implicit solvent, and different protocol details have been proposed (Moreira et al., 2007a,b). While, in general, a better description of the solvent is preferred, supporters of implicit solvent claim that, if the calculation of the solvation term is done with implicit solvent, it is better to perform MD in the same conditions.

In the following section, we will discuss how a single-trajectory CAS on the system VP24–KPNA, using MM–PBSA method to perform free energies calculations, led to the identification of a nonapeptide able to target VP24, potentially inhibiting its interaction with KPNA. MD simulation on VP24–KPNA protein–protein complex was performed using the GROMACS package and Amber99SB-ildn force field. A 50 ns productive-phase MD was carried out, and CAS was performed on the last 20 ns of the dynamics when the system was fully equilibrated. In general a few nanoseconds long trajectories allow a sufficient sampling for an accurate binding free energy calculation with the MM–PBSA approach and consequently for CAS purposes (Hou et al., 2011; Moreira et al., 2006). We can define the protein–protein interface as the ensemble of amino acids, solvent-exposed surface area of which has a nonzero variation upon complex formation, and in the VP24–KPNA complex, it is composed of 56 residues. CAS results are summarized in Table 9.1. A large number of the residues at the interface (66%) are either polar or charged. This fact makes the VP24–KPNA interface rather peculiar, as protein–protein contact surfaces usually show a lower polarity concerning average protein surfaces (Lo Conte et al., 1999). We observed a high degree of electrostatic complementarity between the binding partners, which is mirrored in the number of salt bridges and hydrogen bonds occurring at the interface. In particular, Arg140 and Glu475′ (residues belonging to KPNA are labeled with a prime on their sequence number) form a salt bridge that is populated along 88% of the MD trajectory. And salt bridges between Glu203 and Lys399′, between Asp205 and Arg396′, and between Asp124 and Lys481′ are populated during 85%, 78%, and 64% of the trajectory, respectively (Fig. 9.6). Residue Asp124 is also involved in a hydrogen bond between its carboxyl group and Thr434′ hydroxyl, which is observed during 75% of the simulation, while Arg137 and Asp480′ backbones form a hydrogen bond for 74% of the trajectory. A small hydrophobic cluster, comprising Leu121, Val141, and Tyr477′, Phe484′, and aliphatic portion of the side chain of Lys481′ has been also observed and is steadily formed during the whole simulation (Fig. 9.7).

TABLE 9.1 Computational Alanine Scanning Results With Standard Errors

Mutation	ΔΔG (kcal/mol)	err (kcal/mol)	Mutation	ΔΔG (kcal/mol)	err (kcal/mol)
GLU88	− 10.99	0.79	LEU390	0.10	0.79
GLU113	**6.16**	**0.79**	GLN391	0.41	0.79
LEU121	**6.28**	**0.79**	GLU394	− 13.16	0.79
ASP124	**4.35**	**0.79**	PHE395	1.43	0.76
TRP125	0.96	0.79	**ARG396**	**21.04**	**0.79**
LEU127	0.14	0.79	**ARG398**	**18.80**	**0.74**
THR128	− 0.64	0.79	**LYS399**	**23.07**	**0.84**
THR129	1.53	0.79	**LYS427**	**16.67**	**0.79**
ASN130	0.24	0.79	ASP431	− 14.81	0.79
THR131	− 0.21	0.79	LEU433	0.50	0.79
PHE134	1.84	0.79	**THR434**	**6.09**	**0.79**
ASN135	**3.30**	**0.79**	VAL435	1.19	0.79
MET136	1.62	0.79	**MET436**	**3.63**	**0.79**
ARG137	**23.84**	**0.81**	ASP437	− 13.57	0.74
THR138	− 2.89	0.79	**GLU474**	**7.91**	**0.79**
GLN139	0.29	0.79	GLU475	1.98	0.79
ARG140	**9.94**	**0.79**	**TYR477**	**6.31**	**0.79**
VAL141	**2.65**	**0.79**	LEU479	0.62	0.79
GLN184	− 1.55	0.79	**ASP480**	**17.56**	**0.79**
ASN185	1.15	0.74	**LYS481**	**28.90**	**0.81**
HIS186	0.14	0.79	GLU483	0.79	0.79
LEU201	1.27	0.79	**PHE484**	**6.02**	**0.79**
GLN202	− 0.07	0.79	LEU485	0.55	0.79
GLU203	**10.37**	**0.72**	SER487	− 1.53	0.79
ASP205	**16.31**	**0.76**	HIS488	1.15	0.79
SER207	0.24	0.79	GLU489	− 4.44	0.76
ASN210	− 0.38	0.79	ILE501	− 0.45	0.79
LYS218	**11.13**	**0.79**	PHE505	1.00	0.79

The left column refers to VP24 residues, right column to KPNA residues. Hot spots are reported in bold. KPNA, Karyopherin α; VP24, viral protein 24.

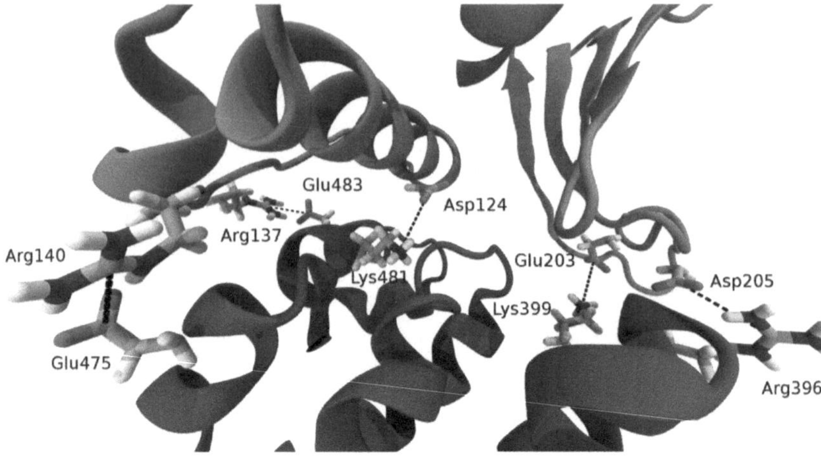

FIGURE 9.6 **Overview of the salt bridge interactions at the VP24–KPNA interface.** Figures were rendered with visual molecular dynamics (VMD) (Humphrey et al., 1996). *KPNA*, Karyopherin α; *VP24*, viral protein 24.

Moreover, few residues exhibit a most negative $\Delta\Delta G$ value, meaning that the mutation of the selected residue into alanine increases the binding between the proteins. Inspecting the electrostatic potential, it can be observed that there is not electrostatic potential complementarity in these regions, suggesting a somehow suboptimal binding between VP24 and KPNA, which deserves further attention from molecular biologists.

Starting from these geometric and energetic features of the interface, we identified two KPNA subsequences comprising a set of hot spots close to one another both in the 3-D structure and in the sequence of KPNA. Peptides sequences corresponding to one of these segments could be able to interfere with VP24–KPNA complex formation. In general the binding capability of an isolated peptide is not guaranteed, even if it contains several hot spots, because it may undergo major structural rearrangements when isolated from the parent protein and because of the complexity of the interaction networks that lead to protein complex formation. For these reasons the subsequences had to be tested with an MD simulation and a subsequent CAS, to verify the interaction with the partner protein, was either lost or retained. The first subsequence we considered, comprising the hot spot residues Arg396′, Arg398′, and Lys399′, ranges from Ala393′ to Glu400′. A 100 ns MD simulation of this peptide in complex with VP24 showed large conformational fluctuations, leading to a partial detachment of the peptide from the protein. In particular the salt bridge interactions involving the former hot spots are completely

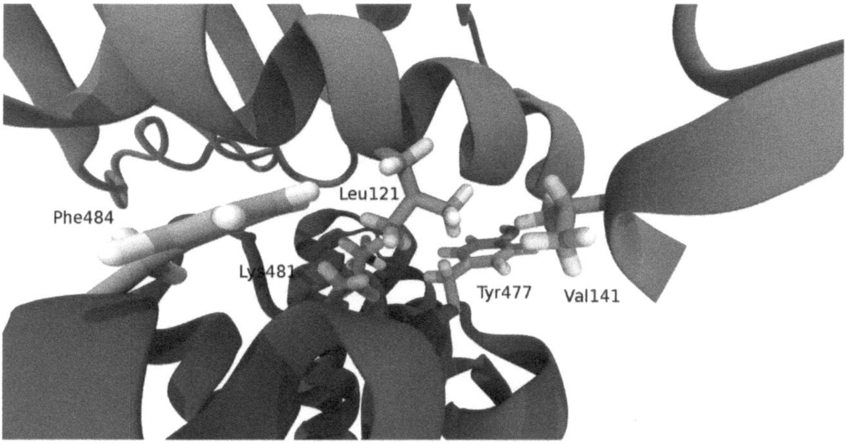

FIGURE 9.7 The hydrophobic cluster at the VP24–KPNA interface. *KPNA*, Karyopherin α; *VP24*, viral protein 24.

lost. For these reasons the peptide seems unable to interact with VP24, thus it was not further considered. The second subsequence comprises residues ranging from Glu474′ to Phe484′. A 100 ns MD simulation and a subsequent CAS on the peptide–VP24 complex were performed, to assay the binding capability of the selected sequence. Both Glu474′ (hot spot in VP24–KPNA complex) and Glu475′ showed negative $\Delta\Delta G$ values, while all the remaining hot spots are conserved. Based on these results, we decided to shorten the sequence discarding Glu474′ and Glu475′, obtaining a peptide with a smaller net charge (−1 instead of −3) and maintaining a very high density of hot spots, namely, Tyr477′, Asp480′, Lys481′, and Phe484′.

The selected peptide (from now on named RS) involves residues ranging from Ala476′ to Phe484′ and has the sequence AYGLDKIEF. The same protocol applied in the two subsequences mentioned above was employed. We ran a 100 ns long MD simulation of the VP24–RS complex, to investigate whether the identified peptide could retain its ability to bind to VP24, even when extracted from its protein environment. In contrast to what happened for the two former sequences, the protein–peptide complex showed good structural stability during the whole simulation, with protein binding site and peptide binding mode well conserved. On the other hand, its helical secondary structure was progressively relaxed and eventually lost during the simulation (Fig. 9.8).

Subsequently, a CAS was carried out to verify if the previously identified KPNA hot spots were conserved also in the VP24–RS complex. Not only all the hot spots are conserved but also a new one, that is, Glu483′, can be identified. This residue forms a salt bridge with Arg137,

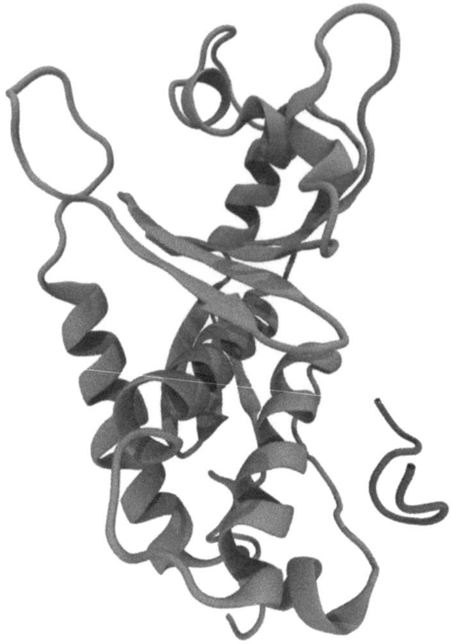

FIGURE 9.8 **VP24 (protein) in complex with the RS peptide (small peptide) extracted from KPNA.** *KPNA*, Karyopherin α; *VP24*, viral protein 24.

which is observed in 90% of the trajectory of the VP24–RS complex. Interestingly, this salt bridge, albeit present also in the VP24–KPNA complex, was observed only in 31% of the trajectory.

The PPIs can be classified based on the strength of interaction. The binding process is regarded as an equilibrium condition that results from a balance between association and dissociation events. The window of biologically relevant K_d values is vast and ranges from high affinity ($K_d <$ nM) to very low affinity ($K_d >$ mM). Strong protein interactions can be addressed experimentally by many techniques, including X-ray crystallography or binding assay, but if the interactions are weak, the vast majority of these approaches fail or become unreliable (Nealon et al., 2017). The nuclear magnetic resonance (NMR) spectroscopy is the only biophysical technique that can detect and quantify weak molecular interactions and at the same time provide detailed structural information of weak protein–ligand complexes with atomic level resolution.

Protein–ligand complexes are dynamic systems, and therefore the rate at which the components of the protein–ligand complex exchange between free and bound forms is central to the NMR method.

Among the ligand-observed NMR techniques that utilize nuclear Overhauser effect (NOE) effects between protein and ligand, saturation transfer difference (STD) can be used as an epitope mapping device to describe the target–ligand interactions (Meyer and Peters, 2003; Potenza et al., 2012).

To experimentally probe the interaction between the VP24 and the nonapeptide RS, we performed STD experiments on a phosphate buffer solution of the VP24–RS complex. Selected regions of ^{1}H-NMR and STD spectra of 1.1 mM peptide in the presence of 22 μM VP24 in phosphate buffer (50:1 ligand–target ratio) are shown in Fig. 9.9.

The STD is based on the transfer of saturation from the protein to the bound peptide, which in turn, by exchange, is moved into solution where it is detected. During the period of saturation the magnetization gradually moves from the protein to the protons of the peptide when the ligand binds to the target. The STD spectrum shows only the signals of peptide, which are in close contact with the protein. The degree of saturation of individual ligand protons (expressed regarding absolute

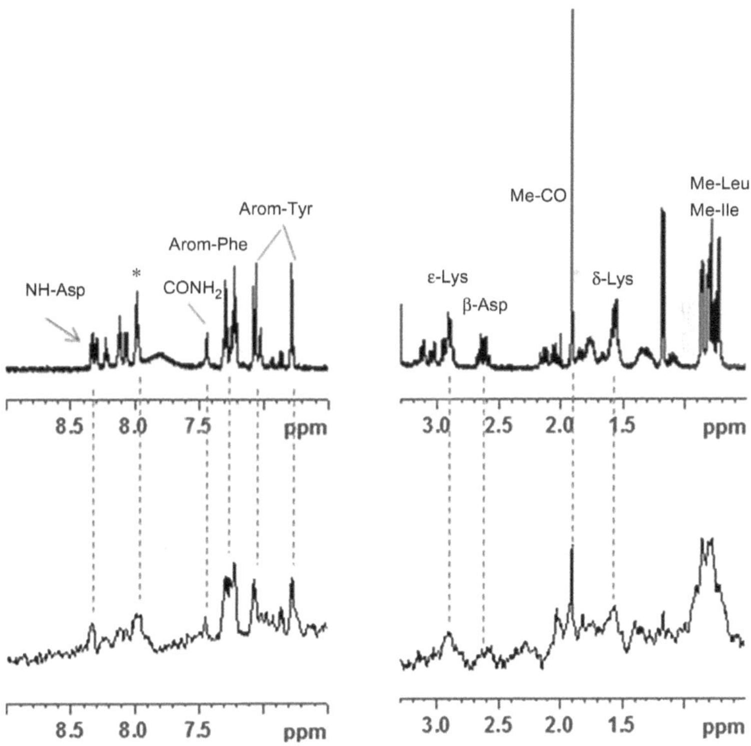

FIGURE 9.9 (Up) Selected regions of ^{1}H-NMR spectrum of 1.1 mM peptide in the presence of 22 μM VP24 in phosphate buffer (50:1 ligand–target ratio). The signal labeled with an asterisk is due to the Lys, Leu, and Ile amide protons, which show the same chemical shift. (Bottom) The same region of the STD spectrum. *NMR*, Nuclear magnetic resonance; *STD*, saturation transfer difference; *VP24*, viral protein 24.

STD percent) reflects the proximity to the receptor surface (Angulo and Nieto, 2011). In fact the ligand protons nearest to the protein are most likely to be saturated to the highest degree and therefore have the strongest signal in the mono-dimensional STD spectrum (Potenza et al., 2012). To better assess the interaction strength the percentage of absolute STD was calculated, and the results, grouped into two intensity ranges for protons which form the epitope, are shown in Fig. 9.10. In particular, we observed a strong interaction of the protein with the C-terminal amide protons (1.6% absolute STD) and the amidic proton of Asp5 (2% absolute STD). Also, the aromatic groups of Tyr2 and Phe9 are in intimate contact with the receptor site (1.5% and 1.6% absolute STD, respectively). Other less intense interactions are observed between protein and the protons of the side chain of Leu4, Asp5, Lys6, and Ile7. Unfortunately, it is not possible to quantify the interactions of the amide protons of Leu 4, Lys 6, and Ile 7 because their chemical shifts are isochronous.

As already specified, the selected peptide RS involves residues ranging from Ala476′ to Phe484′. The results obtained by STD–NMR experiments can in some way be compared with those obtained with CAS analysis even if the two approaches point out a different aspect of PPI. CAS is focused on the energetic of side chain interactions highlighted by their mutation into alanine, while STD supplies information on the proximity of a set of hydrogen atoms of the peptides concerning the protein.

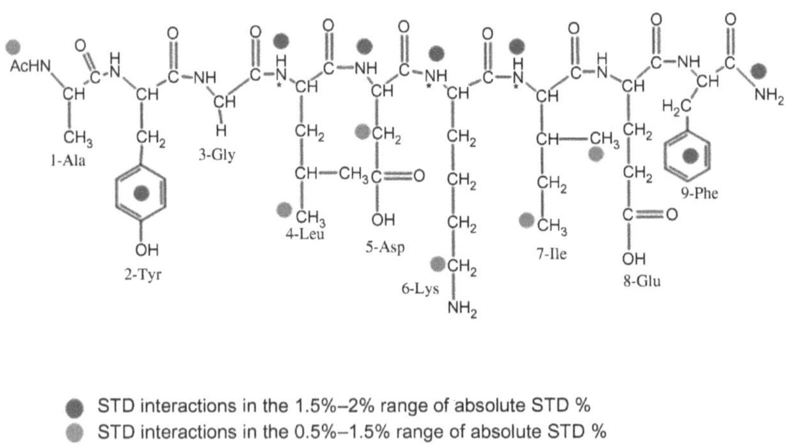

FIGURE 9.10 **The absolute STD percentage (grouped in two intensity ranges) for protons, which forms the epitope that reflects the relative proximity of the atoms to the binding site.** In particular the more strictly interacting moieties are the aromatic protons of Tyr2 and Phe9 (1.5% and 1.6% absolute STD, respectively), the amidic proton of Asp5 (2% absolute STD), and the C-terminal amide protons (1.6% absolute STD). *STD*, Saturation transfer difference.

We may notice that the key role of the hydrophobic interactions, between the Tyr477′ and Phe484′ aromatic residues, is highlighted both by CAS and by NMR analysis. In general, we observed that the two methodologies give mutually compatible and sound results.

SMALL MOLECULES AS INHIBITORS OF VIRAL PROTEIN 24

In this section, we review several attempts to find a small organic molecule that could act as an inhibitor of Ebola's protein, VP24. The virtual screening, an important method that is using often in the process of drug discovery, is appearing as a major procedure in scientific works reviewed in this section. The main task of virtual screening is to identify small molecules, which are plausible to bind a drug target (Lavecchia and Di Giovanni, 2013). The virtual screenings reviewed in this section is related to molecular docking, a method uses the complementary principle to predict the optimal orientation of small molecule within a bigger one, which is typically protein receptor or enzyme. The result of docking is stable complex and the estimated binding affinity between two molecules. The authors form this review recognized the compounds isolated from various herbal plants as a valuable source of drug candidates. Therefore they were searching for VP24 inhibitor in libraries of natural compounds and some of them in the library of FDA-approved drugs. The reader can find the recent reviews of the docking procedures and software in the recommended literature (Grinter and Zou, 2014; Kotev et al., 2016; Pagadala et al., 2017).

In our review, we used Schrödinger's Glide (Friesner et al., 2004, 2006; Halgren et al., 2004; Sherman et al., 2006) for docking simulations and visualizations of the complexes between ligands and VP24. Interaction complexes were produced using Grapheme Toolkit from OpenEye (Grapheme Toolkit 2017. Oct. 1 OpenEye Scientific Software, Santa Fe, NM. http://www.eyesopen.com).

Using in silico experiment, Pleško et al. (2015) suggested that several plant polyphenols may bind to VP24 and thus inhibit the formation of the complex between VP24 and KPNA5. The authors screened against VP24 a cherry-picked library of 18 polyphenols from herbal extracts, such as green tea, pomegranate extract, olive leaves. Numerous scientific reports state that the daily intake of compounds from the library can have a beneficial effect on human health. For example, epigallocatechin gallate (EGCG), which is the main component of green tea, can reduce the risk of Alzheimer's disease, diabetes II, and heart diseases and prevent cancer. Fig. 9.11 shows the structures of four representative compounds from the study.

Fig. 9.12 shows the binding modes of TGG and oleuropein within VP24, along with its docking score. Induced fit docking experiment of TGG resulted in the lowest binding score ($\Delta G = -15.8$ kcal/mol) among all compounds from the study. TGG nicely fits into quite an extensive region of the shallow pocket, which represents KPNA5 recognition domain. Each of four galloyl moieties connected to a glucose scaffold serves as tentacles, which search the optimal accommodation within the binding site of the VP24. The interaction between the ligand and VP24

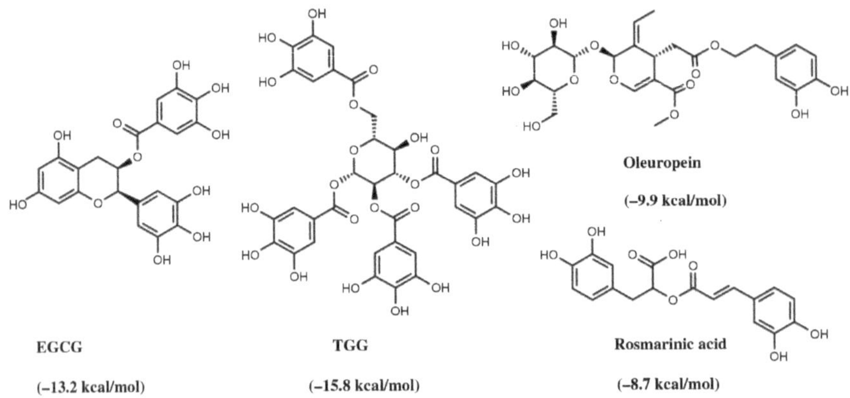

FIGURE 9.11 **Polyphenols extracted from plants as inhibitors of VP24 interferon antagonistic action, along with a docking score obtained with Schrödinger's Induced Fit Protocol.** *VP24*, Viral protein 24.

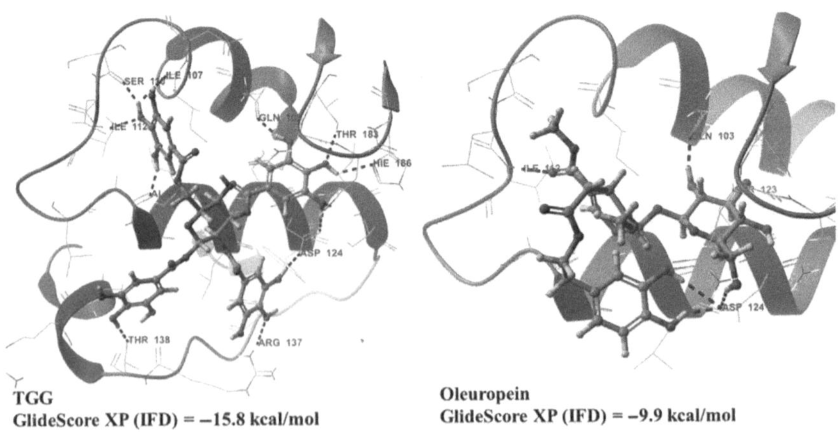

FIGURE 9.12 **Poses of 1,2,3,5-tetragalloyl glucose and oleuropein within VP24, poses and accompanied docking score obtained with Schrodinger's Induced Fit Protocol.** *VP24*, Viral protein 24.

is stabilized by 10 hydrogen bonds, which connects ligand's hydroxyls to protein, amino acids' residues (Ile 107, Ser 110, Ile 112, Ala 116, Asp 124, Arg 137, 2xThr 138, Asn 185).

Oleuropein, kaempferol, and quercetin are active ingredients in olive leaves, according to the in silico study of Kasmi et al., all three compounds have potent inhibitory activity against VP24 and VP30 (Kasmi, 2014). Induced fit docking results of Pleško et al. showed that oleuropein is the best VP24's in silico binder among these three compounds. From Fig. 9.12, it is apparently evident that oleuropein nicely fits into the binding site of VP24. In addition, five hydrogen bonds (Gln103, Ile112, Ser123, $2 \times$ Asp124) stabilize the interaction between oleuropein and VP24. Glide XP score (IFD - Induced fit docking) for the best pose is -9.9 kcal/mol.

We can summarize that Pleško et al. reported several new compounds with decent in silico affinity to inhibit VP24. Despite the promising docking results, it is tough to classify these compounds as promising leads for drug design, because of their questionable ADMET properties.

Sharma et al. (2017) studied the interaction of oseltamivir and its derivates with the model of VP24. They performed in silico–driven experiment, which shows that oseltamivir, developed initially as a neuraminidase inhibitor, binds to the protein model and thus might reduce the suppression of the human immune system induced by VP24. Furthermore, the authors performed also virtual screening using oseltamivir derivates as potential inhibitors; they found some candidates with better in silico performance to bind with VP24. Structures of oseltamivir and some of its derivates along with their Glide docking score are shown in Fig. 9.13. Based on some clinical reports collected in the latest EBOV outbreak, Haque et al. (2015) suggested that oseltamivir in a combination of FTY720, an FDA-approved drug for multiple sclerosis, could provide adequate protection against EBOV infection.

Uzochukwu et al. (2016) explored the binding affinities of the ethnomedicinal compound to five EBOV proteins (VP24, VP30, VP35, VP40, and NP) by molecular docking simulations. They identified five compounds: robustaflavone, cepharanthine, corilagin, hypericin, and theaflavin as potential remedies against Ebola viral infection. Structures of top-scored compounds along with its AutoDock-Vina docking score related to target VP24 (PDB-ID: 4M0Q) are shown in Fig. 9.14. For this review, we docked corilangin to VP24 using Schrödinger Glide docking software, the analysis of a resulted complex between corilangin and VP24 is shown in Fig. 9.15. The authors suggested that treatment with natural remedies could be a beneficiary supplement for anti-Ebola therapy without the time lag. Moreover, the use of the products from natural resources may reduce the cost of medical management of the disease.

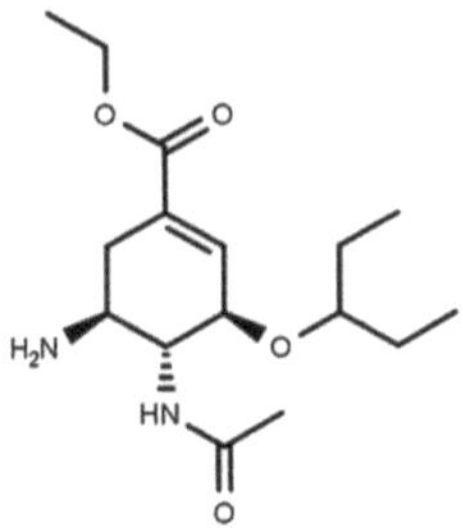

Oseltamivir

–5.8 kcal/mol

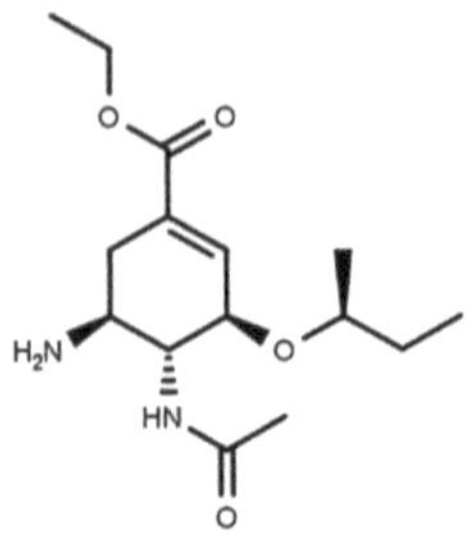

ZINC_77287098

–6.1 kcal/mol

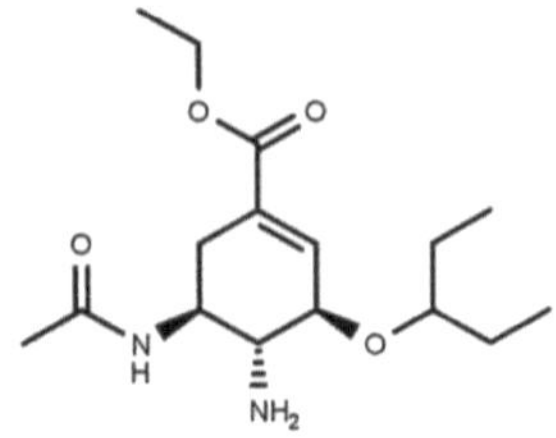

ZINC_77286925

–6.1 kcal/mol

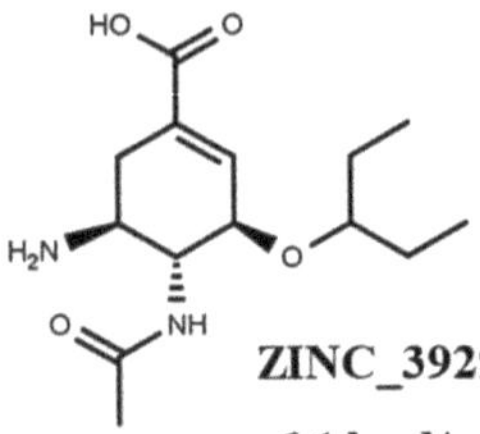

ZINC_3929509

–6.1 kcal/mol

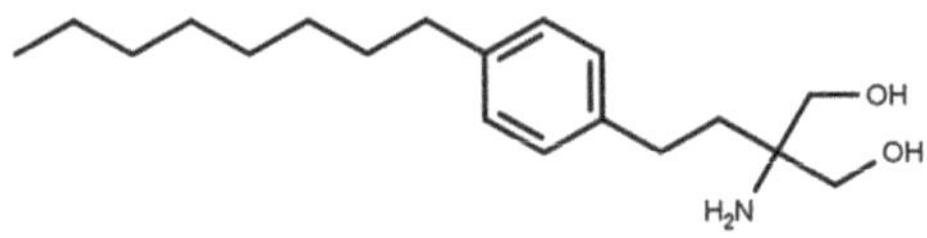

Fingolimod (FTY20)

FIGURE 9.13 Oseltamivir, its derivates, and FTY20 with accompanied Schrodinger's Glide docking score.

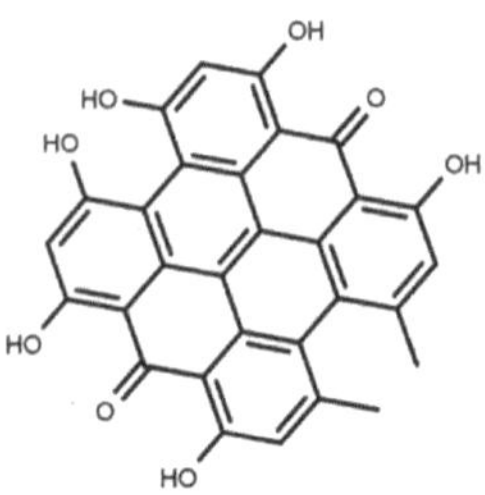

Hypericin

–8.68 kcal/mol

(–10.5 kcal/mol)

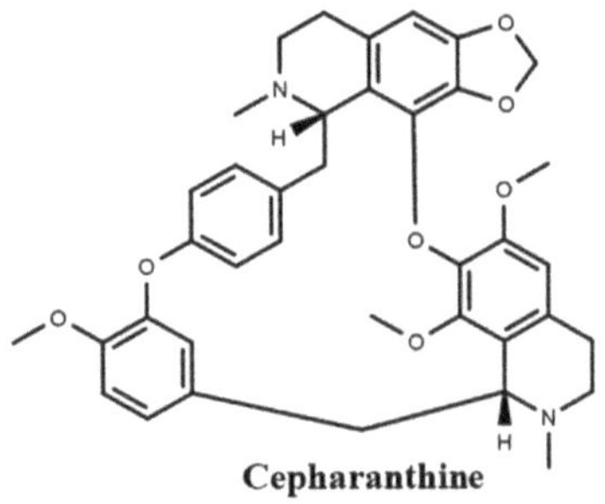

Cepharanthine

–7.95 kcal/mol

(–4.1 kcal/mol)

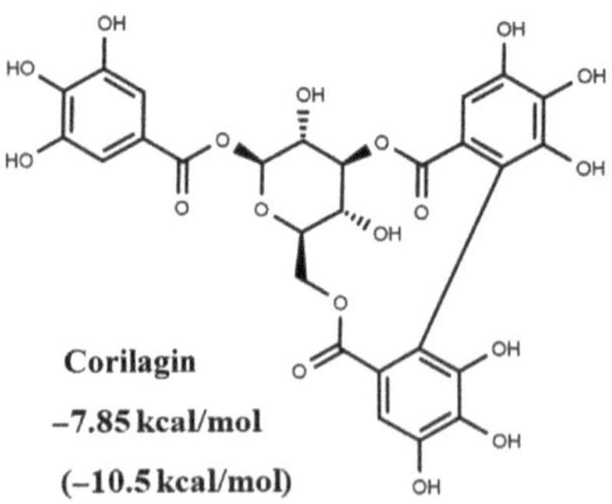

Corilagin

–7.85 kcal/mol

(–10.5 kcal/mol)

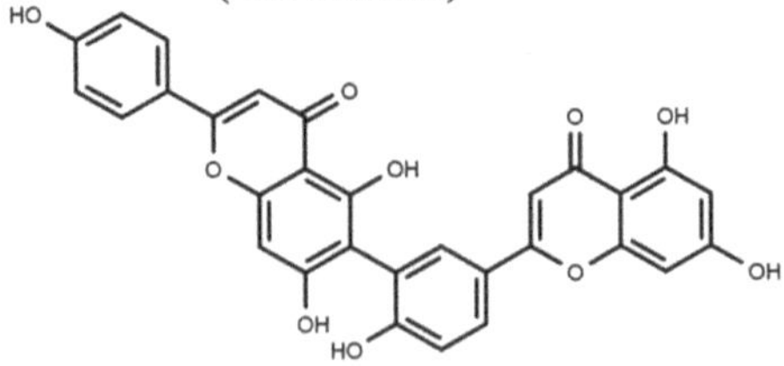

Robustaflavone

–8.48 kcal/mol

(–4.1 kcal/mol)

FIGURE 9.14 Examples of ethnomedicinal compounds studied as a VP24 inhibitor, along with AD-Vina Score and Glide IFD score in brackets. *VP24*, Viral protein 24.

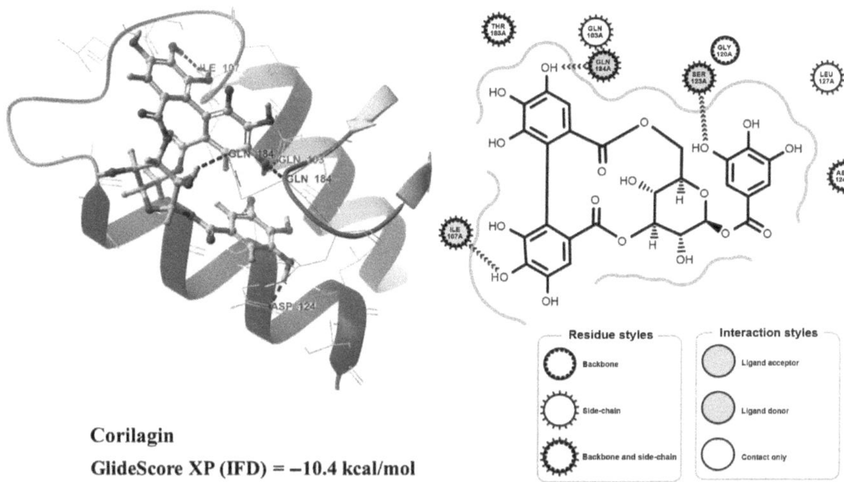

FIGURE 9.15 **Corilagin as an inhibitor of the VP24.** *VP24*, Viral protein 24.

Tambunan et al. performed an exciting study of searching EBOV's VP24 from the pool of natural Indonesian products through in silico drug discovery approach (Tambunan and Nasution, 2017). They collected 2020 Indonesian natural products from various sources, for which they performed initial ADMET screening test. The total number of 301 compounds, which passed initial filter, was subjected to docking protocol based on MOE.2014 software [*Molecular Operating Environment (MOE)*, 2013.08; Chemical Computing Group ULC, 1010 Sherbrooke St. West, Suite #910, Montreal, QC, Canada H3A 2R7, 2018]. Structures of several compounds from their study which theoretically binds to VP24 with decent affinity are shown in Fig. 9.16. The authors expose cycloartocarpin as the best candidate for the VP24 inhibitor since it has the lowest binding energy and decent ADMET properties.

Raj et al. used in silico protocol based on Schrödinger Suite to evaluate the inhibitory activity of flavonoids against the four EBOV proteins: VP40, VP35, VP30, and VP24 (Raj and Varadwaj, 2016). They screened a library of approximately 4000 compounds from extended flavonoid library (http://www.timtec.net/). After ADMET filtering and three-tiered virtual screening, the authors exposed gossypetin and taxifolin, which have the best binding properties surprisingly, to all four targets, among all studied compounds. For our review the induced fit docking of gossypetin and taxifolin to VP24 was performed using Schrödinger protocol. The resulted interaction diagrams (Fig. 9.17) are showing that both compounds share almost the same binding mode to VP24, interacting to protein's residues Gln103, Asp124, Asn185 via hydrogen bonds.

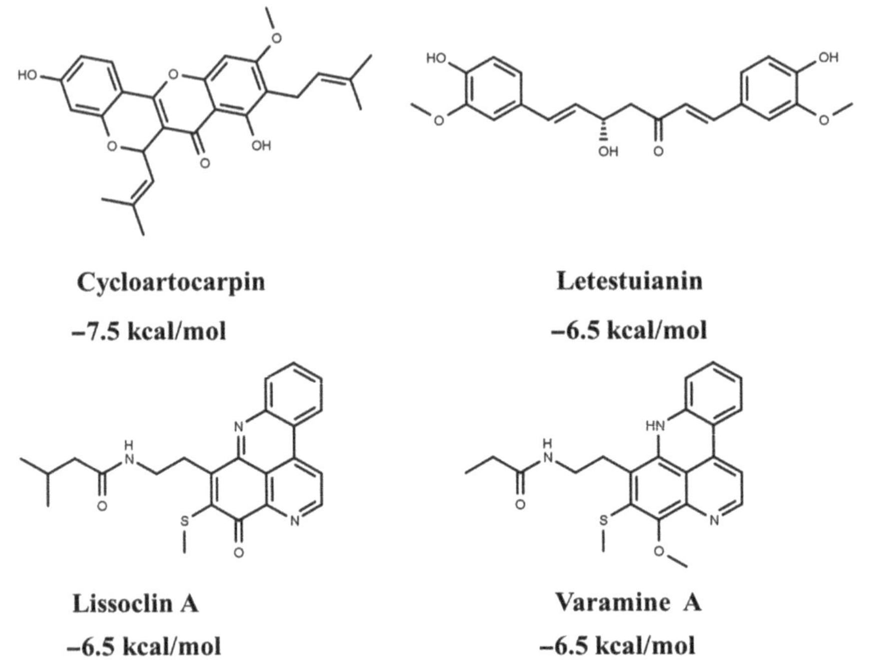

FIGURE 9.16 **Natural Indonesian products as an inhibitor of VP24, along with the binding energy estimated using MOE 2014 software.** *VP24,* Viral protein 24.

Both exposed compounds, gossypetin isolated from the *Hibiscus sabdariffa* and taxifolin isolated from conifers of the different coniferous tree, have low toxicity to human, known antiviral activity and acceptable pharmacokinetic properties. Therefore it can be taken as anti-Ebola therapy without the time lag for extensive clinical trials.

Setlur et al. (2017) performed another in silico study in line. They tried to find bioactive compounds against several EBOV targets, VP24 among them, by docking using AutoDock 4.2 (Morris et al., 2009). Using an extensive literature search, they selected 150 bioactive compounds for virtual screening against EBOV's drugable targets. Limonin, samarcandin, and gummosin, compounds from this study with the highest calculated affinity to VP24, are shown in Fig. 9.18 along with AutoDock docking score and VP24's interaction residues.

Limonin, a compound from the medicinal plants *Syzygium aromaticum,* has the best affinity against VP24 among compounds reported study provided by Setlur et al. (2017). Limonin is also an active component of *Citrus bergamia,* extract of which exhibits antiviral activity against retroviruses such as HIV. Samarcandin and gummosin, other compounds from the study, are active ingredients of *Ferula assa-foetida,*

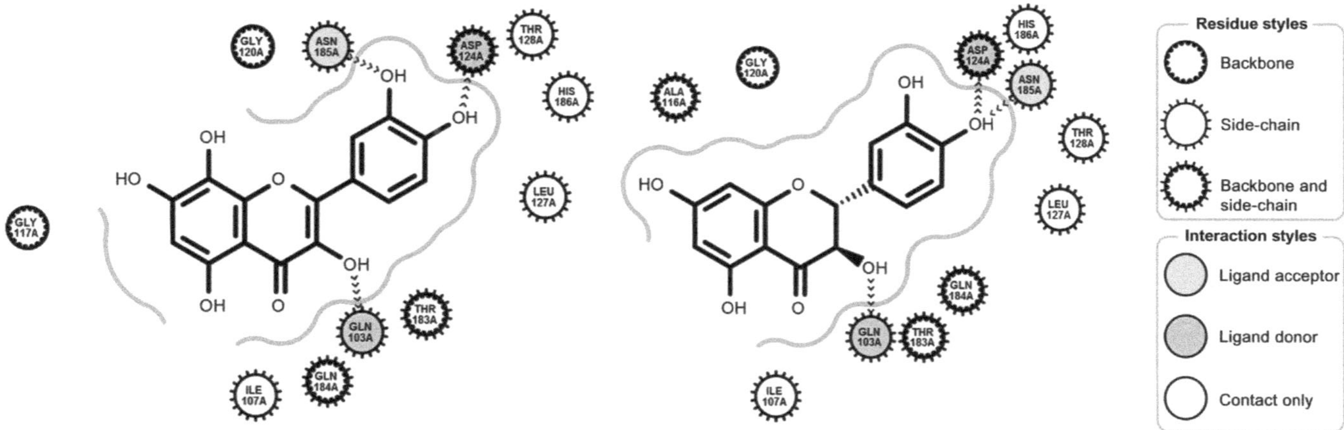

Gossypetin

GlideScore XP (IFD) = –10.6 kcal/mol

Taxifolin

GlideScore XP (IFD) = –10.9 kcal/mol

FIGURE 9.17 **The interaction diagram of gossypetin and taxifolin with VP24, and corresponding docking score.** *VP24*, Viral protein 24.

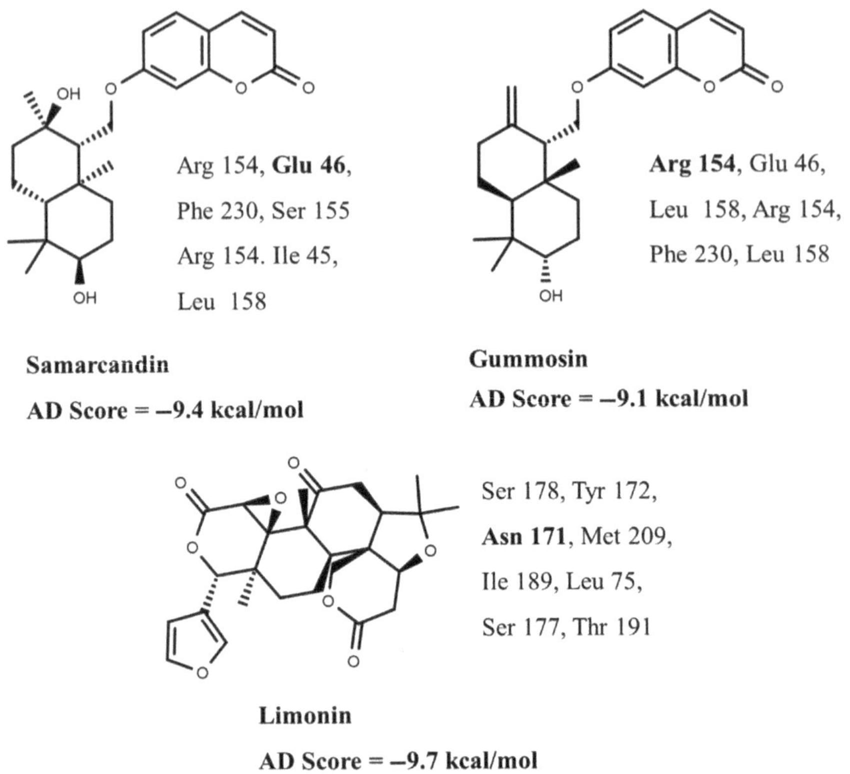

FIGURE 9.18 **Inhibitors of VP24 suggested by Setlur et al. along the VP24's interaction residues, and accompanying AD Score.** *AD Score*, AutoDock Score; *VP24*, viral protein 24.

a spice with known antiinflammatory, antidiabetic, antimutagenic, and antiviral properties. The active compounds from *F. assa-foetida* inhibit H1N1 and several other deleterious viral targets (Lee et al., 2009).

Zhao et al. (2016) used a drug repurposing strategy to find remedies against EVD. Repurposing already approved drugs provides an efficient solution to accelerate the development the therapy for curing EVD, which was already systematically reviewed in several places (Schuler et al., 2017; Sweiti et al., 2017). They were using structural system pharmacology approach for screening 1766 FDA-approved and 259 experimental drugs to identify compounds with the inhibitory activity of the replication and virulence of the virus. Their initial screening has identified numerous promising hits. They exposed indinavir, which is known as HIV protease inhibitor, and may effectively reduce the virulence of EBOV. They also found several antiviral drugs, such as maraviroc, abacavir, telbivudine, and cidofovir, which may inhibit Ebola RNA-directed RNA polymerase. The authors reported VP24 as a target for

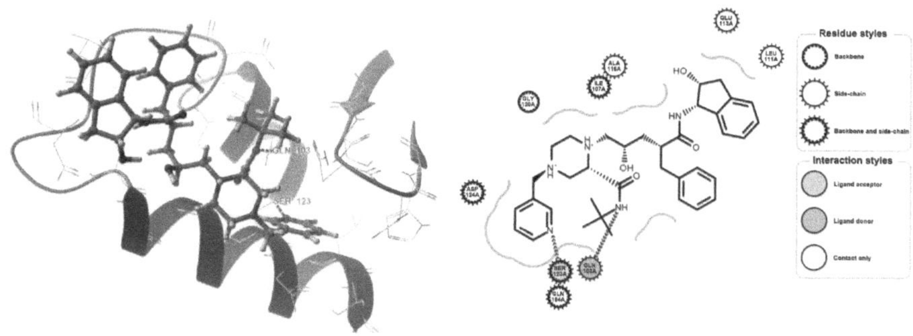

Indinavir ADVina Score = −6.84 kcal/mol

FIGURE 9.19 **Indinavir as an inhibitor of VP24.** *VP24,* Viral protein 24.

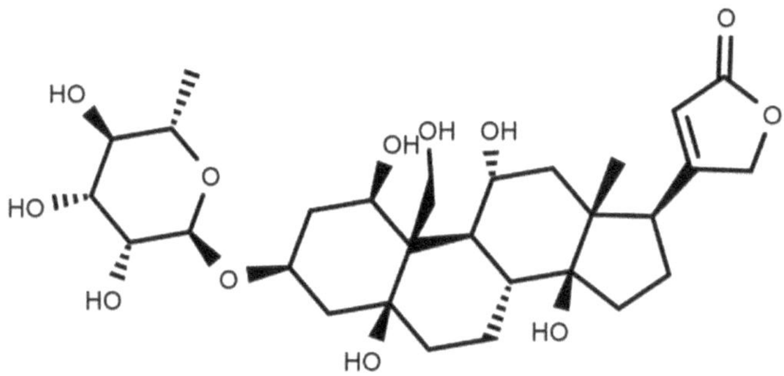

FIGURE 9.20 **Ouabain.**

indinavir. Fig. 9.19 shows the reproduction of the binding mode of the indinavir within VP24's binding pocket; docking experiment was performed with AutoDock/Vina (Trott and Olson, 2010).

ATP1A1 is one of the cellular proteins, which associates with VP24, the function of this protein is related to establishing and maintaining the electrochemical gradients of Na and K cations across the plasma membrane and in cell signaling. Garcia-Dorival et al. (2014) have shown that inhibition the function of ATP1A1 with ouabain (Fig. 9.20) resulting in a decrease in progeny virus. It has also been shown that ouabain can successfully inhibit EBOV replication in human lung cell culture. Unfortunately, and it is not yet FDA approved (Yu et al., 2017) (Fig. 9.20).

> If you know the enemy and know yourself, you need not fear the result of a hundred battles. If you know yourself but not the enemy, for every victory gained you will also suffer a defeat. If you know neither the enemy nor yourself, you will succumb in every battle. ***Sun Tzu, The Art of War.***

CONCLUSIONS

The chapter starts with the brief introduction of *EBOV* in context of its structure and action. The ancient wisdom of Sun Tzu stated that for a successful fight against the enemy, which is in our case *EBOV*, the thorough knowledge about the enemy and about us is extremely important. This is the reason why we started the chapter with a brief introduction of *EBOV* in context of its structure and action.

Recent studies unveiled the multiple strategies of EBOV to surround, attack, and evade the human immune system. Highly lethal EBOV is very adaptive, and it has several mechanisms for disabling the immune system. The viral armamentarium composed of VP24, VP35, and sGP offer the extensive coverage of different pathways by which human's antiviral responses occur. Antagonism of the IFN pathway, which plays a substantial role in the immune response, with VP24 is one of the fundamental mechanisms for devastation action of *EBOV*. In this mechanism, VP24 binds transporter KPNA and thus prevent translocation of transcription factor STAT1 from cytosol to nucleus, which is essential for gene expression related to immune response.

Targeting the protein–protein complex VP24–KPNA could be a promising therapeutic strategy to reduce Ebola virulence. Because of the specific characteristics of protein contact interfaces, which are often large and flat and lacking drugable pockets, the design of inhibitors could be a difficult task (Blundell et al., 2000; Lo Conte et al., 1999). A natural choice for interfering interactions between VP24 and KPNA is the design of oligopeptides, which are mimics of KPNA. Using extensive molecular simulation and NMR methods, Dapiaggi et al. designed nonapeptide with the sequence AYGLDKIEF that binds to VP24 and thus hinders the interaction with human host protein.

Another strategy that we explored was in silico screening for molecules, which may bind to VP24. For this study, we collected data from different sources, mainly virtual screenings. We found some interesting natural compounds with promising in silico activity, such as TGG, oleuropein, corilagin, samarcandin, gossypetin, and indinavir. The advantage of the treatment with natural remedies or with FDA-approved drugs could be a beneficiary supplement for anti-Ebola therapy without the time lag.

Acknowledgments

This work was supported by Slovenian Research Agency (ARRS) through the research program P1-0201. Č.P. express the gratitude to Mr. Sebastian Pleško for fruitful discussion and critical readings of draft. S.P. acknowledges Università degli Studi di Milano for the financial support through the Development Plan of Atheneum grant—line B1. F.D. and S. P. thank Prof. Maurizio Sironi for moral support and useful discussions.

References

Angulo, J., Nieto, P.M., 2011. STD-NMR: application to transient interactions between biomolecules—a quantitative approach. Eur. Biophys. J. 40 (12), 1357–1369. Available from: https://doi.org/10.1007/s00249-011-0749-5.

Arkin, M.R., Randal, M., DeLano, W.L., Hyde, J., Luong, T.N., Oslob, J.D., et al., 2003. Binding of small molecules to an adaptive protein–protein interface. Proc. Natl. Acad. Sci. U.S.A. 100 (4), 1603–1608. Available from: https://doi.org/10.1073/pnas.252756299.

Banadyga, L., Hoenen, T., Ambroggio, X., Dunham, E., Groseth, A., Ebihara, H., 2017. Ebola virus VP24 interacts with NP to facilitate nucleocapsid assembly and genome packaging. Sci. Rep. 7 (1), 7698. Available from: https://doi.org/10.1038/s41598-017-08167-8.

Baseler, L., Chertow, D.S., Johnson, K.M., Feldmann, H., Morens, D.M., 2017. The pathogenesis of Ebola virus disease. Annu. Rev. Pathol. 12, 387–418. Available from: https://doi.org/10.1146/annurev-pathol-052016-100506.

Basu, A., Mills, D.M., Mitchell, D., Ndungo, E., Williams, J.D., Herbert, A.S., et al., 2015. Novel small molecule entry inhibitors of Ebola virus. J. Infect. Dis. 212 (Suppl. 2), S425–S434. Available from: https://doi.org/10.1093/infdis/jiv223.

Berg, T., 2003. Modulation of protein–protein interactions with small organic molecules. Angew. Chem. Int. Ed. Engl. 42 (22), 2462–2481. Available from: https://doi.org/10.1002/anie.200200558.

Berhanu, W.M., Yasar, F., Hansmann, U.H., 2013. In silico cross seeding of Abeta and amylin fibril-like oligomers. ACS Chem. Neurosci. 4 (11), 1488–1500. Available from: https://doi.org/10.1021/cn400141x.

Blundell, T.L., Burke, D.F., Chirgadze, D., Dhanaraj, V., Hyvonen, M., Innis, C.A., et al., 2000. Protein–protein interactions in receptor activation and intracellular signalling. Biol. Chem. 381 (9–10), 955–959. Available from: https://doi.org/10.1515/BC.2000.117.

Bower, H., Smout, E., Bangura, M.S., Kamara, O., Turay, C., Johnson, S., et al., 2016. Deaths, late deaths, and role of infecting dose in Ebola virus disease in Sierra Leone: retrospective cohort study. BMJ 353, i2403. Available from: https://doi.org/10.1136/bmj.i2403.

Chen, F., Liu, H., Sun, H., Pan, P., Li, Y., Li, D., et al., 2016. Assessing the performance of the MM/PBSA and MM/GBSA methods. 6. Capability to predict protein–protein binding free energies and re-rank binding poses generated by protein–protein docking. Phys. Chem. Chem. Phys. 18 (32), 22129–22139. Available from: https://doi.org/10.1039/c6cp03670h.

Chene, P., 2006. Drugs targeting protein–protein interactions. ChemMedChem 1 (4), 400–411. Available from: https://doi.org/10.1002/cmdc.200600004.

Chipot, C., Pohorille, A., 2007. Free Energy Calculations. Springer-Verlag, Berlin Heidelberg.

Coltart, C.E., Lindsey, B., Ghinai, I., Johnson, A.M., Heymann, D.L., 2017. The Ebola outbreak, 2013–2016: old lessons for new epidemics. Philos. Trans. R. Soc. Lond. B Biol. Sci. 372 (1721). Available from: https://doi.org/10.1098/rstb.2016.0297.

Cuendet, M.A., Michielin, O., 2008. Protein–protein interaction investigated by steered molecular dynamics: the TCR-pMHC complex. Biophys. J. 95 (8), 3575–3590. Available from: https://doi.org/10.1529/biophysj.108.131383.

Dapiaggi, F., Pieraccini, S., Potenza, D., Vasile, F., Macut, H., Pellegrino, S., et al., 2017. Computer aided design and NMR characterization of an oligopeptide targeting the Ebola virus VP24 protein. New J. Chem. 41 (11), 4308–4315. Available from: https://doi.org/10.1039/C6NJ04014D.

Fasan, R., Dias, R.L., Moehle, K., Zerbe, O., Vrijbloed, J.W., Obrecht, D., et al., 2004. Using a β-hairpin to mimic an α-helix: cyclic peptidomimetic inhibitors of the p53–HDM2 protein–protein interaction. Angew. Chem. Int. Ed. Engl. 43 (16), 2109–2112. Available from: https://doi.org/10.1002/anie.200353242.

Friesner, R.A., Banks, J.L., Murphy, R.B., Halgren, T.A., Klicic, J.J., Mainz, D.T., et al., 2004. Glide: a new approach for rapid, accurate docking and scoring. 1. Method and assessment of docking accuracy. J. Med. Chem. 47 (7), 1739–1749. Available from: https://doi.org/10.1021/jm0306430.

Friesner, R.A., Murphy, R.B., Repasky, M.P., Frye, L.L., Greenwood, J.R., Halgren, T.A., et al., 2006. Extra precision glide: docking and scoring incorporating a model of hydrophobic enclosure for protein–ligand complexes. J. Med. Chem. 49 (21), 6177–6196. Available from: https://doi.org/10.1021/jm051256o.

Garcia-Dorival, I., Wu, W., Dowall, S., Armstrong, S., Touzelet, O., Wastling, J., et al., 2014. Elucidation of the Ebola virus VP24 cellular interactome and disruption of virus biology through targeted inhibition of host-cell protein function. J. Proteome Res. 13 (11), 5120–5135. Available from: https://doi.org/10.1021/pr500556d.

Goodsell, D.S., Dutta, S., Zardecki, C., Voigt, M., Berman, H.M., Burley, S.K., 2015. The RCSB PDB "Molecule of the Month": inspiring a molecular view of biology. PLoS Biol 13 (5), e1002140. Available from: https://doi.org/10.1371/journal.pbio.1002140.

Grinter, S.Z., Zou, X., 2014. Challenges, applications, and recent advances of protein-ligand docking in structure-based drug design. Molecules 19 (7), 10150–10176. Available from: https://doi.org/10.3390/molecules190710150.

Halgren, T.A., Murphy, R.B., Friesner, R.A., Beard, H.S., Frye, L.L., Pollard, W.T., et al., 2004. Glide: a new approach for rapid, accurate docking and scoring. 2. Enrichment factors in database screening. J. Med. Chem. 47 (7), 1750–1759. Available from: https://doi.org/10.1021/jm030644s.

Han, Z., Boshra, H., Sunyer, J.O., Zwiers, S.H., Paragas, J., Harty, R.N., 2003. Biochemical and functional characterization of the Ebola virus VP24 protein: implications for a role in virus assembly and budding. J. Virol. 77 (3), 1793–1800.

Haque, A., Hober, D., Blondiaux, J., 2015. Addressing therapeutic options for Ebola virus infection in current and future outbreaks. Antimicrob. Agents Chemother. 59 (10), 5892–5902. Available from: https://doi.org/10.1128/AAC.01105-15.

Hou, T., Wang, J., Li, Y., Wang, W., 2011. Assessing the performance of the MM/PBSA and MM/GBSA methods. 1. The accuracy of binding free energy calculations based on molecular dynamics simulations. J. Chem. Inf. Model. 51 (1), 69–82. Available from: https://doi.org/10.1021/ci100275a.

Humphrey, W., Dalke, A., Schulten, K., 1996. VMD: visual molecular dynamics. J. Mol. Graph. 14 (1), 33–38. 27-38.

Huo, S., Massova, I., Kollman, P.A., 2002. Computational alanine scanning of the 1:1 human growth hormone–receptor complex. J. Comput. Chem. 23 (1), 15–27. Available from: https://doi.org/10.1002/jcc.1153.

Janeba, Z., 2015. Development of small-molecule antivirals for Ebola. Med. Res. Rev. 35 (6), 1175–1194. Available from: https://doi.org/10.1002/med.21355.

Kasmi, Y., 2014. Étude in silico des effets inhibitrices des Oleuropein, Kaempferol, et Quercetin sur le Protéine VP30 de Ebola Virus [In silico study of the inhibitory effects of Oleuropein, Kaempferol, and Quercetin on the VP30 protein from Ebola virus]. Int. J. Innov. Appl. Stud. 8 (4), 1566.

Keskin, O., Gursoy, A., Ma, B., Nussinov, R., 2008. Principles of protein–protein interactions: what are the preferred ways for proteins to interact? Chem. Rev. 108 (4), 1225–1244. Available from: https://doi.org/10.1021/cr040409x.

Kotev, M., Soliva, R., Orozco, M., 2016. Challenges of docking in large, flexible and promiscuous binding sites. Bioorg. Med. Chem. 24 (20), 4961–4969. Available from: https://doi.org/10.1016/j.bmc.2016.08.010.

Krieger, E., Vriend, G., 2014. YASARA view—molecular graphics for all devices—from smartphones to workstations. Bioinformatics 30 (20), 2981–2982. Available from: https://doi.org/10.1093/bioinformatics/btu426.

Lavecchia, A., Di Giovanni, C., 2013. Virtual screening strategies in drug discovery: a critical review. Curr. Med. Chem. 20 (23), 2839–2860.

Lee, C.L., Chiang, L.C., Cheng, L.H., Liaw, C.C., Abd El-Razek, M.H., Chang, F.R., et al., 2009. Influenza A (H_1N_1) antiviral and cytotoxic agents from ferula assa-foetida. J. Nat. Prod. 72 (9), 1568–1572. Available from: https://doi.org/10.1021/np900158f.

Lemkul, J.A., Bevan, D.R., 2010. Assessing the stability of Alzheimer's amyloid protofibrils using molecular dynamics. J. Phys. Chem. B 114 (4), 1652–1660. Available from: https://doi.org/10.1021/jp9110794.

Litterman, N., Lipinski, C., Ekins, S., 2015. Small molecules with antiviral activity against the Ebola virus. F1000Res. 4, 38. Available from: https://doi.org/10.12688/f1000research.6120.1.

Lo Conte, L., Chothia, C., Janin, J., 1999. The atomic structure of protein–protein recognition sites. J. Mol. Biol. 285 (5), 2177–2198.

Massova, I., Kollman, P.A., 1999. Computational alanine scanning to probe protein–protein interactions: a novel approach to evaluate binding free energies. J. Am. Chem. Soc. 121 (36), 8133–8143. Available from: https://doi.org/10.1021/Ja990935j.

Massova, I., Kollman, P.A., 2000. Combined molecular mechanical and continuum solvent approach (MM-PBSA/GBSA) to predict ligand binding. Perspect. Drug Discovery Des. 18, 113–135. Available from: https://doi.org/10.1023/A:1008763014207.

Mateo, M., Carbonnelle, C., Martinez, M.J., Reynard, O., Page, A., Volchkova, V.A., et al., 2011. Knockdown of Ebola virus VP24 impairs viral nucleocapsid assembly and prevents virus replication. J. Infect. Dis. 204 (Suppl. 3), S892–S896. Available from: https://doi.org/10.1093/infdis/jir311.

Meyer, B., Peters, T., 2003. NMR spectroscopy techniques for screening and identifying ligand binding to protein receptors. Angew. Chem. Int. Ed. Engl. 42 (8), 864–890. Available from: https://doi.org/10.1002/anie.200390233.

Misasi, J., Gilman, M.S., Kanekiyo, M., Gui, M., Cagigi, A., Mulangu, S., et al., 2016. Structural and molecular basis for Ebola virus neutralization by protective human antibodies. Science 351 (6279), 1343–1346. Available from: https://doi.org/10.1126/science.aad6117.

Moreira, I.S., Fernandes, P.A., Ramos, M.J., 2006. Unraveling the importance of protein–protein interaction: application of a computational alanine-scanning mutagenesis to the study of the IgG1 streptococcal protein G (C2 fragment) complex. J. Phys. Chem. B 110 (22), 10962–10969. Available from: https://doi.org/10.1021/jp054760d.

Moreira, I.S., Fernandes, P.A., Ramos, M.J., 2007a. Computational alanine scanning mutagenesis—an improved methodological approach. J. Comput. Chem. 28 (3), 644–654. Available from: https://doi.org/10.1002/jcc.20566.

Moreira, I.S., Fernandes, P.A., Ramos, M.J., 2007b. Hot spots—a review of the protein–protein interface determinant amino-acid residues. Proteins 68 (4), 803–812. Available from: https://doi.org/10.1002/prot.21396.

Morris, G.M., Huey, R., Lindstrom, W., Sanner, M.F., Belew, R.K., Goodsell, D.S., et al., 2009. AutoDock4 and AutoDockTools4: automated docking with selective receptor flexibility. J. Comput. Chem. 30 (16), 2785–2791. Available from: https://doi.org/10.1002/jcc.21256.

Nealon, J.O., Philomina, L.S., McGuffin, L.J., 2017. Predictive and experimental approaches for elucidating protein–protein interactions and quaternary structures. Int. J. Mol. Sci. 18 (12). Available from: https://doi.org/10.3390/Ijms18122623Artn 2623.

Noda, T., Halfmann, P., Sagara, H., Kawaoka, Y., 2007. Regions in Ebola virus VP24 that are important for nucleocapsid formation. J. Infect. Dis. 196 (Suppl. 2), S247–S250. Available from: https://doi.org/10.1086/520596.

Nyakatura, E.K., Frei, J.C., Lai, J.R., 2015. Chemical and structural aspects of Ebola virus entry inhibitors. ACS Infect. Dis. 1 (1), 42–52. Available from: https://doi.org/10.1021/id500025n.

Okware, S., 2016. Managing Ebola in low-resource settings: experiences from Uganda. In: Podlipnik, Č. (Ed.), Ebola. InTech, Rijeka, Croatia, pp. 3–19.

Pagadala, N.S., Syed, K., Tuszynski, J., 2017. Software for molecular docking: a review. Biophys. Rev. 9 (2), 91–102. Available from: https://doi.org/10.1007/s12551-016-0247-1.

Pandey, A., Atkins, K.E., Medlock, J., Wenzel, N., Townsend, J.P., Childs, J.E., et al., 2014. Strategies for containing Ebola in West Africa. Science 346 (6212), 991–995. Available from: https://doi.org/10.1126/science.1260612.

Picazo, E., Giordanetto, F., 2015. Small molecule inhibitors of Ebola virus infection. Drug Discov. Today 20 (2), 277–286. Available from: https://doi.org/10.1016/j.drudis.2014.12.010.

Pleško, S., Volk, H., Luksic, M., Podlipnik, C., 2015. In silico study of plant polyphenols' interactions with VP24-Ebola virus membrane-associated protein. Acta Chim. Slov. 62 (3), 555–564.

Pleško, S., Podlipnik, Č., 2016. Strategies for the development of small molecule inhibitors of Ebola viral infection. In: Podlipnik, Č. (Ed.), Ebola. InTech, Rijeka, Croatia, pp. 137–162.

Potenza, D., Belvisi, L., Vasile, F., Moroni, E., Cossu, F., Seneci, P., 2012. A NMR and computational study of Smac mimics targeting both the BIR2 and BIR3 domains in XIAP protein. Org. Biomol. Chem. 10 (16), 3278–3287. Available from: https://doi.org/10.1039/c2ob06979b.

Raj, M., Bullock, B.N., Arora, P.S., 2013. Plucking the high hanging fruit: a systematic approach for targeting protein–protein interactions. Bioorg. Med. Chem. 21 (14), 4051–4057. Available from: https://doi.org/10.1016/j.bmc.2012.11.023.

Raj, U., Varadwaj, P.K., 2016. Flavonoids as multi-target inhibitors for proteins associated with Ebola virus: in silico discovery using virtual screening and molecular docking studies. Interdiscip. Sci. 8 (2), 132–141. Available from: https://doi.org/10.1007/s12539-015-0109-8.

Sanchez, A., Kiley, M.P., Holloway, B.P., Auperin, D.D., 1993. Sequence analysis of the Ebola virus genome: organization, genetic elements, and comparison with the genome of Marburg virus. Virus Res. 29 (3), 215–240.

Schafer, A.M., Cheng, H., Lee, C., Du, R., Han, J., Perez, J., et al., 2017. Development of potential small molecule therapeutics for treatment of Ebola virus. Curr. Med. Chem. 25 (38), 5177–5190. Available from: https://doi.org/10.2174/0929867324666171010141416.

Schuler, J., Hudson, M.L., Schwartz, D., Samudrala, R., 2017. A systematic review of computational drug discovery, development, and repurposing for Ebola virus disease treatment. Molecules 22 (10). Available from: https://doi.org/10.3390/molecules22101777.

Schymkowitz, J., Borg, J., Stricher, F., Nys, R., Rousseau, F., Serrano, L., 2005. The FoldX web server: an online force field. Nucleic Acids Res. 33 (Web Server issue), W382–W388. Available from: https://doi.org/10.1093/nar/gki387.

Scott, D.E., Bayly, A.R., Abell, C., Skidmore, J., 2016. Small molecules, big targets: drug discovery faces the protein–protein interaction challenge. Nat. Rev. Drug Discov. 15 (8), 533–550. Available from: https://doi.org/10.1038/nrd.2016.29.

Setlur, A.S., Naik, S.Y., Skariyachan, S., 2017. Herbal lead as ideal bioactive compounds against probable drug targets of Ebola virus in comparison with known chemical analogue: a computational drug discovery perspective. Interdiscip. Sci. 9 (2), 254–277. Available from: https://doi.org/10.1007/s12539-016-0149-8.

Sharma, D., Sharma, N., Pathak, M., Sharma, R., Tyagi, P., Chawla, R., et al., 2017. Homology modeling and docking studies of VP_{24} protein of Ebola virus with an antiviral drug and its derivatives. Chem. Biol. Lett. 4 (1), 27–32.

Sherman, W., Day, T., Jacobson, M.P., Friesner, R.A., Farid, R., 2006. Novel procedure for modeling ligand/receptor induced fit effects. J. Med. Chem. 49 (2), 534–553. Available from: https://doi.org/10.1021/jm050540c.

Sweiti, H., Ekwunife, O., Jaschinski, T., Lhachimi, S.K., 2017. Repurposed therapeutic agents targeting the Ebola virus: a systematic review. Curr. Ther. Res. Clin. Exp. 84, 10–21. Available from: https://doi.org/10.1016/j.curtheres.2017.01.007.

Tambunan, U.S.F., Nasution, M.A.F., 2017. Identification of Novel Ebola Virus (EBOV) VP24 Inhibitor From Indonesian Natural Products Through In Silico Drug Design Approach. AIP Conference Proceedings 1862, 030091. Available from: https://doi.org/10.1063/1.4991195.

Trott, O., Olson, A.J., 2010. Software news and update AutoDock Vina: improving the speed and accuracy of docking with a new scoring function, efficient optimization, and multithreading. J. Comput. Chem. 31 (2), 455–461. Available from: https://doi.org/10.1002/jcc.21334.

Uzochukwu, I.C., Olubiyi, O.O., Ezebuo, F.C., Obinwa, I.C., Ajaegbu, E.E., Eze, P.M., et al., 2016. Ending the Ebola virus scourge: a case for natural products. Open Access J. Pharm. Res. 1 (1), 000105.

Van Durme, J., Delgado, J., Stricher, F., Serrano, L., Schymkowitz, J., Rousseau, F., 2011. A graphical interface for the FoldX forcefield. Bioinformatics 27 (12), 1711–1712. Available from: https://doi.org/10.1093/bioinformatics/btr254.

Wang, S.R., Zhang, Q.Y., Wang, J.Q., Ge, X.Y., Song, Y.Y., Wang, Y.F., et al., 2016. Chemical targeting of a G-Quadruplex RNA in the Ebola virus L gene. Cell Chem. Biol. 23 (9), 1113–1122. Available from: https://doi.org/10.1016/j.chembiol.2016.07.019.

Watanabe, S., Noda, T., Halfmann, P., Jasenosky, L., Kawaoka, Y., 2007. Ebola virus (EBOV) VP24 inhibits transcription and replication of the EBOV genome. J. Infect. Dis. 196 (Suppl. 2), S284–S290. Available from: https://doi.org/10.1086/520582.

Wells, J.A., McClendon, C.L., 2007. Reaching for high-hanging fruit in drug discovery at protein–protein interfaces. Nature 450 (7172), 1001–1009. Available from: https://doi.org/10.1038/nature06526.

Xu, W., Edwards, M.R., Borek, D.M., Feagins, A.R., Mittal, A., Alinger, J.B., et al., 2014. Ebola virus VP24 targets a unique NLS binding site on karyopherin alpha 5 to selectively compete with nuclear import of phosphorylated STAT1. Cell Host Microbe 16 (2), 187–200. Available from: https://doi.org/10.1016/j.chom.2014.07.008.

Yu, D.S., Weng, T.H., Wu, X.X., Wang, F.X.C., Lu, X.Y., Wu, H.B., et al., 2017. The lifecycle of the Ebola virus in host cells. Oncotarget 8 (33), 55750–55759. Available from: https://doi.org/10.18632/oncotarget.18498.

Zhang, A.P., Abelson, D.M., Bornholdt, Z.A., Liu, T., Woods Jr., V.L., Saphire, E.O., 2012. The Ebolavirus VP24 interferon antagonist: know your enemy. Virulence 3 (5), 440–445. Available from: https://doi.org/10.4161/viru.21302.

Zhao, Z., Martin, C., Fan, R., Bourne, P.E., Xie, L., 2016. Drug repurposing to target Ebola virus replication and virulence using structural systems pharmacology. BMC Bioinf. 17, 90. Available from: https://doi.org/10.1186/s12859-016-0941-9.

Zinzalla, G., Thurston, D.E., 2009. Targeting protein–protein interactions for therapeutic intervention: a challenge for the future. Fut. Med. Chem. 1 (1), 65–93. Available from: https://doi.org/10.4155/Fmc.09.12.

Zoete, V., Michielin, O., 2007. Comparison between computational alanine scanning and per-residue binding free energy decomposition for protein–protein association using MM-GBSA: application to the TCR-p-MHC complex. Proteins 67 (4), 1026–1047. Available from: https://doi.org/10.1002/prot.21395.

CHAPTER

10

Application of Nanodiagnostics in Viral Infectious Diseases

Rahma Ait Hammou[1], *Mustapha Benhassou*[1,2,3] *and Moulay Mustapha Ennaji*[1]

[1]Laboratory of Virology, Microbiology, Quality, Biotechnologies/Eco-Toxicology and Biodiversity, Faculty of Sciences and Techniques, Mohammedia, University Hassan II of Casablanca, Casablanca, Morocco [2]Mohammed VI University of Health Sciences of Casablanca, Casablanca, Morocco [3]School of Medicine and Pharmacy of Casablanca, University Hassan II of Casablanca, Casablanca, Morocco

ABBREVIATIONS

BCA	Biobarcode amplification assay
DNA	Deoxyribonucleic acid
ELISA	Enzyme-linked immunosorbent assays
HBV	Hepatitis B virus
HCV	Hepatitis C virus
HIV	Human immunodeficiency virus
NP	Nanoparticles
PCR	Polymerase chain reaction
RNA	Ribonucleic acid

INTRODUCTION

Viruses (Armstead and Li, 2011; Sharma, 2010), bacteria (Kelley et al., 2013; Hyde, 2011), fungi (Pappas et al., 2004), or parasites (Olliaro, 2009) are the main pathogens that cause infectious diseases and are one of the major causes of deaths, accounting for approximately 15 million annual

Emerging and Reemerging Viral Pathogens
DOI: https://doi.org/10.1016/B978-0-12-814966-9.00010-X

deaths worldwide (Von Groote-Bidlingmaier and Diacon, 2011). In spite of the great efforts to enhance and develop effective pharmaceuticals and new technologies to produce drugs at low cost, the remarkable increase in drug resistance of infectious agents prevents efficient treatment of these diseases. Therefore the consequences of drug resistance were higher mortality, increased cost due to the use of more expensive drugs, and increased burden on the public healthcare system.

Consequently, it becomes crucial and mandatory to develop new strategies, pharmaceuticals, and devices to diagnose and treat diseases accurately, easily, and efficiently. Furthermore, in the aim of effective control and suitable infectious diseases, a number of nanotechnology-based materials have been studied (Blecher et al., 2011). These nanotechnologies served to enhance immune responses against antigens for effective vaccination, suitable use of pharmaceuticals in order to achieve the target site and to be released at a controlled rate, and for accurate, rapid, and low-cost detection and diagnosis of infectious diseases. This chapter will discuss the conventional methods of diagnosis and nanotechnologies developed to improve treatment, diagnostics, and prevention of infectious diseases.

CONVENTIONAL DIAGNOSIS FOR INFECTIOUS DISEASES AND LIMITATIONS

Infectious diseases treatment and prevention requires up-to-date diagnostics to be more efficient. Furthermore, monitoring of efficacy of treatment is mandatory during therapy by detection of pathogens. Diverse conventional tools are available for the diagnosis of infectious disease including microscopy, tissue culture, lateral flow immunoassays (also known as dipsticks or immune chromatographic test), enzyme-linked immunosorbent assays (ELISAs), and biochemical tests (Table 10.1). More recently, molecular diagnostic techniques, such as polymerase chain reaction (PCR) and real-time PCR, have been widely used to diagnose and monitor infections such as human immunodeficiency virus (HIV)/AIDS and hepatitis C virus (HCV) because they have a higher specificity and sensitivity than ELISA-based diagnostics. However, the cost of these techniques, time-consuming and require prior sample preparation, that allows their use only in developed countries but are often poorly suited for the developing countries, where infectious diseases are leading causes of morbidity and mortality, because the availability of trained clinical staff and suitable laboratory facilities may be limited (Sosnik and Amiji, 2010). Therefore development of new diagnostic technologies is required. The characteristics of the ideal tool of diagnostic for the developing countries would be a

TABLE 10.1 Current Therapies Against Selected Virus and Their Limitations (Qasim et al., 2014)

Disease name	Causative agent	Current treatment strategies	Limitations in treatment
Hepatitis C	HCV	Combination of interferon and broad spectrum antiviral therapy	• Limited efficacy in patients with HCV genotype 1 • Drug resistance is rapidly emerging • Drug administration by injection over 72 weeks may result in chronic side effects
AIDS	HIV	HAART	Treatment should be continued throughout life • Potential emergence of drug resistance • Complete eradication is not possible • Side effects such as increased rate of heartbeat, diabetes, liver diseases, cancer, and premature aging
Cervical cancer	HPV	Cryosurgery, loop Electrosurgical excision Procedure (LEEP), laser therapy, hysterectomy Vaccine	• No treatment for existing cervical cancer except for the removal of cervix • Vaccine has side effects and is effective only before exposure to the virus
Hepatitis B	HBV	Interferon therapy	• No treatment available for acute hepatitis B • Cold chain issues for vaccine • A booster dose of vaccine is required; therefore, follow-up of patients is a major issue
Poliomyelitis	Polio virus	Vaccine	• No treatments available. Immunization for prevention only • Vaccine is expensive and requires a cold chain for transportation and storage • Issue of OPV degradation

(Continued)

TABLE 10.1 (Continued)

Disease name	Causative agent	Current treatment strategies	Limitations in treatment
Influenza virus	Influenza virus A and B	Vaccine	These drugs have activity only against influenza A strain • Amantadine carries a risk of neuropsychological, atropinic, and dopaminergic adverse effects • Zanamivir carries a risk of life-threatening bronchospasm • Emergence of drug resistance is a crucial problem for the treatment of influenza virus
Mumps, measles, and rubella	Paramyxovirus, mumps virus, and rubella	Vaccine	• No treatment available. Vaccine for prevention only • Potential side effects of vaccine

AIDS, Acquired immunodeficiency syndrome; *HAART*, highly active antiretroviral therapy; *HBV*, hepatitis B virus; *HCV*, hepatitis C virus; *HIV*, human immunodeficiency virus; *HPV*, human papilloma virus; *OPV*, oral polio vaccine.

cost-effective, portable, and point-of-source detection system that is also highly reliable, sensitive, and accurate (Hauck et al., 2010). Furthermore, to consider the technique as an ideal, it should be able to detect multiple pathogens in a single reaction.

Diagnostic is usually realized using conventional molecular diagnostic technologies that are mainly based on the amplification of specific DNA sequences from extracted nucleic acids (DNA or RNA), for example, target amplification (e.g., PCR, reverse transcriptase PCR, and strand displacement amplification), signal amplification (e.g., branched DNA assays and hybrid capture), probe amplification (e.g., ligase chain reaction, cleavage-invader, and cycling probes), or postamplification analysis (e.g., sequencing the amplified products or melting curve analysis), although sensitivity of amplification methods may be false positive due to trace contamination of the specimen or equipment. In addition, these techniques basically depend on enzymatic activity; false negatives can occur when samples contain contaminants that inhibit the enzymes (Hartman et al., 2005).

One of the latest methods for rapid diagnostics of infectious diseases area DNA microarrays or DNA chips DNA microarrays which are essentially a high-throughput version of the Southern blot method (Schena et al., 1995). Each microarray contains a number of different DNA oligonucleotides that recognize specific target genes from a pathogen through complementary DNA–DNA binding. The detection is based on oligonucleotides hybridized to pathogen genes for the diagnosis of infectious diseases.

However, with their potential in diagnostics and practical use in clinical settings, this technique is hampered by several considerations, especially difficulties in the identification of pathogen-unique target genes and in the design of oligonucleotide primers for multiplex PCR. The first difficulty lies in finding a gene that is unique to a particular pathogen among a tremendous number of genes (Janda and Abbott, 2007). Another challenge is the design of oligonucleotides that will amplify particular pathogen genes in multiplex PCR without nonspecific amplifications.

Emerging nanotechnology-based techniques have recently attracted interest as an approach that may overcome the problems of current diagnostic techniques through their specific mode of actions and unique physical properties (i.e., shape, size, surface charge, and dimension) (Daum et al., 2012). These techniques may be applied to develop accurate, reliable, rapid, safe, cost-effective, sensitive, specific, and easily accessible techniques for the detection of pathogens (Rosi and Mirkin, 2005). In the following section, we discuss new nanotechnology-based methods for the treatment, diagnosis, or prevention of infectious diseases.

APPLICATION OF NANOTECHNOLOGY IN INFECTIOUS DISEASE

Nanodiagnostics is defined currently as the use of nanotechnology in diagnostic applications. This new tool has been extensively studied to meet the requirements of clinical diagnostics with high sensitivity and earlier detection of various diseases (Baptista, 2014).

The ability of rapid and real-time detection which can be realized using very small volumes of samples from patients is the specific and unique property of nanomaterials or nanostructures used in nanodiagnostic platforms. Therefore nanodiagnosis represents another interesting property since it has a huge potential to be low-cost, user friendly, and robust systems. Actually, diagnosis and detection of pathogens in infectious diseases and cancer biomarkers are mainly based on nanodiagnostic applications. For the infectious diseases the nanodiagnostic platforms have the ability to achieve reliable and rapid conclusions with simple and portable devices just by using blood, sputum, or urine samples from patients (Kaittanis et al., 2010). In addition, especially in resource-poor areas in the developing countries, it could be suitable to use nanodiagnostics platforms, which has an important potential to be robust, affordable, and reproducible in the diagnosis of infectious diseases.

Nanoparticle-Based Diagnosis for Infectious Diseases

Many nanoparticles (NPs) are currently available for nanodiagnosis, particularly fluorescent NPs, metallic NPs, and magnetic NPs, which have been successfully utilized for the diagnosis of infectious diseases. Sensitivity and photostability of fluorescent NPs allow this category to be used to label many different biological targets. They have been demonstrated as new strategies to perform disease diagnosis in real time through bioimaging or sensing activities.

Furthermore, gold and silver NPs constitute the most used metallic NPs in the diagnosis applications which could emit intense absorption when excited with electromagnetic radiation (Tallury et al., 2010). The gold NPs are the first nanomaterials as nanodiagnostics for the detection of DNA in 1996 (Mirkin et al., 1996). The changes in the color of gold NPs in solution from red to blue have been demonstrated after DNA-guided aggregations, which make them ideal nanomaterials for nanodiagnostics because of their unique color changes and other chemical and physical properties. Thus conjugation of diverse molecules, such as antibodies, antigens, and enzymes with gold particles as electrochemical labels, optical probes, and signal transfer amplifiers, can be used in the diagnosis of various diseases. For instance, the gold nanorods have been used to diagnose HIV through their second-order nonlinear optical properties. A 100 pM target DNA can be recognized by a 145-mer oligonucleotide probe, which was recorded by a hyper-Rayleigh scattering (HRS) spectroscopy with high sensitivity and selectivity. Detection of single-base-mismatch HIV-1 virus DNA through the HRS intensity changes using the gold nanorods was rapid, simple, and efficient (Darbha et al., 2008). In addition, a similar HRS technique with gold NPs has also been developed to detect HCV for infectious diseases.

Ragarding bioimaging, cancer therapy, and nanodiagnostics, magnetic NPs have also been successfully applied in this biomedical applications (Wu et al., 2016; Kolosnjaj-Tabi et al., 2016; Duguet et al., 2006; Gu et al., 2006; Moraes Silva et al., 2016). This diagnosis is based on the enhanced separation and detection of aligned magnetic NPs bonded with targeting agents under an applied magnetic field (Shinde et al., 2012).

For magnetic resonance imaging magnetic NPs, iron oxide NPs, composed of magnetite or maghemite cores, have been used. Therefore modification and conjugation of surface of the iron oxide NPs with antibodies, proteins, and nucleic acids is commonly used in order to detect many different infecting pathogens, such as viruses, bacteria, and parasites. The early and efficient diagnosis of malaria using magnetic NPs with iron oxide core and silver shell is considered a successful example of this method (Yuen and Liu, 2012).

Nanodevice-Based Diagnosis for Infectious Diseases

Simple NP-based nanodiagnostics for infectious diseases have been extensively studied and used but they are still far from meeting the real demands in the clinic. Since many infecting pathogens, such as HIV, always have multiple strain types, and infection may be with multiple pathogens such as HIV and hepatitis B virus (HBV) together for patients. For that, development of complicated diagnosis and more advanced nanodiagnostics of infectious diseases are greatly required. Actually, integration of many techniques with the nanotechnology in order to develop nanodevice-based diagnosis platforms for the diagnosis of infectious diseases is available. Among them, lab-on-a-chip and microfluidics techniques have shown good promise for the detection of infectious diseases (Dixon et al., 2016; Cabibbe et al., 2015).

These integrated techniques represent many advantages that could greatly potentiate to build low-cost and portable devices (Murdock et al., 2017; Sackmann et al., 2014; Li et al., 2016; Su et al., 2015). Since multiple assays could be integrated into one single device (Tay et al., 2016) which can lead to decrease in the volume of samples from infected patients, the consumption of materials, and the analysis time.

A successful integrated nanodevice with high-throughput and multiplexed detecting ability for the most important blood-borne infecting agents such as HIV, HBV, and HCV in serum samples has been developed through the combination of nanotechnology (quantum dots) and microtechnology (microfluidics) (Klostranec et al., 2007). Using this diagnosis, multiple pathogens could be detected precisely using the human serum samples simultaneously.

Nanobiosensors

A biosensor is defined as an analytical device based on conversion of molecular recognition of a target analyte into a measurable signal using a transducer. Glucose sensor is considered the most well-known example in use today, which has had a transformative effect on the management of diabetes since its introduction in the current form 30 years ago. Other example bases on lateral flow assays are widely used such as pregnancy test (Luong et al., 2008; Ngom et al., 2010). Regarding infectious diseases, biosensors have a variety of advantages and offer the possibility of an easy-to-use, are considered as a sensitive and inexpensive technology platform that can identify pathogens rapidly and predict effective treatment (Foudeh et al., 2012; Pejcic et al., 2006; D'orazio, 2011). These advantages include also a small fluid volume manipulation (less reagent and lower cost), short assay time, low energy consumption, high portability, high-throughput, and multiplexing ability (Whitesides, 2006). Development of biosensors with ability to perform complex molecular assays for diagnosis of many infectious diseases is due to current advances in micro- and nanotechnologies. In addition, significant progress has been made in parallel to understand pathogen genomics and proteomics and their interactions with the host (Fournier and Raoult, 2011; Hodges and Connor, 2013; Mairiang et al., 2013). Biosensor-based immunoassays may improve the detection sensitivity of pathogen-specific antigens, while multiplex detection of host immune response antibodies (serology) may improve the overall specificity. To facilitate assay development, further system integration which integrates both pathogen-specific targets and biomarkers representative of host immune responses at different stages of infection may be developed (Mohan et al., 2011).

Label-Free Biosensors

Label-free biosensors are based primarily on changes that occur when target analytes bind to immobilized molecular capture elements on a solid support or cause changes in interfacial capacities or resistance (Hunt and Armani, 2010; Rapp et al., 2010).

Label-free biosensors require only a single recognition element, leading to simplified assay design, decreased assay time, and reduction in reagent costs. This recognition mode is especially appropriate for small molecular targets, which can be buried within the binding pocket of the capturing element, leaving little room for interaction with a detector agent that would be required in a labeled assay. Ability to perform quantitative measurement of molecular interaction in real time constitutes another advantage of label-free method, which allows continuous data recording. Also, target analytes are detected without labeling or chemical modification in their natural and then can be preserved for further analysis. The label-free sensing strategies for various infectious diseases discussed below operate through a binding-event-generated perturbation in optical, electrical, or mechanical signals (Table 10.2).

Optical Transducer

Optical transducers are widely used due to their high sensitivity with several well established optical phenomena such as surface plasmon changes, scattering, and interferometry (Citartan et al., 2013). Surface

TABLE 10.2 Examples of Label-Free Detection Strategies

Label-tree assay	Technology	Advantages	Disadvantages
Optical transducer	Surface plasmon resonance	Real-time detection; possibility of high throughput	Sensitive to sample matrix effects; sensor surface functionalization challenging; bulky optical equipment
Electrical transducer	Redox electrochemistry (amperometric)	Simple sensor design; detection platform amenable to inexpensive and miniaturization	Redox species required to increase current production; no real-time detection; sensitive to sample matrix effects
	Impedance spectroscopy	Simple electrode design; real-time detection	Sensitive to sample matrix effects; bulky equipment; data analysis may not be trivial (theoretical model may be required)
	Potentiometry	Real-time detection; consecutive measurements on different samples are possible	Bulky equipment, sensitive to sample matrix; complicated sample preparation steps; careful control of temperature is essential
	Field effect transistor	Real-time detection; stable sensor response; detection platform amenable to POC system	Sensitive to sample matrix effects; bulky equipment; data analysis may not be trivial (theoretical model may be required)
Mechanical transducer	Microcantilever	Real-time detection; multiplex and high throughput are possible	Sensitive to sample matrix effects; careful control of temperature is essential; bulky equipment
	Quartz crystal microbalance	Simple electrode design; real-time detection; detection platform amenable to POC system	Sensitive to sample matrix effects; careful control of temperature and stress is essential

POC, Point-of-care.

plasmon resonance is the excitation of an electromagnetic wave propagating along the interface of two media with dielectric constants of opposite signs, such as metal and sample buffer, by a specific angle of incident light beam (Guo, 2012). The signal is based on total internal reflection that results in a reduced intensity of the reflected light.

The angle at which the resonance occurs is sensitive to any change at the interface, such as changes in refractive index or formation of a nanoscale film thickness due to surface–molecular interactions. Therefore these changes can be measured by monitoring the light intensity minimum shift over time.

Most label-free optical biosensors require precise alignment of light coupling to the sensing area, which is a major drawback for point-of-care applications. Therefore optical sensing can be significantly improved when this approach is used in an integration scheme. Integrated optics allow several passive and active optical components on the same substrate, allowing flexible development of minimized compact sensing devices, with the possibility of fabrication of multiple sensors on one chip.

This system was used to demonstrate the recognition of small enveloped RNA viruses (vesicular stomatitis virus and pseudotyped ebola) as well as large enveloped DNA viruses (vaccinia virus) at clinically relevant concentrations (Yanik et al., 2010).

Electrical Transducer

With high innate sensitivity and simplicity to be effectively conjugated to miniaturized hardware, electrical analytical methods are considered common sensing approaches. Common types of electrical biosensors that have been applied to infectious disease diagnostics include voltammetric, amperometric, impedance, and potentiometric sensors (Luo and Davis, 2013). Voltammetric and amperometric sensors involve current measurement of an electrolyte with a DC voltage applied across the electrode as a function of voltage and time, respectively. An immunosensor based on the amperometric approach has been developed for the detection of hepatitis B surface antigen, a major index of hepatitis B viruses (Qiu et al., 2011).

Mechanical Transducer

Emergence of micro- and nanoscale mechanical transducers able to detect changes in force, motion, mechanical properties, and mass that comes along with molecular recognition events was facilitated by the advanced development of micro- and nanofabrication technologies (Tamayo et al., 2013; Arlett et al., 2011). Therefore cantilever and quartz crystal microbalances (QCMs) are the main established tools among this category of biosensors. Thus Peduru Hewa et al. (2009) developed a QCM-based immunosensor for the detection of influenza A and B viruses. By conjugating Au NPs to the antiinfluenza A or B monoclonal antibodies, a detection limit of 1×103 pfu/mL for laboratory-cultured preparations and clinical samples (nasal washes) was achieved. In 67 clinical samples the QCM-based immunosensor was comparable with standard methods such as shell vial and cell culture and better than ELISA in terms of sensitivity and specificity.

Generally, this type of sensors does not include signal amplification; improvement of specificity and sensitivity of a given device depends largely on the selection and combination of capturing elements and transducers. Otherwise, advances in biochemistry and molecular biology may ensure the diversity of capturing elements with higher affinity, specificity, and stability. Translation of technologies from detection using laboratories solutions to real-world clinical samples such as serum, blood, or urine represents a real challenge for application of these technologies. The complex sample matrices of clinical samples can lead to nonspecific binding and aberrant signals.

Labeled Biosensors

This class of biosensors is the most common and robust method of biosensing. Classically, in labeled assays, the analyte is sandwiched between the capture and detector agents (Peruski and Peruski, 2003). Electrodes, glass chips, nano- or microparticles constitute solid surfaces on which are immobilized captures agents, while flurophores, enzymes, or NPs considered signaling tags are typically conjugated to detector agents (Ju et al., 2011). As with label-free assays, optical, electrical or mechanical transducers can be coupled to the signaling tag. Optical sensors are used to detect fluorescent (Li et al., 2013), colorimetric (Nam et al., 2007) or luminescent tags (Sapsford et al., 2006), electrochemical sensors used to detect redox reactions from enzyme tags (Mach et al., 2009), and magnetoresistive sensors used to detect magnetic tags (Haun et al., 2010). Furthermore, using these systems, detection of the analyte can be quantitative or semiquantitative by relating the signal generated by the analyte of interest. In general, binding sites are different among capture and detector elements, thereby the specificity increased and the background reduced. However, the multistep protocol can make the assay more costly and complicated. The standard sandwich immunoassay mostly used for infectious disease applications in clinical laboratories is ELISA (Peruski and Peruski, 2003). This technique is based on capture of antibody and a detector antibody modified with an enzyme tag for catalyzing the conversion of chromogenic substrate into colored molecules. Regarding quantitative ELISA, the optical density of the colored product from the sample is compared with a standard serial dilution of a known concentration of the target molecule. Furthermore, nucleic acids can also be detected with sandwich assays. Most well-known commercially available examples include home pregnancy tests and urinalysis strips. Lateral flow assays have been proposed for saliva- or blood-based HIV tests, blood-based malaria antigen test, and

TABLE 10.3 Examples of Label-Free Detection Strategies

Technology	Advantages	Disadvantages
Redox electrochemistry (amperometric)	Detection platform amenable to POC system; easy integration with other electric field-driven modules	No real-time detection; multiple steps assay
Biobarcode	Detection platform amenable to POC system; easily interpreted results	No real-time detection; complicated protocol for probe preparations; multiple steps assay
Metal nanoparticles	Detection platform amenable to POC system; easily interpreted results; multiplex	No real-time detection; temperature fluctuations can affect the results; multiple steps assay

POC, Point-of-care.

serum-based tuberculosis test (Ngom et al., 2010). These assays have wide range of advantages including mainly low cost, minimal to no sample preparation, and straightforward interpretation of the results (Martinez et al., 2010). Moreover, poor sensitivity for many of the clinically relevant targets and qualitative or semiquantitative results are the most important disadvantages of this technique. Thus current efforts have essentially focused on signal amplification to improve the limit of detection. In order to develop the field of nanotechnology, functionalization of NPs with different biological molecules was established over the years, and this made it ideal labels for diverse signal amplification processes in the biosensor platforms. Due to their high surface-to-volume ratio, NPs are attractive means of signal amplification to improve sensitivity and versatility of biosensing devices (D'orazio, 2011; Ju et al., 2011; Wang and Wang, 2014). Labeled biosensors are based essentially on biobarcode, metal NPs, and magnetic NPs (Table 10.3).

Biobarcode

Biobarcode amplification assay (BCA) is one of the most promising NP-based approaches. This technique has the ability to detect both proteins and nucleic acids without enzymatic reactions (Nam et al., 2003, 2007). In addition, BCA is an involvement of sandwich assay with targets captured with micro- or NPs conjugated with oligonucleotides (barcode DNA) using as surrogates for signal amplification. Thus many strands are released with capture of every target for subsequent detection with other means such as electrochemical or optical. This tool was applied recently to detect HIV capsid (p24) antigen, a useful marker for ultimate prediction of CD4 + T-cell decline, disease progression, and early detection of HIV-1 infection (Tang et al., 2007).

Metal Nanoparticles

Based on their unique optical properties, metal NPs are also used as signal amplification labels for bio-recognition processes (Cao et al., 2011; Lin et al., 2013). Gold and silver NPs exhibit plasmon absorbance bands in the visible light spectrum which are determined by the size of the respective particles.

Magnetic Nanoparticles

These particles coupled with detectors for biosensing can serve for signal amplification and are amenable to use in solution phase sandwich assays, such as diagnostic magnetic resonance (Haun et al., 2010; Soelberg et al., 2009). Significantly faster assay times compared to diffusion-dependent surface structure-based assays. Using diagnostic magnetic resonance, the capture and detection agents are both in solution and related to magnetic particles. Therefore the magnetic particles cluster as the antibodies bind to analyte of interest once it is present.

CONCLUSION

This chapter presented current and conventional diagnosis used for infectious diseases and limitations of the current advances facilitated by nanotechnologies to address the limitations. Therefore modern nano-based tools are more reliable than conventional techniques. This field should be developed further for practical use in daily life as it already provides new directions for the advancement of treatments and diagnostics for infectious diseases. Furthermore, in order to control infectious diseases and improve public health, development of this nanotechnology-based therapeutics, vaccines, and diagnostics may foster easy, cheap, safe, and portable use of end products, especially in the developing countries. To ensure more advanced and improved than traditional diagnosis, NPs make a great leap especially in control of many parameters in diagnosis and treatment such as controlled slow release of encapsulated drugs. But, for more development of NPs-based technologies, it is crucial to develop the design of these particles with a targeted function; efficient encapsulation of drugs; specific binding to targets, on-site release, etc.

In addition, infectious diseases are ideal applications for the emerging biosensor technologies. For many infectious diseases, rapid diagnosis and timely initiation of effective treatment can be critical for patient outcome and public health. When integrated with advanced microfluidic systems, biosensor can form the foundation of rapid point-of-care devices with the potential to positively impact patient care.

Acknowledgments

All the authors are thankful to the contributors of the Team of Virology, Oncology, and Medical Biotechnologies of Laboratory of Virology, Microbiology, Quality, and Biotechnologies/ETB and also to the Fondation Lalla Salma de lutte contre le cancer.

References

Arlett, J.L., Myers, E.B., Roukes, M.L., 2011. Comparative advantages of mechanical biosensors. Nat. Nanotechnol. 6 (4), 203–215.

Armstead, A.L., Li, B.Y., 2011. Nanomedicine as an emerging approach against intracellular pathogens. Int. J. Nanomed. 6, 3281–3293.

Baptista, P.V., 2014. Nanodiagnostics: leaving the research lab to enter the clinics? Diagn. 1 (4), 305–309.

Blecher, K., Nasir, A., Friedman, A., 2011. The growing role of nanotechnology in combating infectious disease. Virulence 2 (5), 395–401.

Cabibbe, A.M., Miotto, P., Moure, R., et al., 2015. Lab-on-chip-based platform for fast molecular diagnosis of multidrug-resistant tuberculosis. J. Clin. Microbiol. 53 (12), 3876–3880.

Cao, X., Ye, Y., Liu, S., 2011. Gold nanoparticle-based signal amplification for biosensing. Anal. Biochem. 417 (1), 1–16.

Citartan, M., Gopinath, S.C., Tominaga, J., Tang, T.H., 2013. Label-free methods of reporting biomolecular interactions by optical biosensors. Analyst 138 (13), 3576–3592.

Darbha, G.K., Rai, U.S., Singh, A.K., Ray, P.C., 2008. Gold-nanorod-based sensing of sequence specific HIV-1 virus DNA by using hyper-Rayleigh scattering spectroscopy. Chemistry 14 (13), 3896–3903.

Daum, N., Tscheka, C., Neumeyer, A., Schneider, M., 2012. Novel approaches for drug delivery systems in nanomedicine: effects of particle design and shape. Wiley Interdiscip. Rev. Nanomed. Nanobiotechnol. 4 (1), 52–65.

Dixon, C., Ng, A.H.C., Fobel, R., Miltenburg, M.B., Wheeler, A.R., 2016. An inkjet printed, roll-coated digital microfluidic device for inexpensive, miniaturized diagnostic assays. Lab Chip 16 (23), 4560–4568.

D'orazio, P., 2011. Biosensors in clinical chemistry—2011 update. Clin. Chim. Acta 412 (19–20), 1749–1761.

Duguet, E., Vasseur, S., Mornet, S., Devoisselle, J.M., 2006. Magnetic nanoparticles and their applications in medicine. Nanomedicine 1 (2), 157–168.

Foudeh, A.M., Fatanat Didar, T., Veres, T., Tabrizian, M., 2012. Microfluidic designs and techniques using lab-on-a-chip devices for pathogen detection for point-of-care diagnostics. Lab Chip 12 (18), 3249–3266.

Fournier, P.E., Raoult, D., 2011. Prospects for the future using genomics and proteomics in clinical microbiology. Annu. Rev. Microbiol. 65, 169–188.

Gu, H., Xu, K., Xu, C., Xu, B., 2006. Biofunctional magnetic nanoparticles for protein separation and pathogen detection. Chem. Commun. (9), 941–949.

Guo, X., 2012. Surface plasmon resonance based biosensor technique: a review. J. Biophotonics 5 (7), 483–501.

Hartman, L.J., Coyne, S.R., Norwood, D.A., 2005. Development of a novel internal positive control for Taqman based assays. Mol. Cell. Probes 19 (1), 51–59.

Hauck, T.S., Giri, S., Gao, Y., Chan, W.C., 2010. Nanotechnology diagnostics for infectious diseases prevalent in developing countries. Adv. Drug Delivery Rev. 62 (4–5), 438–448.

Haun, J.B., Yoon, T.J., Lee, H., Weissleder, R., 2010. Magnetic nanoparticle biosensors. Wiley Interdiscip. Rev. Nanomed. Nanobiotechnol. 2 (3), 291–304.

Hodges, E.N., Connor, J.H., 2013. Translational control by negative-strand RNA viruses: methods for the study of a crucial virus/host interaction. Methods 59 (2), 180–187.

Hunt, H.K., Armani, A.M., 2010. Label-free biological and chemical sensors. Nanoscale 2 (9), 1544–1559.

Hyde, R., 2011. Germany reels in the wake of *E. coli* outbreak. Lancet 377 (9782), 1991.

Janda, J.M., Abbott, S.L., 2007. 16S rRNA gene sequencing for bacterial identification in the diagnostic laboratory: pluses, perils, and pitfalls. J. Clin. Microbiol. 45 (9), 2761–2764.

Ju, H., Zhang, X., Wang, J., 2011. NanoBiosensing. Signal Amplification for Nanobiosensing. Springer-Verlag, New York, pp. 39–84, 2011.

Kaittanis, C., Santra, S., Perez, J.M., 2010. Emerging nanotechnology-based strategies for the identification of microbial pathogenesis. Adv. Drug Delivery Rev. 62 (4–5), 408–423.

Kelley, K., Cosman, A., Belgrader, P., Chapman, B., Sullivan, D.C., 2013. Detection of methicillin-resistant *Staphylococcus aureus* by a duplex droplet digital PCR assay. J. Clin. Microbiol. 51 (7), 2033–2039.

Kolosnjaj-Tabi, J., Lartigue, L., Javed, Y., et al., 2016. Biotransformations of magnetic nanoparticles in the body. Nano Today 11 (3), 280–284.

Li, B., Yu, Q., Duan, Y., 2013. Fluorescent labels in biosensors for pathogen detection. Crit. Rev. Biotechnol. 35, 82–93 [Epub ahead of print].

Li, Z., Zhao, Y., Zhang, D., Zhuang, S., Yamaguchi, Y., 2016. The development of a portable buoyancy-driven PCR system and its evaluation by capillary electrophoresis. Sens. Actuators, B: Chem. 230, 779–784.

Lin, M., Pei, H., Yang, F., et al., 2013. Applications of gold nanoparticles in the detection and identification of infectious diseases and biothreats. Adv. Mater. 25 (25), 3490–3496.

Luo, X., Davis, J.J., 2013. Electrical biosensors and the label free detection of protein disease biomarkers. Chem. Soc. Rev. 42 (13), 5944–5962.

Luong, J.H., Male, K.B., Glennon, J.D., 2008. Biosensor technology: technology push versus market pull. Biotechnol. Adv. 26 (5), 492–500.

Mach, K.E., Du, C.B., Phull, H., et al., 2009. Rapid molecular diagnosis of urinary tract infections. J. Urol. 181 (4), 138–139.

Mairiang, D., Zhang, H.M., Sodja, A., et al., 2013. Identification of new protein interactions between dengue fever virus and its hosts, human and mosquito. PLoS One 8 (1), e53535.

Martinez, A.W., Phillips, S.T., Whitesides, G.M., Carrilho, E., 2010. Diagnostics for the developing world: microfluidic paper-based analytical devices. Anal. Chem. 82 (1), 3–10.

Mirkin, C.A., Letsinger, R.L., Mucic, R.C., Storhoff, J.J., 1996. A DNA-based method for rationally assembling nanoparticles into macroscopic materials. Nature 382 (6592), 607–609.

Mohan, R., Mach, K.E., Bercovici, M., et al., 2011. Clinical validation of integrated nucleic acid and protein detection on an electrochemical biosensor array for urinary tract infection diagnosis. PLoS One 6 (10), e26846.

Moraes Silva, S., Tavallaie, R., Sandiford, L., Tilley, R.D., Gooding, J.J., 2016. Gold coated magnetic nanoparticles: from preparation to surface modification for analytical and biomedical applications. Chem. Commun. 52 (48), 7528–7540.

Murdock, R.C., Gallegos, K.M., Hagen, J.A., Kelley-Loughnane, N., Weiss, A.A., Papautsky, I., 2017. Development of a point-of-care diagnostic for influenza detection with antiviral treatment effectiveness indication. Lab Chip 17 (2), 332–340.

Nam, J.M., Thaxton, C.S., Mirkin, C.A., 2003. Nanoparticle-based bio-bar codes for the ultrasensitive detection of proteins. Science 301 (5641), 1884–1886.

Nam, J.M., Jang, K.J., Groves, J.T., 2007. Detection of proteins using a colorimetric bio-barcode assay. Nat. Protoc. 2 (6), 1438–1444.
Ngom, B., Guo, Y., Wang, X., Bi, D., 2010. Development and application of lateral flow test strip technology for detection of infectious agents and chemical contaminants: a review. Anal. Bioanal. Chem. 397 (3), 1113–1135.
Olliaro, P., 2009. Management of fever and malaria—policy and practice. Trop. Med. Int. Health 14 (5), 488–490.
Pappas, P.G., Rex, J.H., Sobel, J.D., Filler, S.G., Walsh, T.J., Edwards, J.E., 2004. Guidelines for treatment of candidiasis. Infect. Dis. Soc. Am. 38 (2), 161–189.
Peduru Hewa, T.M., Tannock, G.A., Mainwaring, D.E., et al., 2009. The detection of influenza A and B viruses in clinical specimens using a quartz crystal microbalance. J. Virol. Methods 162 (1–2), 14–21.
Pejcic, B., De Marco, R., Parkinson, G., 2006. The role of biosensors in the detection of emerging infectious diseases. Analyst 131 (10), 1079–1090.
Peruski, A.H., Peruski, L.F., 2003. Immunological methods for detection and identification of infectious disease and biological warfare agents. Clin. Vaccine Immunol. 10 (4), 506–513.
Qiu, J.D., Huang, H., Liang, R.P., 2011. Biocompatible and label-free amperometric immunosensor for hepatitis B surface antigen using a sensing film composed of poly(allylamine)-branched ferrocene and gold nanoparticles. Microchim. Acta 174 (1–2), 97–105.
Rapp, B.E., Gruhl, F.J., Lange, K., 2010. Biosensors with label-free detection designed for diagnostic applications. Anal. Bioanal. Chem. 398 (6), 2403–2412.
Rosi, N.L., Mirkin, C.A., 2005. Nanostructures in biodiagnostics. Chem. Rev. 105 (4), 1547–1562.
Sackmann, E.K., Fulton, A.L., Beebe, D.J., 2014. The present and future role of microfluidics in biomedical research. Nature 507 (7491), 181–189.
Sapsford, K.E., Pons, T., Medintz, I.L., Mattoussi, H., 2006. Biosensing with luminescent semiconductor quantum dots. Sensors 6 (8), 925–953.
Schena, M., Shalon, D., Davis, R.W., Brown, P.O., 1995. Quantitative monitoring of gene expression patterns with a complementary DNA microarray. Science 270 (5235), 467–470.
Sharma, S.D., 2010. Hepatitis C virus: molecular biology & current therapeutic options. Indian J. Med. Res. 131, 17–34.
Shinde, S.B., Fernandes, C.B., Patravale, V.B., 2012. Recent trends in in-vitro nanodiagnostics for detection of pathogens. J. Controlled Release 159 (2), 164–180.
Soelberg, S.D., Stevens, R.C., Limaye, A.P., Furlong, C.E., 2009. Surface plasmon resonance detection using antibody-linked magnetic nanoparticles for analyte capture, purification, concentration, and signal amplification. Anal. Chem. 81 (6), 2357–2363.
Sosnik, A., Amiji, M., 2010. Nanotechnology solutions for infectious diseases in developing nations. preface. Adv. Drug Delivery Rev. 62 (4-5), 375–377.
Su, W., Gao, X., Jiang, L., Qin, J., 2015. Microfluidic platform towards point-of-care diagnostics in infectious diseases. J. Chromatogr. A 1377, 13–26.
Tallury, P., Malhotra, A., Byrne, L.M., Santra, S., 2010. Nanobioimaging and sensing of infectious diseases. Adv. Drug Delivery Rev. 62 (4–5), 424–437.
Tamayo, J., Kosaka, P.M., Ruz, J.J., et al., 2013. Biosensors based on nanomechanical systems. Chem. Soc. Rev. 42 (3), 1287–1311.
Tang, S.X., Zhao, J.Q., Storhoff, J.J., et al., 2007. Nanoparticle-based biobarcode amplification assay (BCA) for sensitive and early detection of human immunodeficiency type 1 capsid (p24) antigen. J. Acquir. Immune Defic. Syndr. 46 (2), 231–237.
Tay, A., Pavesi, A., Yazdi, S.R., Lim, C.T., Warkiani, M.E., 2016. Advances in microfluidics in combating infectious diseases. Biotechnol. Adv. 34 (4), 404–421.
Von Groote-Bidlingmaier, F., Diacon, A.H., 2011. Medical treatment of tuberculosis—update 2011. Ther. Umsch. 68 (7), 395–401.

Wang, E.C., Wang, A.Z., 2014. Targeted nanoparticles and their applications in biology. Integr. Biol. 6, 9–26.
Whitesides, G.M., 2006. The origins and the future of microfluidics. Nature 442 (7101), 368–373.
Wu, W., Jiang, C.Z., Roy, V.A.L., 2016. Designed synthesis and surface engineering strategies of magnetic iron oxide nanoparticles for biomedical applications. Nanoscale 8 (47), 19421–19474.
Yanik, A.A., Huang, M., Kamohara, O., et al., 2010. An optofluidic nanoplasmonic biosensor for direct detection of live viruses from biological media. Nano Lett. 10, 4962–4969.
Yuen, C., Liu, Q., 2012. Magnetic field enriched surface enhanced resonance Raman spectroscopy for early malaria diagnosis. Biomedo 17 (1), 0170051–0170057.

CHAPTER

11

Baculovirus-Derived Vectors for Immunization and Therapeutic Applications

M. Laura Fabre, Paula N. Arrías, Tomás Masson, Matías L. Pidre and Víctor Romanowski

Laboratory of Molecular Virology, Institute of Biotechnology and Molecular Biology (IBBM), University of La Plata-CONICET, La Plata, Argentina

ABBREVIATION

AAV	adeno-associated viruses
AcMNPV	*Autographa californica* nucleopolyhedrovirus
BEVS	Baculovirus Expression Vector System
BV	budded virus
CMV	cytomegalovirus
DNA	deoxyribonucleic acid
HA	hemagglutinin
ITR	inverted terminal repeats
NC	nucleocapsid
OB	occlusion body
ODV	occlusion derived virus
ORF	open reading frame
PCV	porcine circovirus
POLH	polyhedrin
rAAV	recombinant adeno-associated virus
RNA	ribonucleic acid
VLP	virus-like particle
VSV	vesicular stomatitis virus

Emerging and Reemerging Viral Pathogens
DOI: https://doi.org/10.1016/B978-0-12-814966-9.00011-1

BIOLOGY OF BACULOVIRUSES

Baculoviruses are insect-specific pathogens that have been isolated from about 600 species belonging to the orders Lepidoptera, Hymenoptera, Diptera, Orthoptera, Coleoptera, Neuroptera, Thysaneura, and Trichoptera. Taxonomy of *Baculoviridae* family includes four genera: *Alphabaculovirus* (nucleopolyhedroviruses specific of lepidopterans), *Betabaculovirus* (granuloviruses specific of lepidopterans), *Gammabaculovirus* (nucleopolyhedroviruses specific of hymenopterans), and *Deltabaculovirus* (granuloviruses specific of dipterans) (Jehle et al., 2006; Herniou et al., 2011). These viruses infect arthropod larvae and possess circular double-stranded deoxyribonucleic acid (DNA) genomes that range in size from 80 to 180 kbp. The baculoviral genome is packed inside a distinct rod-shaped nucleocapsid (NC). The life cycle of baculoviruses is biphasic, and each phase is characterized by a different viral phenotype. In the environment, baculoviruses are found as occlusion bodies (OBs). A protein matrix encloses virions and provides protection against harsh environmental conditions such as temperature, dehydration, and ultraviolet light until a susceptible larva ingests the OBs. Briefly, the baculovirus infection cycle of insect hosts is as it follows: OBs are ingested together with foliage by larvae and dissolved in the midgut-releasing occlusion-derived viruses (ODVs), and the damaged peritrophic membrane allows the receptor-mediated endocytosis of the ODV by epithelial cells (Adams and McClintock., 1991). ODVs may contain one or multiple rod-shaped NCs depending on the baculovirus (or less frequently on growth conditions). The NCs enter the cytoplasm and translocate to the nucleus where the viral DNA is transcribed and replicated. Newly synthesized genomic DNA assembles with specific viral proteins to form the NCs in the nucleus. The NCs travel to the plasma membrane where budding takes place to yield the budded viruses (BVs), the second viral phenotype, responsible for the systemic infection of the larvae.

Gene expression is temporally regulated; early genes are transcribed directly by host ribonucleic acid (RNA) polymerase II, whereas late and very-late gene transcription is dependent on viral RNA polymerase. Late genes code mostly for structural proteins characteristic of the beginning of systemic infection while very-late genes code for most ODV structural proteins including the major OB matrix protein (Horton and Burand, 1993; Haas-Stapleton et al., 2004; Au et al., 2013). BVs can infect both in vivo tissues and insect cell cultures (Laakkonen et al., 2008; Long et al., 2006a,b). The possibility to propagate these viruses in cell culture has made baculoviruses an accessible tool for genetic engineering (Goodwin et al., 1970), both for the study of their biology and for the development of a recombinant protein expression system (*BEVS: baculovirus expression vector system*) (Smith et al., 1983). This system is

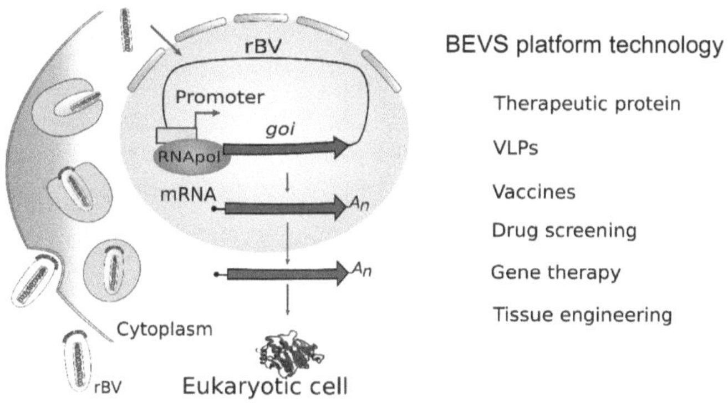

FIGURE 11.1 **Baculovirus expression vector system (BEVS).** The recombinant baculovirus (rBV) enters the cell via endocytosis. Following this event, the nucleocapsid (NC) is released into the cytoplasm and translocated into the nucleus, where viral DNA will be expressed. A simplified scheme of the viral DNA is presented, highlighting the promoter and the coding sequence of the gene of interest (goi). Alternative promoters: Immediate early protein ie1, polh, p10, gp64, p6.9, vp39 promoters from baculovirus genomes are used for protein expression in insect cells and promoters such as Simian Virus SV40, immediate early protein ie1 cytomegalovirus (CMV), elongation factor (EF)-1α, hybrid CAG promoter, human ubiquitin C, U6, Pol III H1, and *Drosophila melanogaster* hsp70, among others, are commonly used for expression in mammalian cells. On the left of the schematic: different applications of the baculovirus expression system.

considered a useful, efficient, and robust tool in many areas, including but not limited to biotechnology, clinical biochemistry, and molecular biology (Fig. 11.1).

GENETIC ENGINEERING OF BACULOVIRUSES

Even though in the last three decades, different strategies have been developed for the generation of recombinant baculoviruses, two remain the most widely used. The more traditional approach consists of cotransfecting insect cells with both a baculoviral genome and a transfer plasmid. The transfer plasmid carries the genes of interest flanked by two regions containing sequences for homologous recombination; meanwhile, the viral genome provides the genetic information of a wild-type virus. Isolation of the recombinant event was made by selecting the OB-phenotype (nonoccluded viruses) observed by microscopy. However, due to a low recombination frequency ranging from 0.1% to 1%, this original technique yielded a high proportion of nonrecombinant wild-type virus (Fig. 11.2A) (O'Reilly and Miller, 1991). This contamination with parental virus makes isolation and purification of recombinant viruses laborious since it involves multiple plaque

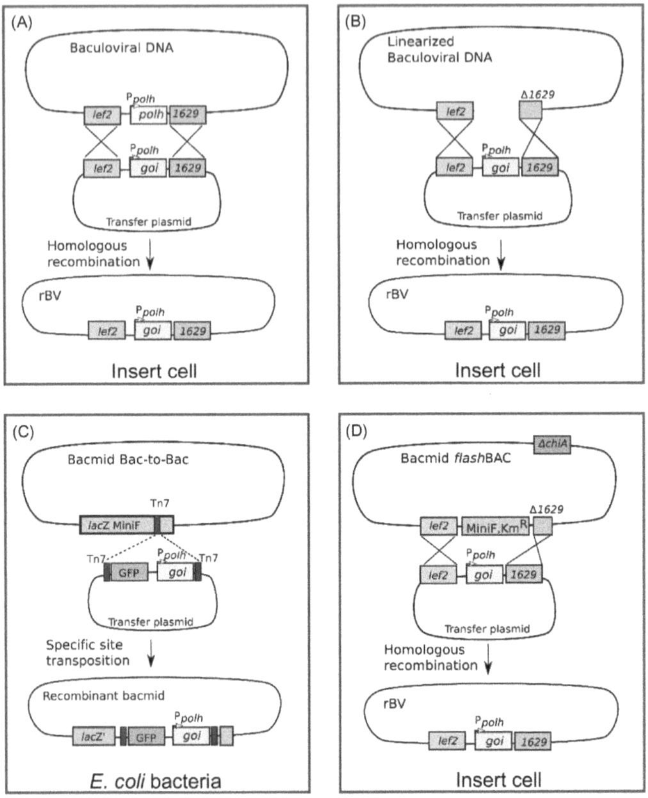

FIGURE 11.2 Methods for the generation of recombinant baculovirus (rBV). (A) Classic methodology. rBV are generated by homologous recombination between a baculoviral wild-type genome and a transfer plasmid containing the gene of interest (goi) in insect cells; (B) Linearized viral genome approach. The viral genome is linearized and the essential gene orf1629 is truncated. This approach facilitates the recombinant virus' selection steps. Homologous recombination reconstitutes orf1629; (C) Bac-to-Bac transposition-based technology. The goi is inserted by site-specific transposition into a bacmid inside *Escherichia coli* cells; (D) *Flash*BAC system. Homologous recombination occurs between a bacmid and a transfer plasmid in insect cells. Bacterial sequences are eliminated after the recombination event.

selection steps that are time consuming. In order to reduce the parental virus progeny, the baculoviral genome was linearized in a unique restriction site (Bsu36I) in the accessory gene *polh*, which codes for the major protein of the nucleopolyhedroviruses' OB, polyhedrin (POLH) and has a strong baculoviral promoter for high expression in insect cells. Since linear genome cannot replicate (noninfective), only the genomes that become circular after homologous recombination with the transfer vector are able to produce progeny. This modification to the system provides a recombination frequency near 30%, but the possibility of isolating wild-type virus still exists due to genetic repair events or incomplete digestion (Fig. 11.2B) (Possee et al., 1991). Further modifications have been made to this system, by inserting additional unique restriction sites, one of which is inserted in the essential gene *orf1629*, which will be disrupted during linearization of the viral DNA by restriction enzyme digestion. This open reading frame (ORF) codes for a phosphoprotein that is a key component of the NC and its deletion renders undesired recircularization events nonviable (Kitts and Possee, 1993).

The second and most widely used methodology of site-specific transposition of Tn7 transposon into a bacmid containing the baculoviral genome (Fig. 11.2C). This system is commercially available as Bac-to-Bac Expression System (Invitrogen, Inc.). Shortly, this technique for generating recombinant baculoviruses consists in the site-specific transposition of a transfer vector (pFastBac) that carries the goi under a strong baculoviral promoter, and a bacmid. The bacmid contains Tn7 sites and a bacterial origin of replication (*mini-F*) and is propagated in DH10Bac *Escherichia coli*. This *E. coli* strain contains a helper plasmid that codes for Tn7 transposase. When both bacmid and transfer plasmid are transformed into bacteria containing the helper plasmid, the site-specific transposition takes place generating a recombinant baculoviral genome. The selection of clones is made by white and blue colony screening. The bacmid can be purified from bacterial cultures and transfected into Sf9 cells inside of which it can replicate, thus generating recombinant baculoviruses. The downside to this system when developing therapeutic products for both veterinary and human use is the presence of bacterial genomic sequences and selection markers such as antibiotic resistance.

Oxford Expression Technologies has made a safer approach with its *flash*BAC platform, which combines the two most widely used methodologies for the generation of recombinant baculoviruses (Fig. 11.2D). This platform is based on a bacmid where an essential gene such as *orf1629* has been partially deleted (other accessory genes such as *p74*, *p10*, *p26* have been also knocked out) and homologous recombination that occurs in insect cells with transfer plasmid containing part of *orf1629* restores the complete sequence of this essential gene, while eliminating bacterial sequences (Je et al., 2003; Hitchman et al., 2010). The *flash*BAC platform also offers variants to increase yields for specific target proteins (GOLD, ULTRA, and PRIME). More recently, a new platform known as MultiBac was developed (Geneva Biotech) for the expression of multiple polypeptides, especially suitable for the production of eukaryotic multiprotein complexes (Berger et al., 2013). This system allows the insertion of foreign genes into two different *loci* of an engineered baculoviral genome: *polyhedrin* locus and *chiA*/*v-cath* locus.

The advantages of MultiBac are the tandem recombining that permits expression of proteins with various subunits and the reduction of proteolytic activity in the cell.

EXPRESSION OF HETEROLOGOUS PROTEINS FOR DIAGNOSIS AND IMMUNIZATION

Expression systems for the production of therapeutic proteins have been established using recombinant DNA technology in mammalian, plant, insect, fungal, and bacterial cells, as well as transgenic multicellular organisms (Wu et al., 2002; Steinbach et al., 2002; Chen et al., 2003; Katagiri et al., 2003; Hammonds et al., 2007).

The BEVS can be divided into two stages: the first is the generation and isolation of a recombinant baculovirus carrying the goi and the second is the infection of a susceptible host (cultured cells or larvae) with this recombinant virus (O'Reilly and Miller, 1991). The very first recombinant protein that was produced using BEVS was human interferon beta (Smith et al., 1983), but to date thousands of proteins have been expressed using this system. By far the most widely used virus is *Autographa californica* nucleopolyhedrovirus (AcMNPV); however, *Bombyx mori* nucleopolyhedrovirus is also used due to the higher protein expression using silkworm larvae or its pupae (Motohashi et al., 2005). The most commonly used promoters for the expression of heterologous proteins in BEVS are the POLH promoter (polh) and p10 promoter (p10). These promoters have a high expression capacity and they become active following the late stages of infection.

In terms of cell lines, those derived from *Spodoptera frugiperda* ovary (Sf9, Sf21, and ExpressSF +) and *Trichoplusia ni* (HighFive, BTI-TN-5B1-4, Invitrogen, Carlsbad, CA, United States) are the ones most frequently chosen by the users of BEVS. All these lines are able to grow both adherent or in suspension, in serum-free media and without requirements of CO_2 (Grace, 1962; Granados, 1994; Vaughn et al., 1997; Lynn, 2001; Ikonomou et al., 2003; Agathos, 2010). It has been reported that HighFive cells have the highest heterologous protein expression yield (Wang et al., 2011, 1992; Yamaji et al., 2006). Larvae belonging to the order Lepidoptera, such as the silkworm, can also be used for the expression, with similar protein yields, the higher rate of production and, lower costs compared to those of cell culture. However, the requirements of Good Manufacturing Practice and quality control using larvae are more difficult to achieve, which makes the use of these insects not as common. Although insect cell lines are more secure than mammalian ones they have to be screened for adventitious agents that pose a risk to human and veterinary health, such as *Alphanodavirus* and *Rhabdovirus* that have been isolated from *Helicoverpa armigera* and *S. frugiperda* cells, respectively (Bai et al, 2011; Ma et al., 2014). A disadvantage associated with the use of both cell culture and larvae is that after the first round of infection, cells lyse, and the entire culture has to be replaced, which does not allow a continuous expression of the goi. However, the use of low-cost serum-free media has somewhat helped overcome this disadvantage.

The recombinant protein production has also been improved by the deletion of baculoviral accessory genes such as chitinase and cathepsin that had a positive effect on the integrity of both the intracellular and secreted recombinant protein (Kaba et al., 2004; Hitchman et al., 2010).

There are many advantages of BEVS over other expression systems. It provides a eukaryotic platform for adequate folding, disulfide bridge establishment, oligomerization, and other posttranslational modifications needed for biological activity of proteins. Moreover, high protein yields can be obtained using strong viral promoters, and the system allows the simultaneous expression of multiple genes due to the baculoviral genome's capacity to insert DNA fragments up to 34 kbp (O'Reilly et al, 1994). It is important to mention that the production of recombinant proteins using this system has great scale-up potential and low-associated cost; low biosecurity requirements are needed since baculoviruses have a very limited host range and are nonpathogenic to humans.

In terms of posttranslational modifications of recombinant proteins, such as the eukaryotic glycosylation, BEVS presents simple protein glycosylation patterns. However, the *N*-glycosylation of proteins in insect cells does not resemble fully that of mammalian cell culture, while insect *N*-glycans have terminal mannose residues, and mammalian *N*-glycans often have terminal sialic acid residues. This difference may result in a lower or even total loss of biological function of the protein of interest (Harrison and Jarvis, 2006, 2016). In order to overcome this problem, humanized insect cells have been developed that express the mammalian glycosylating enzymes (Harrison and Jarvis, 2006; Jarvis et al., 1998; Mabashi-Asazuma et al., 2014; Okada et al., 2010). Another strategy consists in modifying the baculoviral genome in order to express the goi along with these glycosylating enzymes (SweetBac) (Palmberger et al., 2012).

Product contamination with baculoviruses might be a problem when using BEVS for development of human and veterinary vaccines. Marek et al. have developed a system called Bacfree. This methodology uses *vp80* deficient baculovirus that can only be propagated in cells that provides VP80 in *trans*. Deletion of this essential gene prevents the assembly of the viral phenotypes BV and ODV (Marek et al., 2011).

The rapid, scalable, and low-cost production using this technology is particularly important in terms of confronting sanitary emergencies such as the influenza pandemic of 2009. Fedson and Dunnill (2007) reported that if just 25% of global bioreactor capacity was used in the generation of anti-flu vaccines with the BEVS, 425 million vials of 10 μg/dose could be produced in only 1 month, as opposed to the classic platform used for this purpose, which involves embryonated chicken eggs and takes about 6 months (Wong and Webby, 2013). This means that nowadays, the production of influenza vaccines begins months before the exact prevalent strain of influenza is know for certain. If BEVS platform was used instead, the time required for the production would be just 45 days (Cox, 2012).

Over the last 30 years, the scientific community has used this platform for the production of recombinant proteins at a laboratory scale. However, in the last decade, the BEVS has started to gain priority in the biotechnological industry. To date, five veterinary vaccines and four products for human use exist on the market including, vaccines, immunotherapy, and gene therapy vectors (Table 11.1).

FluBlok is the first recombinant vaccine to be licensed by the US Food and Drug Administration for the prevention of seasonal influenza and is manufactured using BEVS in expresSF+ insect cells. It is a trivalent recombinant hemagglutinin (HA) vaccine, produced by Protein Sciences Corporation, that contains HA antigen derived from three influenza virus strains, selected for inclusion in the annual influenza vaccine by the World Health Organization and updated on an annual basis (Holtz et al., 2003; Wang et al., 2006). Since in this system, there cannot be replication, subsequent inactivation steps of the live influenza virus that are crucial in the licensed egg-based vaccines are not needed using this alternative.

Diamyd is a therapeutic vaccine produced using BEVS in Sf9 cells aimed to treat type 1 diabetes. Diamyd is a recombinant 65 kDa glutamate decarboxylase corresponding to human glutamate decarboxylase 65 that is the major autoantigen in type 1 diabetes. Diamyd is developed by Diamyd Medical and produced by Protein Sciences Corporation (Hinke, 2008).

In summary, the BEVS is a valuable platform for the manufacture of a great variety of biotechnological products since with just one insect cell line and a single method for the construction of recombinant baculoviruses it is possible to express a large variety of products.

TABLE 11.1 Commercial Vaccines and Therapies Based on Baculovirus Expression Technology for Veterinary or Human Use

Pathogen of disease	Expressed product	Vaccine type	Commercial name	Company	Year	References
VETERINARY USE						
Classical swine fever	E2 glycoprotein	Subunit/ marker	Porcilis Pesti	Merck	1998	Kirnbauer et al. (1992)
Classical swine fever	E2 glycoprotein	Subunit/ marker	Bayovac CSF E21	Bayer AG	2001	Small et al. (2000)
Porcine circovirus type 2	PCV ORF2	VLP	Circumvent PCV	MSD Animal Health	2005	Blanchard et al. (2003)
Porcine circovirus type 2	PCV ORF2	VLP	Ingelvac CircoFLEX	Boehringer Ingelheim	2008	Fachinger et al. (2008)
Porcine circovirus type 2	PCV ORF2	VLP	Porcilis PCV	Merck	2009	van Aarle (2003)
HUMAN USE						
HPV	HPV L1 protein (strain 16 and 18)	VLP	Cervarix	GlaxoSmithKline	2007	Muñoz et al. (2003)
Prostate cancer	PAP–GM–CSF[a] AAV	Immunotherapy	Provenge	Dendreon	2010	Cheever and Higano (2011)
Familial lipoprotein lipase deficiency	AAV vector with lipoprotein lipase transgene[b]	Gene therapy	Glybera	uniQure	2012	Ferreira et al. (2014)
Avian influenza virus	HA	Subunit	Flublok	Protein Sciences Corporation	2013	Holtz et al. (2003)

[a]*Prostatic acid phosphatase coupled to granulocyte macrophage colony-stimulating factor.*
[b]*Discontinued.*
AAV, Adeno-associated viruses; *HA*, hemagglutinin; *HPV*, human papillomavirus; *VLPs*, virus-like particles.

THE BACULOVIRUS EXPRESSION VECTOR SYSTEM FOR THE GENERATION OF VIRUS-LIKE PARTICLES

The baculovirus/insect cell expression system has been widely used for the generation of antigens and the development of viral vaccines both for human and veterinary use (Cox, 2012; Mena and Kamen, 2011; van Oers, 2006; Yamaji, 2014). Classic viral vaccines (first-generation vaccines), such as those for measles, rubella, rabies, and yellow fever, use attenuated viral strains or inactivated viruses. Attenuation corresponds with limited viral replication after vaccination, and although the immune response induced by this kind of vaccines is similar to natural infection adverse, reactions cannot be dismissed. Conversely, inactivated strains cannot replicate since the genetic material has been destroyed, making them more secure than the attenuated ones. However, the immune response generated by these formulations is not as efficient, and three to five doses are generally required to attain optimal antibody titers; in addition, risks of incomplete inactivation can lead to events such as the Cutter incident, in which about 40,000 polio cases were caused by a defective polio vaccine in the United States (Fitzpatrick, 2006).

An alternative that is set to outclass this type of vaccines consists of subunit or recombinant protein formulations, which include virus-like particles (VLPs). These recombinant vaccines carry a limited number of viral proteins, either complete or truncated, in a nonpathogenic viral vector. VLPs are self-assembling, complex protein structures that resemble wild-type viral capsids or enveloped virions. The difference with the latter is that VLPs do not contain pathogenic (replicative) genomes. Vaccines formulated with these structures combine the advantages of attenuated vaccines with those of subunit vaccines since the epitopes can be recognized by host's immune system and efficiently stimulate the adaptive immune response while avoiding the possible reversion to virulent phenotypes.

VLPs are classified in two types based on structure: enveloped VLPs, that include a host-derived lipid membrane and naked or nonenveloped (simply referred to as VLPs). Likewise, they are also classified as simple or multiple VLPs depending on the number of different recombinant proteins included in the structure (Noad and Roy, 2003; Roldão et al., 2011). Some difficulties may be associated with VLP production: for example, protein stoichiometry can be a problem since the incorporation of adequate proportions of each protein may not occur (Latham and Galarza, 2001). However, even if the VLPs do not fully resemble the wild-type virus, they can stimulate the host immune system to an extent high enough to meet the requirements of a vaccine candidate.

VLPs represent a significant advance in terms of developing preventive tools against viruses that cannot be propagated in cell lines or ex vivo cultures, such as the hepatitis B virus and enteric adenoviruses types 40 and 41.

VLP production using BEVS can be accomplished by one of two alternatives, or by the combination of both. The first consists of the coinfection of insect cells with as many recombinant viruses as proteins the manufacturer wants to assemble in the VLP structure (monocistronic baculoviruses) or with a single polycistronic virus that codes for all the proteins. Complex multiprotein VLPs present difficulties for their production since the infection with multiple recombinant baculoviruses are not homogenous. Vieiria et al. compared both strategies for the production of rotavirus VLPs containing VP2, VP6, and VP7. They observed that the coexpression strategy was superior to the coinfection one; meanwhile Roldão et al. for the same VLPs found the opposite (Vieira et al., 2005; Roldão et al., 2006). To this day, a vast array of taxonomically and structurally different VLPs has been produced using BEVS. Many of these already are in preclinical and clinical trial stages, and seven VLP-based vaccines have already been approved and are currently being commercialized (Table 11.2). One of the most promising products already on the market is Cervarix, a VLP-based vaccine against human papillomaviruses 16 and 18, which are responsible for 70% of cervical cancer cases, produced by GlaxoSmithKline (Muñoz et al., 2003; Smith et al., 2007).

BACULOVIRUS SURFACE DISPLAY

As mentioned earlier, the BEVS can be used to produce substantial amounts of heterologous proteins for their use as immunogens. This strategy, although key to subunit vaccines development, can be replaced by peptide display on virion surfaces. Baculovirus surface display refers to the display of heterologous peptides or proteins on the surface of BVs by different strategies all of which involve a translational fusion of a peptide to GP64 or a portion of it (Kost et al., 2005; Xu et al., 2011). GP64 is the major surface protein and a class III integral membrane glycoprotein of budded virions that is able to mediate membrane fusion under acidic pH conditions. AcMNPV GP64 is a 512-amino acid long phosphoglycoprotein with 15 cysteine residues that participate in disulfide bridge formation required for stability and biological activity (Hefferon et al., 1999; Garry and Garry, 2008).

The surface display of the heterologous polypeptide can be achieved by constructing a chimera with complete wild-type GP64 or only the transmembrane, multimerization, and carboxy-terminal (cytoplasmic tail) domains to the target protein, or by expression of a recombinant

TABLE 11.2 Generation of Virus-Like Particles (VLPs) Using Baculovirus Expression Vector System

Virus	Type of VLP[a]	Cells	Recombinant proteins	Expression strategies	References
Feline leukemia virus	eVLP	*Sf9*	Gp85 and Gag	Coinfection	Thomsen et al. (1992)
Norwalk virus	VLP	*Sf9*	Capsid protein	Single infection	Jiang et al. (1992)
Porcine parvovirus	VLP	*Sf9*	VP2	Single infection	Martinez et al. (1992)
Bluetongue virus	VLP	*Sf21*	VP2, VP6, VP7, and NS1	Coexpression	Belyaev and Roy (1993)
Poliovirus	VLP	*Sf21*	VP0, VP3, and VP1	Coinfection	Bräutigam et al. (1993)
HPV	eVLP	*Sf9*	L1 and L2	Coexpression	Kirnbauer et al. (1992), Boxus et al. (2016), Huber et al. (2017)
Herpes simplex virus	VLP	*Sf21*	VP23, VP5, VP22,[a] VP21–VP24, VP26–VP19c	Coinfection	Tatman et al. (1994)
Rotavirus	VLP	*Sf9*	VP2, VP6, VP7, and VP4	Coinfection	Crawford et al. (1994)
Porcine parvovirus (LCMV)[b]	VLP	*Sf9*	VP2 containing LCMV epitope	Single infection	Sedlik et al. (1997)
African horse sickness virus	VLP	*Sf9*	VP3 and VP4	Coinfection	Maree et al. (1998)
Human immunodeficiency virus	eVLP	*Sf9*	Gag and Gp120	Coexpression	Buonaguro et al. (2001)
Human severe acute respiratory syndrome coronavirus	eVLP	*Sf21*	Spike, membrane and envelope proteins	Coinfection	Ho et al. (2004)
Human astrovirus	VLP	*Sf9*	ORF2	Single infection	Caballero et al. (2004)
Enterovirus-71	VLP	*Sf9*	P1 and 3CD	Coexpression and coinfection	Chung et al. (2006)

(*Continued*)

TABLE 11.2 (Continued)

Virus	Type of VLP[a]	Cells	Recombinant proteins	Expression strategies	References
Feline calicivirus	VLP	*Sf9*	Capsid	Single infection	Di Martino et al. (2007)
Simian virus 40	VLP	*Sf9*	VP1, VP2, and VP3	Single infection and coinfection	Kosukegawa et al. (1996)
Rift Valley fever virus	eVLP	*Sf9*	Gn, Gc, and N	Coexpression	Liu et al. (2008b)
Porcine circovirus	VLP	*Tn*	VP2	Single infection	Liu et al. (2008a)
Avian influenza virus	eVLP	*Sf9*	HA, NA, and others	Coexpression	Tao et al. (2009), Pushko et al. (2016)
Encephalomyocarditis virus	VLP	*Sf9*	P1-2A-3C	Single infection	Jeoung et al. (2011)
Rous sarcoma virus	eVLP	*Tn* and *Sf9*	Gag and Gp120	Single infection	Deo et al. (2011)
Japanese encephalitis virus	eVLP	*Sf9*	prM and gE	Coinfection	Yamaji et al. (2006)
Coxsackievirus A16	VLP	*Sf9*	P1 and 3CD	Coexpression	Liu et al. (2012)
Foot-and-mouth disease virus	VLP	*Sf9*	P1-2A-3C	Single infection	Mohana Subramanian et al. (2012)
RHDV	VLP	*SL*	Capsid protein	Single infection	Zheng et al. (2016)
PRRS	eVLP	*Sf9*	Gp2, Gp3, Gp4, Gp5, E, and M	Single infection and coinfection	García Durán et al. (2016), Nam et al. (2013)
Chikungunya virus	eVLP	*Sf21*	Structural polyprotein	Single infection	Metz et al. (2013)
Ebola virus	eVLP	*Tn* and *Sf9*	GP, NP, and VP40	Coexpression and coinfection	Sun et al. (2009), Ye et al. (2006), Warfield et al. (2007, 2015)
Marburg virus	eVLP	*Tn* and *Sf9*	GP, NP, and VP40	Coexpression and coinfection	Warfield et al. (2007)

(*Continued*)

TABLE 11.2 (Continued)

Virus	Type of VLP[a]	Cells	Recombinant proteins	Expression strategies	References
Sudan virus	eVLP	*Tn* and *Sf9*	GP, NP, and VP40	Coexpression and coinfection	Warfield et al. (2015)
Duck hepatitis A virus	VLP	*Sf9*	P1 and 3CD	Coexpression	Wang et al. (2018)

[a]*Classification of VLP: VLP (nonenveloped VLP), eVLP (enveloped VLP).*
[b]*LCMV, lymphocytic choriomeningitis virus;* Sf, Spodoptera frugiperda; Tn, Trichoplusia ni; *SL, Silkworm larvae. HPV,* Human papillomavirus; *PRRS,* porcine reproductive and respiratory syndrome; *RHDV,* rabbit hemorrhagic disease virus.
This table has been modified and expanded on the basis of Liu, F., Wu, X., Li, L., Liu, Z., Wang, Z., 2013. Use of baculovirus expression system for generation of virus-like particles: successes and challenges. Protein Expr. Purif. 90, 104–116.

GP64 fused to a short or long heterologous peptide (Grabherr and Ernst, 2010).

Baculovirus display of heterologous proteins by fusion to GP64 was demonstrated to be very effective immunogens, successfully able to induce antibody response to a variety of displayed proteins (Kost et al., 2005). In addition, baculoviruses by themselves are able to trigger a robust innate immune response by activation of professional antigen-presenting cells.

Baculoviruses displaying heterologous envelope proteins, such as vesicular stomatitis virus (VSV)-G protein, have also been constructed. Several studies have shown that VSV-G is capable of improving transduction efficiency of baculovirus in vertebrate cells (Barsoum et al., 1997). Different strategies have been developed to increase the effectiveness of virus display (Chapple and Jones 2002). The polypeptides displayed by VSV-G exhibited higher densities of recombinant proteins than GP64 fusions because this protein is homogenous distributed in the BV compared with GP64 that is mostly apical.

In contrast to surface display, heterologous protein can be displayed on the NC of the virion. The NC core is composed of a circular, covalently closed dsDNA genome and the major basic core protein P6.9, and the NC sheath is formed by stacked ring-like subunits consisting of one major polypeptide, VP39, and several minor proteins. The NC display strategy involves the translational fusion of a peptide to VP39 (Kukkonen et al., 2003; Molinari et al., 2011; Pidre et al., 2013).

Depending on the host immune response that is targeted, one strategy might be chosen over the other. Antigen displayed on the NC should be able to reach the cytosol and preferentially trigger major histocompatibility complex class I presentation pathway and mount a strong CD8 + T-cell response (Molinari et al., 2011). In turn, the BV displaying antigens on the surface elicit a humoral response (Gronowski et al., 1999; Abe et al., 2005).

Several studies have shown that baculovirus display can generate high titers of specific antibodies that protect against different pathogens such as Japanese encephalitis virus, avian reovirus, human enterovirus, influenza, and malaria (Xu et al., 2011; Lin et al., 2014; Marek et al., 2011; Jin et al., 2008; Madhan et al., 2010).

BACULOVIRUS EXPRESSION VECTOR SYSTEM PLATFORM FOR THE PRODUCTION OF ADENO-ASSOCIATED VIRUSES

Adeno-associated viruses (AAV) have become relevant for in vivo gene-therapy (Nayerossadat et al., 2012; Flotte, 2013; Wang and Gao, 2014). This therapy involves the incorporation of genes into cells. AAV used as viral a vector is the tool that allows the safe, efficient, and specific incorporation of the gene inside the organism. The first group of hereditary diseases for which gene-therapy was investigated included hemophilia A and B, Leber's macular degeneration, and Duchenne's muscular dystrophy, but nowadays there are many studies focused on central nervous system diseases such as Parkinson's, different types of cancers, cardiovascular afflictions, and bone regeneration (Flotte et al., 1996; Kay et al., 2000; Kaplitt et al. 2007; Jiang et al., 2011). An estimate of 70%–80% of human population has been exposed to at least one event of infection with AAV, and no adverse consequences have been associated to the infection and latency of these viruses (Boutin et al., 2010; Erles et al., 1999; Tobiasch et al., 1998). AAVs have a single-stranded DNA genome and are nonenveloped viruses of 18–25 nm of diameter. They belong to the Parvovirus family and are classified as members of the *Dependovirus* genus. Twenty human and nonhuman primate strains have been isolated, and all serotypes except for serotype 5 share a very similar capsid structure, genome size, and organization. AAV genome has a size of approximately 4.7 kb and contains two ORFs that code for four regulatory proteins. These ORFs are flanked by two inverted terminal repeats (ITRs) that contain *cis*-regulatory sequences needed for viral infection. Major proteins Rep78 and Rep68 are involved in genome excision, rescue, replication, and integration. Integration into the host genome is site-specific and takes place in chromosome 19, more specifically into *locus* AAVS1 of the human genome, without adverse consequences. Minor proteins Rep52 and Rep40 are responsible for the accumulation of DNA and packaging. VP1, VP2, and VP3 are structural proteins encoded in the Cap ORF.

Recombinant AAVs are constructed by replacement of *rep* and *cap* genes with the goi flanked by ITR regions. This viral vector is able to transduce both active division and quiescent cells. It is very important to highlight that, in contrast to wt AAV, the recombinant AAV has no site-specific integration capacity since *rep* genes are absent; thus the integration occurs at random. However, the estimated integration frequency of these recombinant adeno-associated viruses (rAAVs) is 0.1%–0.5% in mammalian cell lines (McCarty et al., 2004). In view of this, to establish stable and persistent expression in patients, an alternative to genome integration has been developed consisting in double-stranded DNA that persists as an episome inside the cells (Yang et al., 1999). To date, adeno-associated vectors have been modified to be tissue-specific, the preferred organs being those that are immune-privileged such as the eye or the brain.

The classic technique for the production of AAVs involves HEK293, A549, or HeLa cells cotransfected with two plasmids, one containing the recombinant DNA (recombinant adeno-associated virus plasmid) and the other the *rep* and *cap* genes (pHelper). The cells are then infected with a helper virus such as an adenovirus or a herpes simplex virus to allow the replication of the recombinant AAV (Samulski et al., 1987). The downside to this methodology is the possibility of undesired recombination events between plasmids and that purification of the helper virus is time consuming, and the scale-up process is difficult to achieve. A good alternative has been the use of BEVS platform [revised in Galibert and Merten (2011)], which involves the use of three recombinant baculoviruses and Sf9 cells. Each of the three recombinant baculoviruses contains one of the following genes: *rep*, *cap*, and the *construct of interest* flanked by the ITRs. Alternatives involving a polycistronic baculovirus containing *rep* and *cap* genes are also used (Smith et al., 2009). The OneBac system involves transgenic insect cell lines that express the genes *rep* and *cap* of the AAV and only one recombinant baculovirus. The efficiency of this technique and the scale-up feasibility are higher than the previous alternatives, and it provides a more flexible strategy since it allows the production of other AAV serotypes by simply changing cell lines (Mietzsch, 2014). Such as other products derived from BEVS, the contamination of rAAVs with baculovirus particles has to be reduced. There is already one rAAV produced by BEVS on the market commercialized as Glybera (alipogene tiparvovec) developed and marketed by uniQure that compensates for lipoprotein lipase deficiency, which received regulatory approval by the European Medicines Agency in 2012, although now it has been discontinued due to lack of demand.

BACULOVIRUSES AS MAMMALIAN TRANSDUCTION VECTORS

It was demonstrated in the 1990s that the baculoviral genome was capable of transducing mammalian cells and that if heterologous genes were under control of adequate mammalian promoters, the ORFs carried by this baculoviral genome were transiently expressed in these cells. The viral promoters such as Simian Viruses SV40, *immediate early protein* ie1 CMV, elongation factor-1α, hybrid CAG promoter, human ubiquitin C, U6, Pol III H1, and *Drosophila melanogaster* hsp70 among others are commonly used for expression in mammalian cells. Following the first days, the baculoviral genome is degraded and unable of insertion in the mammalian genome in the absence of selection pressure (Volkman and Goldsmith, 1983; Carbonell et al., 1985; Boyce and Bucher, 1996). At present, the baculoviral transduction ability has been documented both in vitro and in vivo for a variety of cell types including embryonic stem cells and pluripotent stem cells (Kost et al., 2005; Ho et al., 2005; Airenne et al., 2010; Chen et al., 2011). It is suggested that the mechanism by which the baculoviruses ingress mammalian cells is clathrin and dynamin-dependent endocytosis and macropinocytosis. Makkonen showed that this process also requires the host cell proteins heparan sulfate proteoglycans and syndecan-1 that serves as a receptor (Long et al., 2006a,b; Kataoka et al., 2012; Makkonen et al., 2013, 2014; Nasimuzzaman, 2014). Once inside the cell, the baculovirus virions are transported into the endosome, where acidic conditions promote membrane fusion mediated by GP64, and the genetic material is thus liberated into the cytoplasm. Actin cytoskeleton reorganization promotes translocation of the baculoviral genome into the nucleus where transcription occurs by host machinery.

One of the advantages of using baculovirus as vectors for mammalian cell transduction is that in contrast to other viruses used to the same end, they are not human pathogens, they do not replicate in mammalian cells, no integration in the genome has been documented, and there is no preexisting immunity to baculoviruses (Strauss et al., 2007; Jin et al., 2008). A human study was made in which volunteers consumed OBs in their diet; the results were promising since no inflammatory response, allergies, or adverse effects were observed in a 5-day follow-up (Heimpel et al., 1973). These studies have been extended to animal models to evaluate possible side-effects such as skin irritation and cytotoxicity, all of which have been negative (Kost and Condreay, 2002; Jin et al., 2008).

Even though a downside to the use of baculovirus as viral vectors for gene delivery is the lack of long-term expression of the transgene, alternatives to overcome this have been developed. For example, Lo et al. engineered a baculovirus vector system with Cre/loxP sites. In their approach, the construct is excised and remains as an episome in the cells that prolong expression (Lo et al., 2017).

All this evidence, together with the fact that baculoviruses infect no other organisms besides those of the phylum *Arthropoda*, shows they are safe vectors for developing DNA vaccines, immunotherapy, and gene therapy. A reflection of this is the increasing number of peer-reviewed publications that use recombinant baculoviruses over other viral vectors. For example, Swift et al. (2013) used a baculovirus with a prodrug approach to mediate cell-death of prostate cancer cells with success, even observing the good capacity of cell-layer penetration compared to other viral vectors. In the last few years, gene silencing by RNA interference has become more relevant for gene regulation. Baculoviral vectors have been engineered to deliver microribonucleic acid and short hairpin ribonucleic acid into insect and mammalian cells successfully. Gottardo and Pidre showed that baculovirus-mediated gene silencing of human in increased survival and reduced the progression of pituitary tumor growth in vivo (Gottardo et al., 2018). This type of strategy can be also applied to interfere with viral pathogens infecting humans and other animals (Makkonen et al., 2015).

DISCUSSION

For over 30 years, baculoviruses have been exploited in molecular biology laboratories, and the trend has never stopped increasing. Since they are a highly versatile tool, with low-associated costs and scalable production, pharmaceutical giants such as Merck are taking advantage of them. Baculoviruses are not mere biological control agents anymore but are also key in the state of the art of gene-therapy and immunization departments. One of the most relevant advantages of BEVS is the rapid response under an epidemiological emergency such as an influenza pandemic. Meanwhile, the classical first-generation, egg-based approach for developing flu vaccines can be nonsuitable for allergic users and takes about six months to develop; the BEVS recombinant technology approach can give a fully functional vaccine within the first 45 days after the outbreak. Using the BEVS is not only inexpensive and direct, but it also provides unique plasticity. Together with the fact that almost no special biosafety measures have to be taken into account for its application, it makes an excellent platform when compared to other systems.

References

Abe, T., Hemmi, H., Miyamoto, H., Moriishi, K., Tamura, S., Takaku, H., et al., 2005. Involvement of the Toll-like receptor 9 signaling pathway in the induction of innate immunity by baculovirus. J. Virol. 79, 2847–2858.

Adams, J.R., McClintock, J.T., 1991. Baculoviridae, nuclear polyhedrosis viruses Part 1. Nuclear polyhedrosis viruses of insects. In: Adams, J.R., Bonami, J.R. (Eds.), Atlas Invertebr Viruses. CRC Press, Boca Raton, FL, pp. 87–180.

Agathos, S.N., 2010. Insect cell culture. In: Baltz, R.H., Davies, J.E., Demain, A.L. (Eds.), Manual of Industrial Microbiology and Biotechnology. ASM Press, Washington, DC, pp. 212–222.

Airenne, K.J., Makkonen, K.-E., Mähönen, A.J., Ylä-Herttuala, S., 2010. In vivo application and tracking of baculovirus. Curr. Gene Ther. 10, 187–194.

Au, S., Wu, W., Panté, N., 2013. Baculovirus nuclear import: open, nuclear pore complex (NPC) sesame. Viruses 5, 1885–1900.

Bai, H., Wang, Y., Li, X., Mao, H., Li, Y., Han, S., et al., 2011. Isolation and characterization of a novel *Alphanodavirus*. Virol. J. 8, 311.

Barsoum, J., Brown, R., McKee, M., Boyce, F.M., 1997. Efficient transduction of mammalian cells by a recombinant baculovirus having the vesicular stomatitis virus G glycoprotein. Hum. Gene Ther. 8, 2011–2018.

Belyaev, A.S., Roy, P., 1993. Development of baculovirus triple and quadruple expression vectors: co-expression of three or four bluetongue virus proteins and the synthesis of bluetongue virus-like particles in insect cells. Nucleic Acids Res. 21, 1219–1223.

Berger, I., Garzoni, F., Chaillet, M., Haffke, M., Gupta, K., Aubert, A., 2013. The multiBac protein complex production platform at the EMBL. J. Vis. Exp. 77, e50159.

Blanchard, P., Mahé, D., Cariolet, R., Keranflec'h, A., Baudouard, M.A., Cordioli, P., et al., 2003. Protection of swine against post-weaning multisystemic wasting syndrome (PMWS) by porcine circovirus type 2 (PCV2) proteins. Vaccine 21, 4565–4575.

Boutin, S., Monteilhet, V., Veron, P., Leborgne, C., Benveniste, O., Montus, M.F., et al., 2010. Prevalence of serum IgG and neutralizing factors against adeno-associated virus (AAV) types 1, 2, 5, 6, 8, and 9 in the healthy population: implications for gene therapy using AAV vectors. Hum. Gene Ther. 21, 704–712.

Boxus, M., Fochesato, M., Mertens, E., Dendouga, N., Brendle, S., Balogh, K.K., et al., 2016. Broad cross-protection is induced in preclinical models by a human. J. Virol. 90, 6314–6325. Available from: https://doi.org/10.1128/JVI.00449-16.

Boyce, F.M., Bucher, N.L., 1996. Baculovirus-mediated gene transfer into mammalian cells. Proc. Natl Acad. Sci. U.S.A. 93, 2348–2352.

Bräutigam, S., Snezhkov, E., Bishop, D.H., 1993. Formation of poliovirus-like particles by recombinant baculoviruses expressing the individual VP0, VP3, and VP1 proteins by comparison to particles derived from the expressed poliovirus polyprotein. Virology 2, 512–524.

Buonaguro, L., Buonaguro, F.M., Tornesello, M.L., Mantas, D., Beth-Giraldo, E., Wagner, R., et al., 2001. High efficient production of Pr55gag virus-like particles expressing multiple HIV-1 epitopes, including a gp120 protein derived from an Ugandan HIV-1 isolate of subtype A. Antiviral. Res. 49, 35–47.

Caballero, S., Guix, S., Ribes, E., Bosch, A., Pintó, R.M., Pinto, R.M., 2004. Structural requirements of astrovirus virus-like particles assembled in insect cells. J. Virol. 78, 13285–13292.

Carbonell, L.F., Klowden, M.J., Miller, L.K., 1985. Baculovirus-mediated expression of bacterial genes in dipteran and mammalian cells. J. Virol. 56, 153–160.

Chapple, S.D., Jones, I.M., 2002. Non-polar distribution of green fluorescent protein on the surface of *Autographa californica* nucleopolyhedrovirus using a heterologous membrane anchor. J. Biotechnol. 95, 269–275.

Cheever, M.A., Higano, C.S., 2011. PROVENGE (sipuleucel-T) in prostate cancer: the first FDA-approved therapeutic cancer vaccine. Clin. Cancer Res. 17, 3520–3526.

Chen, J., Duncan, M.B., Carrick, K., Pope, R.M., Liu, J., 2003. Biosynthesis of 3-*O*-sulfated heparan sulfate: unique substrate specificity of heparan sulfate 3-*O*-sulfotransferase isoform 5. Glycobiology 13, 785–794.

Chen, C.-Y., Wu, H.-H., Chen, C.-P., Chern, S.-R., Hwang, S.-M., Huang, S.-F., et al., 2011. Biosafety assessment of human mesenchymal stem cells engineered by hybrid baculovirus vectors. Mol. Pharm. 8 (5), 1505–1514.

Chung, Y.C., Huang, J.H., Lai, C.W., Sheng, H.C., Shih, S.R., Ho, M.S., et al., 2006. Expression, purification and characterization of enterovirus-71 virus-like particles. World J. Gastroenterol. 12, 921–927.

Cox, M.M.J., 2012. Recombinant protein vaccines produced in insect cells. Vaccine 30, 1759–1766.

Crawford, S.E., Labbé, M., Cohen, J., Burroughs, M.H., Zhou, Y.-J., Estes, M.K., 1994. Characterization of virus-like particles produced by the expression of rotavirus capsid proteins in insect cells. J. Virol. 68, 5945–5952.

Deo, V.K., Tsuji, Y., Yasuda, T., Kato, T., Sakamoto, N., Suzuki, H., et al., 2011. Expression of an RSV-gag virus-like particle in insect cell lines and silkworm larvae. J. Virol. Methods 177, 147–152.

Di Martino, B., Marsilio, F., Roy, P., 2007. Assembly of feline calicivirus-like particle and its immunogenicity. Vet. Microbiol. 120, 173–178.

Erles, K., Sebokova, P., Schlehofer, J.R., 1999. Update on the prevalence of serum antibodies (IgG and IgM) to adeno-associated virus (AAV). J. Med. Virol. 59, 406–411.

Fachinger, V., Bischoff, R., Jedidia, S.B., Saalmüller, A., Elbers, K., 2008. The effect of vaccination against porcine circovirus type 2 in pigs suffering from porcine respiratory disease complex. Vaccine 26, 1488–1499.

Fedson, D.S., Dunnill, P.M., 2007. New approaches to confronting an imminent influenza pandemic. Perm. J. 11, 63–69.

Ferreira, V., Petry, H., Salmon, F., 2014. Immune responses to AAV-vectors, the Glybera example from bench to bedside. Front. Immunol. 5, 1–15.

Fitzpatrick, M., 2006. The cutter incident: how America's first polio vaccine led to the growing vaccine crisis. J. R. Soc. Med. 99 (7543), 156.

Flotte, T.R., 2013. Birth of a new therapeutic platform: 47 years of adeno-associated virus biology from virus discovery to licensed gene therapy. Mol. Ther. 21 (11), 1976–1981.

Flotte, T., Carter, B., Conrad, C., Guggino, W., Reynolds, T., Rosenstein, B., et al., 1996. A phase I study of an adeno-associated virus-CFTR gene vector in adult CF patients with mild lung disease. Hum. Gene Ther. 7 (9), 1145–1159.

Galibert, L., Merten, O.W., 2011. Latest developments in the large-scale production of adeno-associated virus vectors in insect cells toward the treatment of neuromuscular diseases. J. Invertebr. Pathol. 107, S80–S93.

García Durán, M., Costa, S., Sarraseca, J., de la Roja, N., García, J., García, I., et al., 2016. Generation of porcine reproductive and respiratory syndrome (PRRS) virus-like-particles (VLPs) with different protein composition. J. Virol. Methods 236, 77–86.

Garry, C.E., Garry, R.F., 2008. Proteomics computational analyses suggest that baculovirus GP64 superfamily proteins are class III penetrenes. Virol. J. 5, 28.

Goodwin, R.H., Vaughn, J.L., Adams, J.R., Louloudes, S.J., 1970. Replication of a nuclear polyhedrosis virus in an established insect cell line. J. Invertebr. Pathol. 16, 284–288.

Gottardo, M.F., Pidre, M.L., Zuccato, C., Asad, A.S., Imsen, M., Jaita, G., et al., 2018. Baculovirus-based gene silencing of Humanin for the treatment of pituitary tumors. Apoptosis 23, 1–9.

Grace, T.D.C., 1962. Establishment of four strains of cells from insect tissues grown in vitro. Nature 195, 788–789.

Grabherr, R., Ernst, W., 2010. Baculovirus for eukaryotic protein display. Curr. Gene Ther. 10, 195–200.

Granados, R.R., 1994. *Trichoplusia ni* cell line which supports replication of baculoviruses. United States Patent: US005300435A. Boyce Thompson Institute for Plant Research, Ithaca, NY.

Gronowski, A.M., Hilbert, D.M., Sheehan, K.C., Garotta, G., Schreiber, R.D., 1999. Baculovirus stimulates antiviral effects in mammalian cells. J. Virol. 73, 9944–9951.

Haas-Stapleton, J.O., Washburn, J.O., Volkman, L.E., 2004. P74 mediates specific binding of *Autographa californica* multiple nucleopolyhedrovirus occlusion-derived virus to primary cellular targets in the midgut epithelia of *Heliothis virescens* larvae. Virology 78, 6786–6791.

Hammonds, J., Chen, X., Zhang, X., Lee, F., Spearman, P., 2007. Advances in methods for the production, purification, and characterization of HIV-1 Gag-Env pseudovirion vaccines. Vaccine 25, 8036–8048.

Harrison, R.L., Jarvis, D.L., 2006. Protein *N*-glycosylation in the baculovirus-insect cell expression system and engineering of insect cells to produce 'Mammalianized' recombinant glycoproteins. Adv. Virus Res. 68, 159–191.

Harrison, R.L., Jarvis, D.L., 2016. Transforming lepidopteran insect cells for continuous recombinant protein expression. Methods Mol. Biol 1350, 329–348. Available from: https://doi.org/10.1007/978-1-4939-3043-2_16.

Hefferon, K.L., Oomens, A.G., Monsma, S.A., Finnerty, C.M., Blissard, G.W., 1999. Host cell receptor binding by baculovirus GP64 and kinetics of virion entry. Virology 258, 455–468.

Heimpel, A.M., Thomas, E.D., Adams, J.R., Smith, L.J., 1973. The presence of nuclear polyhedrosis viruses of Trichoplusia ni on cabbage from the market shelf. Environ. Entomol. 2, 72–75.

Herniou, B.M., Becnel, J.J., Blissard, G.W., Bonning, B., Harrison, R., Jehle, J.A., et al., 2011. Baculoviridae. In: King, A.M.Q., Adams, M.J., Carstens, E.B., Lefkowitz, E.J. (Eds.), Virus Taxonomy: Classification and Nomenclature of Viruses: Ninth Report of the International Committee on Taxonomy of Viruses. Elsevier Acad Press, San Diego, CA, pp. 163–173.

Hinke, S.A., 2008. Diamyd, an alum-formulated recombinant human GAD65 for diabetes and the prevention of autoimmune diabetes. Curr. Opin. Mol. Ther. 12 (6), 724–733.

Hitchman, R.B., Possee, R.D., Crombie, A.T., Chambers, A., Ho, K., Siaterli, E., et al., 2010. Genetic modification of a baculovirus vector for increased expression in insect cells. Cell Biol. Toxicol. 26, 57–68.

Ho, Y., Lin, P.-H., Liu, C.Y.Y., Lee, S.-P., Chao, Y.-C., 2004. Assembly of human severe acute respiratory syndrome coronavirus-like particles. Biochem. Biophys. Res. Commun. 318, 833–838.

Ho, Y.-C., Chung, Y.-C., Hwang, S.-M., Wang, K.-C., Hu, Y.-C., 2005. Transgene expression and differentiation of baculovirus-transduced human mesenchymal stem cells. J. Gene Med. 7, 860–868.

Holtz, K., Anderson, K., Cox, M.J., 2003. Production of a recombinant influenza vaccine using the baculovirus expression vector system. Bioprocessing 2, 25–32.

Horton, H.M., Burand, J.P., 1993. Saturable attachment sites for polyhedron-derived baculovirus on insect cells and evidence for entry via direct membrane fusion. J. Virol. 67, 1860–1868.

Huber, B., Schellenbacher, C., Shafti-Keramat, S., Jindra, C., Christensen, N., Kirnbauer, R., 2017. Chimeric L2-based virus-like particle (VLP) vaccines targeting cutaneous human papillomaviruses (HPV). PLoS One 12, 1–27. Available from: https://doi.org/10.1371/journal.pone.0169533.

Ikonomou, L., Schneider, Y.-J., Agathos, S.N., 2003. Insect cell culture for industrial production of recombinant proteins. Appl. Microbiol. Biotechnol. 62 (1), 1–20.

Jarvis, D.L., Kawar, Z.S., Hollister, J.R., 1998. Engineering *N*-glycosylation pathways in the baculovirus-insect cell system. Curr. Opin. Biotechnol. 9, 528–533.

Je, Y.H., Jin, B.R., Park, H.W., Roh, J.Y., Chang, J.H., Seo, S.J., et al., 2003. Baculovirus expression vectors that incorporate the foreign protein into viral occlusion bodies. Biotechniques 34, 81–87.

Jehle, J.A., Blissard, G.W., Bonning, B.C., Cory, J.S., Herniou, E.A., Rohrmann, G.F., et al., 2006. On the classification and nomenclature of baculoviruses: a proposal for revision. Arch. Virol. 151, 1257–1266.

Jeoung, H.-Y., Lee, W.-H., Jeong, W., Shin, B.-H., Choi, H.-W., Lee, H.S., et al., 2011. Immunogenicity and safety of virus-like particle of the porcine encephalomyocarditis virus in pig. Virol. J. 8, 170.

Jiang, X., Wang, M., Graham, D.Y., Estes, M.K., 1992. Expression, self-assembly, and antigenicity of the Norwalk virus capsid protein. J. Virol. 66, 6527–6532.

Jiang, M., Liu, Z., Xiang, Y., Ma, H., Liu, S., Liu, Y., et al., 2011. Synergistic antitumor effect of AAV-mediated TRAIL expression combined with cisplatin on head and neck squamous cell carcinoma. BMC Cancer 11, 54.

Jin, R., Lv, Z., Chen, Q., Quan, Y., Zhang, H., Li, S., et al., 2008. Safety and Immunogenicity of H5N1 influenza vaccine based on baculovirus surface display system of *Bombyx mori*. PLoS One 3, 1–7.

Kaba, S.A., Salcedo, A.M., Wafula, P.O., Vlak, J.M., Van Oers, M.M., 2004. Development of a chitinase and v-cathepsin negative bacmid for improved integrity of secreted recombinant proteins. J. Virol. Methods 122, 113–118.

Kaplitt, M.G., Feigin, A., Tang, C., Fitzsimons, H.L., Mattis, P., Lawlor, P.A., et al., 2007. Safety and tolerability of gene therapy with an adeno-associated virus (AAV) borne GAD gene for Parkinson's disease: an open label, phase I trial. Lancet 369 (9579), 2097–2105.

Katagiri, Y., Brew, S.A., Ingham, K.C., 2003. All six modules of the gelatin-binding domain of fibronectin are required for full affinity. J. Biol. Chem. 278, 11897–11902.

Kataoka, C., Kaname, Y., Taguwa, S., Abe, T., Fukuhara, T., Tani, H., et al., 2012. Baculovirus GP64-mediated entry into mammalian cells. J. Virol. 86 (5), 2610–2620.

Kay, M.A., Manno, C.S., Ragni, M.V., Larson, P.J., Couto, L.B., McClelland, A., et al., 2000. Evidence for gene transfer and expression of factor IX in haemophilia B patients treated with an AAV vector. Nat. Genet. 24 (3), 257–261.

Kirnbauer, R., Booy, F., Cheng, N., Lowy, D., Schiller, J., 1992. Papillomavirus L1 major capsid protein self-assembles into virus-like particles that are highly immunogenic. Proc. Natl Acad. Sci. U.S.A. 89, 12180–12184.

Kitts, P.A., Possee, R.D., 1993. A method for producing recombinant baculovirus expression vectors at high frequency. Biotechniques 14, 810–817.

Kost, T.A., Condreay, J.P., 2002. Recombinant baculoviruses as mammalian cell gene-delivery vectors. Trends Biotechnol. 20, 173–180.

Kost, T.A., Condreay, J.P., Jarvis, D.L., 2005. Baculovirus as versatile vectors for protein expression in insect and mammalian cells. Nat. Biotechnol. 23, 567–575.

Kosukegawa, A., Arisaka, F., Takayama, M., Yajima, H., Kaidow, A., Handa, H., 1996. Purification and characterization of virus-like particles and pentamers produced by the expression of SV40 capsid proteins in insect cells. Biochim. Biophys. Acta 1, 37–45.

Kukkonen, S.P., Airenne, K.J., Marjomaki, V., Laitinen, O.H., Lehtolainen, P., Kankaanpaa, P., et al., 2003. Baculovirus capsid display: a novel tool for transduction imaging. Mol. Ther. 8, 853–862.

Laakkonen, J.P., Kaikkonen, M.U., Ronkainen, P.H., Ihalainen, T.O., Niskanen, E.A., Häkkinen, M., et al., 2008. Baculovirus-mediated immediate-early gene expression and nuclear reorganization in human cells. Cell Microbiol. 2, 667–681.

Latham, T., Galarza, J.M., 2001. Formation of wild-type and chimeric influenza virus-like particles following simultaneous expression of only four structural proteins. J. Virol. 75, 6154–6165. Available from: https://doi.org/10.1128/JVI.75.13.6154-6165.2001.

Lin, S.Y., Chung, Y.C., Hu, Y.C., 2014. Update on baculovirus as an expression and/or delivery vehicle for vaccine antigens. Expert Rev. Vaccines 13, 1501–1521.

Liu, L.J., Suzuki, T., Tsunemitsu, H., Kataoka, M., Ngata, N., Takeda, N., et al., 2008a. Efficient production of type 2 porcine circovirus-like particles by a recombinant baculovirus. Arch. Virol. 153, 2291–2295.

Liu, L., Celma, C.C.P., Roy, P., 2008b. Rift valley fever virus structural proteins: expression, characterization and assembly of recombinant proteins. Virol. J. 5, 82.

Liu, Q., Yan, K., Feng, Y., Huang, X., Ku, Z., Cai, Y., et al., 2012. A virus-like particle vaccine for coxsackievirus A16 potently elicits neutralizing antibodies that protect mice against lethal challenge. Vaccine 30, 6642–6648.

Liu, F., Wu, X., Li, L., Liu, Z., Wang, Z., 2013. Use of baculovirus expression system for generation of virus-like particles: successes and challenges. Protein Expr. Purif. 90, 104–116.

Lo, S.C., Li, K.C., Chang, Y.H., Hsu, M.N., Sung, L.Y., Vu, T.A., et al., 2017. Enhanced critical-size calcarial bone healing by ASCs engineered with Cre/loxP-based hybrid baculovirus. Biomaterials 124, 1–11.

Long, G., Westenberg, M., Wang, H., Vlak, J.M., Hu, Z., 2006a. Function, oligomerization and *N*-linked glycosylation of the *Helicoverpa armigera* single nucleopolyhedrovirus envelope fusion protein. J. Gen. Virol. 87, 839–846.

Long, G., Pan, X., Kormelink, R., Vlak, J.M., 2006b. Functional entry of baculovirus into insect and mammalian cells is dependent on clathrin-mediated endocytosis. J. Virol. 80, 8830–8833.

Lynn, D.E., 2001. Novel techniques to establish new insect cell lines. In Vitro Cell Dev. Biol. Anim. 37, 319–321.

Ma, H., Galvin, T.A., Glasner, D.R., Shaheduzzaman, S., Khan, A.S., 2014. Identification of a novel rhabdovirus in *Spodoptera frugiperda* cell lines. J. Virol. 88, 6576–6585.

Mabashi-Asazuma, H., Kuo, C.W., Khoo, K.H., Jarvis, D.L., 2014. A novel baculovirus vector for the production of non fucosylated recombinant glycoproteins in insect cells. Glycobiology 24, 325–340.

Madhan, S., Prabakaran, M., Kwang, J., 2010. Baculovirus as vaccine vectors. Curr. Gene Ther. 10, 201–213.

Makkonen, K.-E., Turkki, P., Laakkonen, J.P., Yla-Herttuala, S., Marjomaki, V., Airenne, K. J., 2013. 6-*O*- and *N*-sulfated syndecan-1 promotes baculovirus binding and entry into mammalian cells. J. Virol. 87, 11148–11159.

Makkonen, K.-E., Turkki, P., Laakkonen, J.P., Ylä-Herttuala, S., Marjomäki, V., Airenne, K. J., 2014. Reply to "Heparan sulfate in baculovirus binding and entry of mammalian cells". J. Virol. 88, 4609–4610.

Makkonen, K.-E., Airenne, K., Ylä-Herttulala, S., 2015. Baculovirus-mediated gene delivery and RNAi applications. Viruses 7 (4), 2099–2125.

Maree, S., Durbach, S., Maree, F.F., Vreede, F., Huismans, H., 1998. Expression of the major core structural proteins VP3 and VP7 of African horse sickness virus, and production of core-like particles. Arch. Virol. Suppl. 14, 203–209.

Marek, M., van Oers, M.M., Devaraj, F.F., Vlak, J.M., Merten, O.W., 2011. Engineering of baculovirus vectors for the manufacture of virion-free biopharmaceuticals. Biotechnol. Bioeng. 108, 1056–1067.

Martinez, C., Dalsgaard, K., Lopez de Turiso, J.A., Cortés, E., Vela, C., Casal, J.I., 1992. Production of porcine parvovirus empty capsids with high immunogenic activity. Vaccine 10, 684–690.

McCarty, D.M., Young, S.M., Samulski, R.J., 2004. Integration of adeno-associated virus (AAV) and recombinant AAV vectors. Annu. Rev. Genet. 38, 819–845.

Mena, J.A., Kamen, A.A., 2011. Insect cell technology is a versatile and robust vaccine manufacturing platform. Expert Rev. Vaccines 10, 1063–1081.

Metz, S.W., Gardner, J., Geertsema, C., Le, T.T., Goh, L., Vlak, J.M., et al., 2013. Effective chikungunya virus-like particle vaccine produced in insect cells. PLoS Negl. Trop. Dis. 7, e2124.

Mietzsch, M., 2014. Adeno-Associated Virus Vectors for Gene Therapy: From Scalable Production Systems via Serotype-Specific Glycan Interaction Patterns to Gene Transfer Applications. Submitted to the Department of Biology, Chemistry and Pharmacy of the Freie Universität Ber.

Mohana Subramanian, B., Madhanmohan, M., Sriraman, R., Chandrasekhar Reddy, R.V., Yuvaraj, S., Manikumar, K., et al., 2012. Development of foot-and-mouth disease virus (FMDV) serotype O virus-like-particles (VLPs) vaccine and evaluation of its potency. Antivir. Res. 96, 288–295.

Molinari, P., Crespo, M.I., Gravisaco, M.J., Taboga, O., Moron, G., 2011. Baculovirus capsid display potentiates OVA cytotoxic and innate immune responses. PLoS One 6, e24108.

Motohashi, T., Shimojima, T., Fukagawa, T., Maenaka, K., Park, E.Y., 2005. Efficient large-scale protein production of larvae and pupae of silkworm by *Bombyx mori* nuclear polyhedrosis virus bacmid system. Biochem. Biophys. Res. Commun. 326, 564–569.

Muñoz, N., Bosch, F., de Sanjose, S., Herrero, R., Castellsague, X., Shah, K., et al., 2003. Epidemiologic classification of human papillomavirus types associated with cervical cancer. N. Engl. J. Med. 348, 518–527.

Nam, H.M., Chae, K.S., Song, Y.J., Lee, N.H., Lee, J.B., Park, S.Y., et al., 2013. Immune responses in mice vaccinated with virus-like particles composed of the GP5 and M proteins of porcine reproductive and respiratory syndrome virus. Arch. Virol. 158 (6), 1275–1285.

Nasimuzzaman, M., 2014. Heparan sulfate in baculovirus binding and entry of mammalian cells. J. Virol. 88 (8), 4607–4608.

Nayerossadat, N., Maedeh, T., Ali, P.A., 2012. Viral and nonviral delivery systems for gene delivery. Adv. Biomed. Res. 1, 27.

Noad, R., Roy, P., 2003. Virus-like particles as immunogens. Trends Microbiol. 11, 438–444.

Okada, T., Ihara, H., Ito, R., Nakano, M., Matsumoto, K., Yamaguchi, Y., et al., 2010. *N*-Glycosylation engineering of lepidopteran insect cells by the introduction of the β1,4-*N*-acetylglucosaminyltransferase III gene. Glycobiology 20, 1147–1159.

O'Reilly, D.R., Miller, L.K., 1991. Improvement of a baculovirus pesticide by deletion of the egt gene. Nat. Biotechnol. 9, 1086–1089.

O'Reilly, D.R., Miller, L.K., Luckow, V.A., 1994. Baculovirus Expression Vectors: A Laboratory Manual. Oxford University Press, 347 pp.

Palmberger, D., Wilson, I.B.H., Berger, I., Grabherr, R., Rendic, D., 2012. Sweetbac: a new approach for the production of mammalianised glycoproteins in insect cells. PLoS One 7, 1–8.

Pidre, M.L., Ferrelli, M.L., Haase, S., Romanowski, V., 2013. Baculovirus display: a novel tool for vaccination. Current Issues in Molecular Virology—Viral Genetics and Biotechnological Applications, Intech, pp. 137–164.

Possee, R.D., Sun, T.P., Howard, S.C., Ayres, M.D., Hill-Perkins, M., Gearing, K.L., 1991. Nucleotide sequence of the *Autographa californica* nuclear polyhedrosis 9.4 kbp EcoRI-I and -R (polyhedrin gene) region. Virology 185, 229–241.

Pushko, P., Sun, X., Irina, T., Rachmat, H., Pulit-Penaloza, J.A., Belser, J.A., et al., 2016. Mono- and quadri-subtype virus-like particles (VLPs) containing H10 subtype elicit protective immunity to H10 influenza in a ferret challenge model. Vaccine 531, 390–394.

Roldão, A., Vieira, H.L.A., Alves, P.M., Oliveira, R., Carrondo, M.J.T., 2006. Intracellular dynamics in rotavirus-like particles production: evaluation of multigene and monocistronic infection strategies. Process Biochem. 41, 2188–2199.

Roldão, A., Vicente, T., Peixoto, C., Carrondo, M.J.T., Alves, P.M., 2011. Quality control and analytical methods for baculovirus-based products. J. Invertebr. Pathol. 107, 94–105.

Samulski, R.J., Chang, L.S., Shenk, T., 1987. A recombinant plasmid from which an infectious adeno-associated virus genome can be excised in vitro and its use to study viral replication. J. Virol. 61, 3096–3101.

Sedlik, C., Saron, M., Sarraseca, J., Casal, I., Leclerc, C., 1997. Recombinant parvovirus-like particles as an antigen carrier: a novel nonreplicative exogenous antigen to elicit protective antiviral cytotoxic T cells. Proc. Natl Acad. Sci. U.S.A. 94, 7503–7508.

Small, E., Fratesi, P., Reese, D., Strang, G., Laus, R., Peshwa, M., 2000. Immunotherapy of hormone-refractory prostate cancer with antigen-loaded dendritic cells. J. Clin. Oncol. 18, 3894–3903.

Smith, G.E., Summers, M.D., Fraser, M.J., 1983. Production of human beta interferon in insect cells infected with a baculovirus expression vector. Mol. Cell. Biol. 3, 2156–2165.

Smith, J.S., Lindsay, L., Hoots, B., Keys, J., Franceschi, S., Winer, R., et al., 2007. Human papillomavirus type distribution in invasive cervical cancer and high-grade cervical lesions: a meta-analysis update. Int. J. Cancer 121, 621–632.

Smith, R.H., Levy, J.R., Kotin, R.M., 2009. A simplified baculovirus-AAV expression vector system coupled with one-step affinity purification yields high-titer rAAV stocks from insect cells. Mol. Ther. 17, 1888–1896.

Steinbach, F., Mauel, S., Beier, I., 2002. Recombinant equine interferons: expression cloning and biological activity. Vet. Immunol. Immunopathol. 84, 84–95.

Strauss, R., Hüser, A., Ni, S., Tuve, S., Kiviat, N., Sow, P.S., et al., 2007. Baculovirus-based vaccination vectors allow for efficient induction of immune responses against plasmodium falciparum circumsporozoite protein. Mol. Ther. 15, 193–202.

Sun, Y., Carrion, R., Ye, L., Wen, Z., Ro, Y.T., Brasky, K., et al., 2009. Protection against lethal challenge by Ebola virus-like particles produced in insect cells. Virology 383, 12–21.

Swift, S.L., Rivera, G.C., Dussupt, V., Leadley, R.M., Hudson, L.C., MA de Ridder, C., et al., 2013. Evaluating baculovirus as a vector for human prostate cancer gene therapy. PLoS One 8 (6), e65557.

Tao, P., Luo, M., Zhu, D., Qu, S., Yang, Z., Gao, M., et al., 2009. Virus-like particle vaccine comprised of the HA, NA, and M1 proteins of an avian isolated H5N1 influenza virus induces protective immunity against homologous and heterologous strains in mice. Viral Immunol. 22, 273–281.

Tatman, J.D., Preston, V.G., Nicholson, P., Elliott, R.M., Rixon, F.J., 1994. Assembly of herpes-simplex virus type-1 capsids using a panel of recombinant baculoviruses. J. Gen. Virol. 75, 1101–1113.

Thomsen, D.R., Meyer, A.L., Post, L.E., 1992. Expression of feline leukaemia virus gp85 and gag proteins and assembly into virus-like particles using the baculovirus expression vector system. J. Gen. Virol. 73, 1819–1824.

Tobiasch, E., Burguete, T., Klein-Bauernschmitt, P., Heilbronn, R., Schlehofer, J.R., 1998. Discrimination between different types of human adeno-associated viruses in clinical samples by PCR. J. Virol. Methods 71, 17–25.

van Aarle, P., 2003. Suitability of an E2 subunit vaccine of classical swine fever in combination with the E(rns)-marker-test for eradication through vaccination. Dev. Biol. 114, 193–200.

van Oers, M.M., 2006. Vaccines for viral and parasitic diseases produced with baculovirus vectors. Adv. Virus Res. 68, 193–253.

Vaughn, J.L., Goodwin, R.H., Tompkins, G.J., McCawley, P., 1997. The establishment of two cell lines from the insect *Spodoptera frugiperda* (Lepidoptera; Noctuidae). In Vitro 13, 213–217.

Vieira, H.L.A., Estêvão, C., Roldão, A., Peixoto, C.C., Sousa, M.F.Q., Cruz, P.E., et al., 2005. Triple layered rotavirus VLP production: kinetics of vector replication, mRNA stability and recombinant protein production. J. Biotechnol. 120, 72–82.

Volkman, L.E., Goldsmith, P.A., 1983. In vitro survey of *Autographa californica* nuclear polyhedrosis virus interaction with nontarget vertebrate host cells. Appl. Environ. Microbiol. 45, 1085–1093.

Wang, D., Gao, G., 2014. State-of-the-art human gene therapy: Part I. Gene delivery technologies. Discov. Med. 18 (97), 67–77.

Wang, P., Granados, R.R., Shuler, M.L., 1992. Studies on serum-free culture of insect cells for virus propagation and recombinant protein production. J. Invertebr. Pathol. 59, 46–53.

Wang, K., Holtz, K., Anderson, K., Chubet, R., Mahmoud, W., Cox, M., 2006. Expression and purification of an influenza hemagglutinin—one step closer to a recombinant protein-based influenza vaccine. Vaccine 24, 2176–2185.

Wang, J., Faust, S.M., Rabinowitz, J.E., 2011. The next step in gene delivery: molecular engineering of adeno-associated virus serotypes. J. Mol. Cell. Cardiol. 50, 793–802.

Wang, A., Gu, L., Wu, S., Zhu, S., 2018. Duck hepatitis A virus structural proteins expressed in insect cells self-assemble into virus-like particles with strong immunogenicity in ducklings. Vet. Microbiol. 215, 23–28. Available from: https://doi.org/10.1016/j.vetmic.2017.12.020.

Warfield, K.L., Posten, N.A., Swenson, D.L., Olinger, G.G., Esposito, D., Gillette, W.K., et al., 2007. Filovirus-like particles produced in insect cells: immunogenicity and protection in rodents. J. Infect. Dis. 196 (s2), S421–S429.

Warfield, K.L., Dye, J.M., Wells, J.B., Unfer, R.C., Holtsberg, F.W., Shulenin, S., et al., 2015. Homologous and heterologous protection of nonhuman primates by ebola and sudan virus-like particles. PLoS One 10 (3), 1–16.

Wong, S.-S., Webby, R.J., 2013. Traditional and new influenza vaccines. Clin. Microbiol. Rev. 26, 476–492.

Wu, D., Murakami, K., Liu, N., Inoshima, Y., Yokoyama, T., Kokuho, T., et al., 2002. Expression of biologically active recombinant equine interferon-gamma by two different baculovirus gene expression systems using insect cells and silkworm larvae. Cytokine 20 (2), 63–69.

Xu, X.-G., Wang, Z.-S., Zhang, Q., Li, Z.-C., Zhao, H.-N., Li, W., et al., 2011. Baculovirus surface display of E envelope glycoprotein of Japanese encephalitis virus and its immunogenicity of the displayed proteins in mouse and swine models. Vaccine 29 (4), 636–643.

Yamaji, H., 2014. Suitability and perspectives on using recombinant insect cells for the production of virus-like particles. Appl. Microbiol. Biotechnol. 98, 1963–1970.

Yamaji, H., Manabe, T., Kitaura, A., Izumoto, E., Fukuda, H., 2006. Efficient production of recombinant protein in immobilized insect cell culture using serum-free basal media after baculovirus infection. Biochem. Eng. J. 28, 67–72.

Yang, J., Zhou, W., Zhang, Y., Zidon, T., Ritchie, T., Engelhardt, J.F., 1999. Concatamerization of adeno-associated virus circular genomes occurs through intermolecular recombination. J. Virol. 73, 9468–9477.

Ye, L., Lin, J., Sun, Y., Bennouna, S., Lo, M., Wu, Q., et al., 2006. Ebola virus-like particles produced in insect cells exhibit dendritic cell stimulating activity and induce neutralizing antibodies. Virology 351 (2), 260–270.

Zheng, X., Wang, S., Zhang, W., Liu, X., Yi, Y., Yang, S., et al., 2016. Development of a VLP-based vaccine in silkworm pupae against rabbit hemorrhagic disease virus. Int. Immunopharmacol. 40, 164–169.

Further Reading

Au, S., Wu, W., Zhou, L., Theilmann, D.A., Panté, N., 2016. A novel mechanism for nuclear import by actin-based propulsion used by the baculovirus nucleocapsid. JCS Adv. Online Artic. 129, 2905–2911. https://doi.org/10.1242/jcs.191668.

Aucoin, M.G., Perrier, M., Kamen, A.A., 2006. Production of adeno-associated viral vectors in insect cells using triple infection: optimization of baculovirus concentration ratios. Biotechnol. Bioeng. 95, 1081–1092.

Aucoin, M.G., Mena, J.A., Kamen, A.A., 2010. Bioprocessing of baculovirus vectors: a review. Curr. Gene. Ther. 10, 174–186.

Bouma, A., De Smit, A.J., De Kluijver, E.P., Terpstra, C., Moormann, R.J.M., 1999. Efficacy and stability of a subunit vaccine based on glycoprotein E2 of classical swine fever virus. Vet. Microbiol. 66, 101–114.

Buckland, B., Boulanger, R., Fino, M., Srivastava, I., Holtz, K., Khramtsov, N., et al., 2014. Technology transfer and scale-up of the Flublok recombinant hemagglutinin (HA) influenza vaccine manufacturing process. Vaccine 32, 5496–5502.

Grimm, D., Zhou, S., Nakai, H., Thomas, C.E., Storm, T.A., Fuess, S., et al., 2003. Preclinical in vivo evaluation of pseudotyped adeno-associated virus vectors for liver gene therapy. Blood 102, 2412–2419.

Kaeppel, C., Beattie, S.G., Fronza, R., van Logtenstein, R., Salmon, F., Schmidt, S., et al., 2013. A largely random AAV integration profile after LPLD gene therapy. Nat. Med. 19, 889–891.

Kitts, P.A., Ayres, M.D., Possee, R.D., 1990. Linearization of baculovirus DNA enhances the recovery of recombinant virus expression vectors. Nucleic Acids Res. 18, 5667–5672.

Kool, M., Voncken, J.W., van Lier, F.L., Tramper, J., Voncken, J.W., 1991. Detection and analysis of *Autographa californica* nuclear polyhedrosis virus mutants with defective interfering properties. Virology 746, 739–746.

Kotterman, M.A., Schaffer, D.V., 2014. Engineering adeno-associated viruses for clinical gene therapy. Nat. Rev. Genet. 15, 445–451.

Lee, H., Krell, P.J., 1994. Reiterated DNA fragments in defective genomes of *Autographa californica* nuclear polyhedrosis virus are competent for AcMNPV-dependent DNA replication. Virology 202, 418–429.

Liebert, M.A., Carter, B.J., 2005. Adeno-associated virus vectors in clinical trials. Hum. Gene Ther. 16, 541–550.

Lin, C.-Y., Lu, C.-H., Luo, W.-Y., Chang, Y.-H., Sung, L.-Y., Chiu, H.-Y., et al., 2010. Baculovirus as a gene delivery vector for cartilage and bone tissue engineering. Curr. Gene. Ther. 10, 242–254.

Matilainen, H., Rinne, J., Gilbert, L., Reunanen, H., Oker-blom, C., Marjoma, V., 2005. Baculovirus entry into human hepatoma cells baculovirus entry into human hepatoma cells. J. Virol. 79, 15452–15459.

Mietzsch, M., Grasse, S., Zurawski, C., Weger, S., Bennett, A., Agbandje-McKenna, M., et al., 2014. OneBac: platform for scalable and high-titer production of adeno-associated virus serotype 1–12 vectors for gene therapy. Hum. Gene Ther. 25, 212–222.

Negrete, A., Kotin, R.M., 2009. Technology. Blood 7, 303–311.

Pijlman, G.P., 2015. Enveloped virus-like particles as vaccines against pathogenic arboviruses. Biotechnol. J. 10, 659–670.

Pijlman, G.P., van den Born, E., Martens, D.E., Vlak, J.M., 2001. *Autographa californica* baculoviruses with large genomic deletions are rapidly generated in infected insect cells. Virology 283, 132–138.

Salmon, F., Grosios, K., Petry, H., 2014. Safety profile of recombinant adeno-associated viral vectors: focus on alipogene tiparvovec (Glybera(®)). Expert Rev. Clin. Pharmacol. 7, 53–65.

Santiago-Ortiz, J.L., Schaffer, D.V., 2016. Adeno-associated virus (AAV) vectors in cancer gene therapy. J. Controlled Release 240, 287–301.
Sokolenko, S., George, S., Wagner, A., Tuladhar, A., Andrich, J.M.S., Aucoin, M.G., 2012. Co-expression vs. co-infection using baculovirus expression vectors in insect cell culture: benefits and drawbacks. Biotechnol. Adv. 30, 766–781.
Urabe, M., Ding, C., Kotin, R.M., 2002. Insect cells as a factory to produce adeno-associated virus type 2 vectors. Hum. Gene Ther. 13, 1935–1943.
Wu, Z., Asokan, A., Samulski, R., 2006. Adeno-associated virus serotypes: vector toolkit for human gene therapy. Mol. Ther. 14, 316–327.

CHAPTER

12

Recombinant Veterinary Vaccines Against Rabies: State of Art and Perspectives

María Paula Del Médico Zajac[1], *Débora Garanzini*[1,2], *Oscar Ramón Pérez*[2] *and Gabriela Calamante*[1]

[1]National Institute of Agricultural Technology (INTA), CONICET, Institute of Agrobiotechnology and Molecular Biology (IABiMo), Buenos Aires, Argentina [2]National Administration of Laboratories and Health Institutes Dr C.G. Malbrán, National Institute of Biological Production, Antirabic Vaccine Service, Buenos Aires, Argentina

ABBREVIATIONS

ALVAC-RG	canarypox virus–expressing rabies glycoprotein
CAV	canine adenovirus
CDV	canine distemper virus
FAO	Food and Agriculture Organization of the United Nations
GARC	Global Alliance for Rabies Control
i.m.	intramuscular
i.n.	intranasal
MVA	modified vaccinia virus Ankara
NA	neutralizing antibodies
OIE	World Organization for Animal Health
ORV	oral rabies vaccines
RABV	rabies virus
RCN	racoonpox virus
s.c.	subcutaneous
V-RG	vaccinia virus–expressing rabies glycoprotein
WHO	World Health Organization

Emerging and Reemerging Viral Pathogens
DOI: https://doi.org/10.1016/B978-0-12-814966-9.00012-3

INTRODUCTION

Rabies is a zoonotic viral disease transmitted through bites or scratches from an infected animal (usually via saliva). It affects a wide range companion animals, livestock and wild mammals, including bats. The disease is almost always fatal following the onset of clinical symptoms. Rabies is one of the 18 neglected tropical diseases listed by the World Health Organization (WHO) that affect poor and vulnerable populations (http://www.who.int/neglected_diseases/diseases/en). An estimated 59,000 people die of rabies every year, primarily in Africa and Asia (Hampson et al., 2015).

Rabies virus (RABV) belongs to the Rhabdoviridae family, genus *Lyssavirus* (King et al., 2012). It has a single-chain negative-sense RNA genome of 12 kb encoding five structural proteins: the nucleoprotein (N), phosphoprotein (P), matrix protein (M), glycoprotein (G), and RNA-dependent RNA polymerase (L) (Wunner et al., 1988). The G protein is the surface protein of rabies virion and is capable of inducing neutralizing antibodies (NA) (VNA) against RABV (Benmansour et al., 1991).

Human deaths from rabies are 100% preventable. There are effective human vaccines to protect people who are at risk of exposure to rabies (preexposure vaccination) and to prevent the development of clinical rabies after exposure has occurred (postexposure prophylaxis). The vaccines are applied subcutaneously or intramuscularly in the deltoid region. In addition, heterologous and homologous rabies immune globulin is used for increasing effectiveness in postexposure treatments. These immunoglobulins are inoculated at a point other than vaccination site.

Considering that in up to 99% of cases, dogs are responsible for RABV transmission to humans (World Health Organization, 2013), it is proposed that with a minimum of 70% dog vaccination coverage, countries can effectively end dog-transmitted rabies (World Health Organization, 2007). However, other domestic animals (such as cats and livestock), wildlife, and bats are susceptible to RABV infection and constitute a possible source of the virus. Rabies, as a zoonotic disease, should be managed with a *One Health* approach where human, animal, and environmental health are connected. Control and/or elimination of rabies require cross-sector efforts including human and veterinary health governmental agencies and professionals, educators, scientists, and community groups. In this context, vaccination of rabies natural reservoirs (companion animals, livestock and wildlife) is the most cost-effective strategy over the long term to eliminate this disease in animals and, subsequently, in humans.

Therefore the World Organization for Animal Health (OIE) recommends the vaccination of susceptible animals in the areas where rabies is endemic. Inactivated and live-attenuated vaccines against rabies are immunogenic and effective. However, they present several disadvantages such as uncertain antigen composition, manipulation of large quantity of the pathogen during the vaccine manufacturing, need of cold chain during storage and transportation, and inability to differentiate vaccinated from infected animals, requirement of laboratories with high biosecurity level (BSL-3) and trained employees.

In order to overcome this inconveniences, there is a need to develop new vaccines that are immunogenic and safe, which required only one or two administrations, and which are suitable to be administered by parenteral and oral routes. Besides, it would be desirable to avoid manipulation of the infectious RABV during vaccine production. In this context, for developing new vaccines against rabies, a lot of research is being done using different recombinant immunogens (viral-vectored, subunit, and nucleic acid–based vaccines) (Yang et al., 2013). Viral-vectored vaccines are composed of genetically modified viruses that encode a foreign gene in its genome and express the antigen in the infected cell. According to the selected virus used as vector, they could be replicative or nonreplicative since they produce or not the viral progeny in the vaccinated organism, respectively. In addition, these viral-vectored vaccines are safe and highly immunogenic in the target species; they enable the differentiation between infected and vaccinated animals, and they are innocuous to the environment.

Currently, available commercial vaccines include both traditional inactivated or attenuated RABV vaccines and also viral-vectored vaccines based on pox- and adenovirus. Indeed, humans and domestic animals (companion animals and livestock) are mostly vaccinated with inactivated RABV vaccines while wildlife is immunized using baits containing attenuated strains of RABV or viral vector–based vaccines. The latter include recombinant vaccinia virus or adenovirus codifying the rabies glycoprotein.

In order to establish the best vaccine to be used and the vaccination strategy to perform, it is necessary to first define (1) the target of vaccination (humans, domestic animals, bats, or wildlife); (2) the economic resources available in the region/country/state where the vaccine will be applied; and (3) the environmental conditions of the area (such as coexistence or/not of humans with animal sources, different animal species, main form of transmission of the disease).

This chapter will focus on the description of viral-vectored vaccines against rabies tested in domestic animals and wildlife.

VIRAL-VECTORED VACCINES AGAINST RABIES

Viral-Vectored Vaccines for Companion Animals

Among companion animals, dogs, cats, and ferrets are susceptible to RABV infection. According to governmental legislations and guidelines for responsible pet ownership, companion animals should be vaccinated against rabies at 12 weeks old and revaccinated every year.

RABV-inactivated vaccines are highly effective in pets. However, the presence of an adjuvant in the vaccine formulation has been associated with the development of fibrosarcoma in genetically predisposed cats (Backer, 1998). In order to improve the safety of vaccines by reducing local inflammation at injection sites, the development of nonadjuvanted vaccines for cats gains importance.

Currently, there is a commercial vaccine against *feline rabies* based on a canarypox virus (CNPV) (PUREVAX Feline Rabies, Merial, Boehringer Ingelheim). The vCP65 strain encoding the rabies G protein [canarypox virus–expressing rabies glycoprotein (ALVAC-RG)] was developed by Taylor et al. (1991). This vaccine induced high levels of NA and protection against rabies challenge after subcutaneous (s.c.) immunization of dogs and cats (Taylor et al., 1991). It also showed a safe profile in target and nontarget species (Taylor et al., 1991, 1995). The ALVAC-RG vaccine was approved for 1-year duration of immunity, and it has been licensed for its use in cats in the United States and Canada for at least 20 years.

Another way to reduce the potential adverse reactions of frequent boosters is extending the revaccination intervals (Day, 2006; Gaskell et al., 2006; Schultz, 2006) or the coadministration with other antigens. The vaccination scheme of cats with the ALVAC-RG vaccine includes the administration of boosters first after 1 year and then every 3 years. This regime induces NA response and full protection against rabies challenge performed after 3 years of the booster (Jas et al., 2012). Most of the veterinary vaccines are multivalent where a single dose contains several immunogens against different pathogens. The vectorized vaccine vCP65 also provided full protection against rabies when coadministered with other conventional feline vaccines [feline calici- (FCV), herpes- (FHV-1), parvo- (FPV) virus and *Chlamydophila felis*] (Jas et al., 2012).

Other poxviruses have been tested as potential vector vaccines in cats against rabies. In 1988, Taylor et al. (1988) demonstrate that a fowlpox virus (FWPV) expressing RABV G protein administered by s.c. route was protective against rabies. On the other hand, other two vectors based on raccoonpox (RCN-G) and parapox viruses (ORF-RG) induced specific humoral response against rabies but the protection has not been evaluated. In the first case, the RCN-G was evaluated as a mucosal vaccine. The administration by intranasal (i.n.) route induced the highest level of NA against rabies and this response remained for 6 months. In addition to that, it has been demonstrated that RCN-G had limited replication after i.n. immunization remarking its safe profile for cats (Osorio et al., 2003). In the last case, ORF-RG administered subcutaneously induced specific NA of the same magnitude of the commercial ALVAC-RG vaccine (Amann et al., 2013).

Other viruses have been also evaluated as vector vaccines in cats. Canine adenovirus (CAV) 2 (CAV-2) does not naturally infect cats, thus, it constitutes a good candidate for feline vaccines. Remarkably, intramuscular (i.m.) administration of a single dose of CAV-2-E3Δ-RGP induced a long-lasting antibody response (52 weeks) and protection against rabies in cats (Hu et al., 2007). A recombinant Newcastle disease virus (NDV) expressing rabies glycoprotein was evaluated in cats. All animals seroconverted to rabies NA following i.m. vaccination and showed a 10-fold increase after the third dose applied 60 weeks after the first immunization (Ge et al., 2011).

As stated before, in order to eliminate *dog-transmitted rabies*, the WHO recommends that 70% of dogs population needs to be vaccinated, and this level of herd immunity must be maintained for 3–7 years (World Health Organization, 2007). Political commitment at local, national, regional, and international levels is essential in order to allocate the needed resources both to motivate responsible dog ownership and to perform mass dog vaccination campaigns.

Indeed, Latin American countries achieved a 98% reduction in rabies cases in dogs that contributed to a 96% reduction in human rabies (in the period 2005–15) due to large-scale dog vaccination campaigns, control of their free-ranging dog populations, and enforcement of the legislation for responsible pet ownership (Velasco-Villa et al., 2017).

In contrast, most developing countries in Africa and Asia present several barriers that preclude successful dog vaccination campaigns including lack of funding, infrastructure and political will, poorly organized campaigns, owner's inability to control their dogs, vaccinator's inability to reach dogs without extraordinary effort, and large proportions of free-roaming dogs (Castillo-Neyra et al., 2017; Jackman and Rowan, 2007; Muthiani et al., 2015; Wallace et al., 2017; World Health Organization, 1988).

Dog vaccination could be achieved by central point, door-to-door, oral rabies vaccines (ORV), and capture-vaccinate-release strategies. Conventional and recombinant vaccines suitable for parenteral and oral administration are available. Oral vaccination of dogs against rabies may be useful in combination with parenteral vaccines to improve dog population vaccination coverage, especially in areas having large populations of nonaccessible dogs (Cliquet et al., 2008).

On the whole, the vaccination strategy (including vaccination method, type of vaccine, and route of administration) should be designed for each particular target region considering the economic resources and the dog population accessibility.

Several viruses have been engineered to express rabies glycoprotein and evaluated as potential rabies vectored vaccines for dogs (Table 12.1).

TABLE 12.1 Viral Vector–Based Vaccine Evaluated in Dogs

Viral vector[a]	Immunization route[b]	Seroconversion[c]	Efficacy[d]	Reference
AdHu5	s.c.	100	NT	Prevec et al. (1990)
	i.n.	100	NT	
	p.o.	0	NT	Vos et al. (2001)
	p.o./i.m.	100	NT	
	s.c.	100	NT	Tims et al. (2000)
	i.m.	100	NT	
CAV-2	s.c.	100	100	Hu et al. (2006)
	oral	87	100	Zhang et al. (2008)
	i.n.	87	NT	
	p.o.	0	0	Wright et al. (2013)
	p.o. (two doses)	50	50	
CDV	i.m.	100	NT	Wang et al. (2012)
CNPV	s.c.	100	100	Taylor et al. (1991)
FWPV	s.c.	100	100	Taylor et al. (1988)
MVA	i.m.	100	NT	Weyer et al. (2007)
	p.o.	0	NT	
NDV	i.m.	100	100	Ge et al. (2011)
ORFV	i.m.	100	NT	Amann et al. (2013)
	s.c.	100	NT	
PrV	p.o.	100	NT	Yuan et al. (2008)
VV	p.o.	66	83	Rupprecht et al. (2005)
	Square bait	22	71	Cliquet et al. (2008)
	Rectangular bait	11	89	
SINV	i.d.	100	NT	Gupta et al. (2009)

[a]AdHu5, *Human adenovirus serotype 5;* CAV-2, *canine adenovirus type 2;* CDV, *canine distemper virus;* CNPV, *canarypox virus;* FWPV, *fowlpox virus;* MVA, *modified vaccinia virus Ankara;* NDV, *Newcastle disease virus;* ORFV, *Orf Virus;* PrV, *pseudorabies virus;* VV, *vaccinia virus;* SINV, *Sindbis virus.*
[b]s.c., *Subcutaneous;* i.m., *intramuscular;* p.o., *per os;* i.n., *intranasal;* i.d., *intradermal.*
[c]*Expressed as percentage of animals that seroconverted (specific rabies NA) after immunization.*
[d]*Expressed as percentage of protected animals after rabies challenge.*
NT, Not tested.

Dogs routinely receive canine distemper virus (CDV) live vaccines at 6 weeks of age and a booster dose every 6 months. In this context, Wang et al. (2012) proposed CDV as a feasible bivalent vector against CDV and rabies. Indeed, they showed a long-lasting immunity (at least 70 weeks) in dogs against both pathogens after i.m. administration of recombinant CDV expressing rabies glycoprotein (rCDV-RG). Efficacy against CDV and rabies remains to be studied. CAV was also proposed as bivalent vector vaccine against CAV (strains 1 and 2) and rabies since s.c. administration of CAV2-RG induced antibodies against both CAV and rabies and protected against rabies challenge (Hu et al., 2006; Zhang et al., 2008). However, Wright et al. (2013) report that preexisting immunity against the viral vector impairs efficacy against rabies.

Nonreplicating viruses in dogs have also been tested as vaccine vectors for rabies. Several poxviruses [CNPV, FWPV, orf virus (ORFV), modified vaccinia virus ankara (MVA)], pseudorabies virus (herpesvirus), and NDV (paramyxovirus) induced specific NA and protection against rabies challenge (for details see Table 12.1).

Recombinant vaccines have relatively low manufacturing costs as they require only minimal downstream processing, and no adjuvants are used in their formulation (Brun et al., 2008). Although several viral-vectored vaccines induced NA and protection against rabies challenge in dogs, none of them are currently commercially available. Indeed, massive dog vaccination is still performed using inactivated RABV vaccines.

In relation to *oral dog vaccination*, poor information is available and little research has been done. As already mentioned, this strategy offers new approaches promising a significant increase in the dog vaccination coverage (especially of free-roaming and poorly supervised dogs) both when applied exclusively or in combination with parenteral vaccination (Fishbein et al., 1992; Matter and Fico, 1992). However, special attention must be addressed on the safety of the selected vaccine in nontarget species, especially humans. The WHO published a document detailing a number of requirements regarding the safety of candidate oral vaccines for dogs and safety, efficacy, and economics of bait delivery (World Health Organization, 2007).

Up to the moment, only vaccinia virus expressing rabies glycoprotein (V-RG) have been tested as an oral vectored vaccine for dogs (Cliquet et al., 2008; Rupprecht et al., 2005). In addition, two different types of baits were equally accepted by dogs and induced 71.4% and 88.8% of protection against rabies challenge (square and rectangular baits, respectively). However, the vaccine delivery system is used in the United States for raccoons (see Rabies in Wildlife: Control Strategies), but it has not been adopted as ORV for dogs.

Thus it is quite important to continue the evaluation of promising viral vectors with a high safety profile (some of the outlined in Table 12.1 or new ones) as potential oral vaccines for dogs.

Domestic *ferrets* (*Mustela putorius furo*) must be vaccinated against CDV and rabies at 12 weeks old. Since ferrets seem to have bad reactions to vaccination most frequently than dogs or cats, it is strongly recommended to use monovalent vaccines (https://www.ferret.org/read/vaccinations.html). However, there are no specific vaccines for ferrets, and multivalent dog's vaccines are commonly used. Further research is necessary in order to test safety and efficacy of available viral-vectored vaccines against rabies in domestic ferrets in order to identify a vaccine candidate that diminishes the number of vaccination events and consequently the adverse reactions. For example, as described earlier, CDV-RG seems to be a good candidate to be evaluated in ferrets since it induces long-lasting humoral immune response both for CDV and rabies in dogs.

Rabies in Wildlife: Control Strategies

A sylvatic cycle of rabies also exists, with wild animals (see Table 12.2) serving as the maintenance host of the virus. Several factors, such as climate change, globalization, demographic evolution, and linked new social behaviors, increase the circulation of infectious agents between animal populations and between animals and humans. Indeed, domestic animal grazing areas abut or overlap with wildlife reserves, farmed wildlife (such as deer and elk); national and international relocation of wildlife, and the encroachment of humans in to formerly remote habitats and environments constitute additional issues to consider (OIE wildlife 2015).

In this context, vaccination against rabies of susceptible wild species should be mandatory. Oral vaccination with edible baits is the only feasible alternative to control rabies in wildlife. A vaccine-filled sachet is enveloped by a bait casing typically consisting of fishmeal, fat, and paraffin. Animals will chew the bait and release the vaccine into the mouth, which will orally vaccinate the animal against rabies. Vaccine baits need to be deposited throughout all potential habitats of the target species.

The distribution of the vaccine baits is differently achieved according to the target territories. To cover large unoccupied areas of the aerial distribution, by aircraft or helicopter, is the most effective way. On the other hand, in urban and suburban regions the manual distribution of vaccine baits is the strategy chosen.

Two viral-vectored vaccines are licensed for use in wildlife: a recombinant human adenovirus serotype 5 vector containing the

TABLE 12.2 Typical Carnivore Host Reservoirs for Rabies Virus

Region	Species
Africa	Domestic dog (*Canis lupus familiaris*)
	Jackals (*Canis adustus and Canis mesomelas*)
	Mongoose (*Herpestes* spp.)
Middle East and Asia	Domestic dog (*C. lupus familiaris*)
	Red fox (*Vulpes vulpes*)
	Ferret badger (*Melogale moschata*)
	Golden jackals (*Canis aureus*)
Europe	Red fox (*V. vulpes*)
	Raccoon dog (*Nyctereutes procyonoides*)
North America	Raccoon (*Procyon lotor*)
	Gray fox (*Urocyon cinereoargenteus*)
	Striped skunk (*Mephitis mephitis*)
	Coyote (*Canis latrans*)
South America	Domestic dog (*C. l. familiaris*), crab-eating fox (*Cerdocyon thous*)
	Marmoset (*Callithrix jacchus*)
Caribbean islands	Domestic dog (*C. l. familiaris*), small Indian mongoose (*Herpestes auropunctatus*)
Eurasian and American arctic and subarctic regions	Arctic fox (*Alopex lagopus*)

Evelyn-Rokitnicki-Abelseth (ERA) strain RABV glycoprotein gene (AdRG1.3) (Yarosh et al., 1996) (trade name ONRAB, Artemis Technologies Inc., Guelph, Canada) and a recombinant vaccinia virus Copenhagen strain expressing the ERA strain rabies glycoprotein (V-RG) (Kieny et al., 1984; Wiktor et al., 1984) (trade name Raboral V-RG, Merial, Boehringer Ingelheim, United States).

The ONRAB vaccine is used in Ontario, Canada, to control rabies in foxes, raccoons, and skunks (https://www.ontario.ca/page/rabies-wildlife). Meanwhile, RABORAL V-RG is extensively distributed in the United States to control rabies in coyotes and raccoons (http://www.raboral.com/). On the other hand, in most parts of Western and Central Europe rabies has been successfully controlled and eradicated (mostly from red fox population) using attenuated live RABV or V-RG vaccines (for an exhaustive analysis of ORV in Europe see Müller et al., 2015).

Both recombinant vaccines are safe in target and as well as nontarget species (Brochier et al., 1989a; Fry et al., 2013; Knowles et al., 2009).

The success of ORV to eradicate or diminish wildlife rabies depends on various factors such as distribution pattern and vaccine bait density, target species population and its habitat features, and the availability of competing foods. Based on these facts, Fehlner-Gardiner et al. (2012) performed a field study in which these covariates were restricted as much as possible and compared the performance of ONRAB and RABORAL V-RG in stripped skunks and raccoons. The authors reported no significant difference between ONRAB and RABORAL V-RG in the proportion of antibody-positive striped skunks, while the proportion of antibody-positive raccoons was significantly higher in the ONRAB- versus the RABORAL V-RG–baited areas (74% vs 30%, respectively). The low response of raccoons to RABORAL V-RG could be related to preexisting immunity against racoonpox virus as described by Root et al. (2008). This interference was solved using a mucosal adjuvant (*N,N,N*-trimethylated chitosan), which increased the viscosity of RABORAL V-RG vaccine and augmented the number of responding raccoons after oral administration (Fry et al., 2012).

V-RG was also evaluated in other target species showing variable results. Raccoon dogs were fully protected after bait consuming (Cliquet et al., 2008), while 67% of stripped skunks were protected after oral instillation of the vaccine (no protection was observed using baits) (Grosenbaugh et al., 2007). In addition, the recombinant vaccine administered orally was not effective inducing rabies NA in mongoose (Blanton et al., 2006), and it was poorly immunogenic in European badger (Brochier et al., 1989b). These results remark the importance of studying the optimal vaccine dose for protection in each target species, in addition to the eating habits, the appropriate bait shape, and the cover material to ensure vaccine uptake.

As state earlier, current oral vaccines are not efficacious for all wild species. Alternative viral-vectored vaccines with high safety profiles have been tested in wild reservoirs. Oral administration of CAV2-RG stimulates protective immunity against rabies in ferrets (Zhao et al., 2014), raccoons, and skunks (Henderson et al., 2009). In addition, baits containing RCN-RG induced NA and protection against rabies in raccoons (Esposito et al., 1988), although oral inoculation of this viral-vectored vaccine was partially efficacious in skunks (Fekadu et al., 1991). Finally, MVA-RG induced anamnestic responses in raccoons when administered by i.m. route but not when administered orally (Weyer et al., 2007).

Nowadays, there are no approved parenteral vaccines for use in captive wildlife or exotic mammals. Thus these animals are routinely vaccinated using inactivated monovalent rabies vaccines (Bush et al., 1985).

However, some animals develop undesirable effects following routine (re)vaccination due to the presence of adjuvants in the formulations. Indeed, the Smithsonian Conservation Biology Institute (Washington, United States) detected adverse reactions after vaccination of a herd of the endangered Eld's deer (*Rucervus eldi thamin*) with an inactivated rabies vaccine (Marrow et al., 2014). The authors suggested a vaccination scheme combining the inactivated vaccine (prime) and the CNPV-based vaccine ALVAC-RG (booster doses) in order to both arise acceptable antirabies NA and reduce adverse effects due to repeated administration of adjuvanted vaccines.

A special case of wild rabies is related to bats. These animals play an important role in the dispersion of seeds, the pollination of plants, and the reduction of night-flying insects such as mosquitoes, but they are also frequent lyssavirus hosts (Walker, 2001). In North, Central, and South America bat-mediated rabies involved both insectivorous and vampire bats. Indeed, bat rabies antigenic variants have been found in skunks and gray foxes in North America and in domestic carnivores (dogs and cats) in several countries of Latin America (Escobar et al., 2015). In addition, the common vampire (*Desmodus rotundus*), has been associated with rabies in livestock (Mayen, 2003).

In the 1960s the strategy used to diminish bat-mediated rabies was the reduction in bat population by capturing and poisoning them, destroying roosts and caves with explosives. However, this method attempted against enormous quantities of bats of ecological importance (diversity).

Outbreaks of bat-mediated rabies are associated with habitat disturbance and ecosystem alterations. Risk areas need to be identified and constantly monitored. An integrated approach should be performed in these areas combining public health, agriculture, and biodiversity conservation interests.

Bat populations could be vaccinated in regions of coexistence of bats with humans or livestock. In this sense, up to now there are only two studies reporting the evaluation of poxvirus-based vaccines in bats (Aguilar-Setién et al., 2002; Stading et al., 2016).

The first study evaluated the V-RG vaccine in vampire bats administrated by i.m., scarification, oral or aerosol routes. All the animals vaccinated intramuscularly and by scarification seroconverted, while 37.5% and 50% of rabies NA was induced by oral and aerosol routes, respectively. After rabies challenge, 100% of protection was recorded in i.m., scarification and oral routes; meanwhile, only one animal out of eight aerosol vaccinated succumbed (Aguilar-Setién et al., 2002).

More recently, MVA and RCN poxvirus were evaluated in the Brazilian free-tailed bat (*Tadarida brasiliensis*). First, the authors evaluated safety and replication of the viral vectors using recombinant viruses expressing the luciferase gene (MVAluc, RCNluc). After oronasal (o.n.) and i.m. administration no clinical illness was observed with both vectors; but RCNluc construct showed higher and longer levels of expression of luciferase than MVAluc recombinant virus. Once selected the viral vector, the authors demonstrated that o.n. administration of RCN-RG induced rabies NA in immunized bats (Stading et al., 2016).

Further research is necessary in order to contribute with more information about potential vaccines useful for preventing rabies in bats.

Rabies Vaccines for Livestock

RABV also infects calves, horses, and domestic livestock such as camels, goats, pigs, llamas, and alpacas, resulting in economic losses and impacting food security. Besides, it would be a contagion focus for field workers. Animals can be infected with the virus through the bite of a rabid primary reservoir hosts, such as raccoon, fox, skunk, vampire bat, or roaming dog, depending on their geographical location.

Livestock and horses, which have frequent contact with humans, should be vaccinated on a regular basis against rabies. This would include all horses, exhibition animals, livestock "pets," petting zoo animals, among others. Although, OIE recommends prophylactic vaccination against rabies for cattle in rabies endemic areas, licensed rabies vaccines recommendations vary by manufacturer, and in some countries there are not licensed livestock vaccines. In addition, no rabies vaccines are currently licensed worldwide for goats, pigs, llamas, or alpacas.

Taking into account that in many rabies endemic countries actual location of livestock grazing abut or overlap wildlife areas, the better cost-effective strategy to control rabies is vaccination of both livestock and wildlife. Notwithstanding, vaccination in those regions is mostly voluntary.

Cattle, horses, and sheep are commonly vaccinated with inactivated rabies vaccines. Even though classical vaccines seem to be effective in these animals, they cannot be used in countries with rabies-free status or in eradication campaigns. In this sense, viral-vectored vaccines may be used in differentiating infected and vaccinated animals (DIVA). However, as detailed later, little research is done in obtaining and evaluating viral-vectored vaccines against rabies for livestock.

Vectored vaccines derived from CAV type 2 have been shown to be safe and protective against rabies in susceptible mammals and reservoirs species, such as dogs, raccoons, and skunks (Henderson et al., 2009; Hu et al., 2006; Liu et al., 2008; Zhao et al., 2014). Concerning farm animals, swine vaccinated with one dose of CAV2-RG intramuscularly produced long-term (for at least 28 weeks) specific rabies NA (Liu et al., 2008).

In addition, this viral vector showed promising results in sheep after simultaneous i.m. and s.c. immunization because it induced a protective level of RABV NA (> 0.5 IU/mL; Bouet-Cararo et al., 2011).

Poxvirus-based vectors have been also evaluated as vaccines for livestock. First, a recombinant RCN-RG was evaluated in a small number of sheep by i.m., s.c. and oral routes. The authors reported the induction of rabies NA using the parenteral routes and also the harmless of the viral vector after bait ingestion (De Martini et al., 1993).

More recently, Martins et al. (2017) showed that i.m. administration of two doses of a recombinant ORFV, which expresses rabies glycoprotein induced high level of RABV NA in cattle and pigs.

Finally, lumpy skin disease virus (LSDV) was evaluated as a dual vaccine against lumpy skin disease and rabies in cattle. A vaccination scheme combining i.m. and s.c. routes induced good levels of rabies NA that were maintained almost for 1 year. Also, NA against LSDV were detected (Aspden et al., 2002). LSDV-RG could also be evaluated as a dual vaccine in small ruminants since LSDV protects against sheep pox virus and goat pox virus, which are antigenically related to the vector (Boshra et al., 2013; Diallo et al., 2002; Ma et al., 2014).

Nevertheless, in order to suggest the use of these recombinant viruses as potential vaccines against rabies in livestock, efficacy trials should be performed.

CONCLUDING REMARKS

The control of rabies by prophylactic vaccination is the most cost-effective strategy. Political commitment and funding have to be increased to end human deaths from dog-transmitted rabies by 2030, as proposed by several international organizations (GARC, WHO, OIE, and FAO).

It should be considered that a case of human rabies represents a strong indicator of the weakness of the public health system because numerous tools are currently available to prevent this disease. Surveillance actions should be intensified according to the appropriate knowledge of the local epidemiological cycles (aerial and/or terrestrial). This identification is essential for applying an adequate immunization strategy in people and animal populations at risk as well as in wild animals that serve as maintenance host of the virus.

The administration routes for veterinary vaccines, especially those used in small animals (high value pets) have to avoid adverse effects generally caused by adjuvants, or by the use of replicative viral vector. In contrast, to prevent rabies in wildlife a replicative competent virus is generally selected as viral vector oral vaccine. However, the latter has to be based in an attenuated strain, or it has to spread in a limited way in the vaccinated animal.

At the moment, most of the veterinary vaccines used worldwide to prevent rabies are based on inactivated virus because they can be easily accepted by regulatory authorities. Nevertheless, at least three viral vectors based antirabies vaccines are commercially available at this moment (PUREVAX, RABORAL V-RG, and ONRAB). These are examples that safety issues can be solved, and viral vectors vaccine platforms are a reliable alternative to develop safety and effective vaccines to prevent rabies in companion animals, livestock and wildlife contributing to the One Health concept.

References

Aguilar-Setién, A., Campos, Y.L., Cruz, E.T., Kretschmer, R., Brochier, B., Pastoret, P.-P., 2002. Vaccination of vampire bats using recombinant vaccinia-rabies virus. J. Wildl. Dis. 38 (3), 539–544.

Amann, R., Rohde, J., Wulle, U., Conlee, D., Raue, R., Martinon, O., et al., 2013. A new Rabies vaccine based on a recombinant Orf virus (parapoxvirus) expressing the rabies virus glycoprotein. J. Virol. 87 (3), 1618–1630.

Aspden, K., van Dijk, A.A., Bingham, J., Coxa, D., Passmore, J.-A., Williamson, A.-L., 2002. Immunogenicity of a recombinant lumpy skin disease virus (neethling vaccine strain) expressing the rabies virus glycoprotein in cattle. Vaccine 20, 2693–2701.

Backer, R.J., 1998. Feline fibrosarcomas in vaccination sites. Feline. Pract. 26, 18–20.

Benmansour, A., Leblois, H., Coulon, P., Tuffereau, C., Gaudin, Y., Flamand, A., et al., 1991. Antigenicity of rabies virus glycoprotein. J. Virol. 65 (8), 4198–4203.

Blanton, J.D., Meadows, A., Murphy, S.M., Manangan, J., Hanlon, C.A., Faber, M.-L., et al., 2006. Vaccination of Small Asian Mongoose (*Herpestes javanicus*) against rabies. J. Wildl. Dis. 42 (3), 663–666.

Boshra, H., Truong, T., Nfon, C., Gerdts, V., Tikoo, S., Babiuk, L.A., et al., 2013. Capripoxvirus-vectored vaccines against livestock diseases in Africa. Antiviral Res. 98 (2), 217–227.

Bouet-Cararo, C., Contreras, V., Fournier, A., Jallet, C., Guibert, J.M., Dubois, E., et al., 2011. Canine adenoviruses elicit both humoral and cell-mediated immune responses against rabies following immunisation of sheep. Vaccine 29 (6), 1304–1310.

Brochier, B., Blancou, J., Thomas, I., Languet, B., Artois, M., Kieny, M.P., et al., 1989a. Use of recombinant vaccinia-rabies glycoprotein virus for oral vaccination of wildlife against rabies: innocuity to several non-target bait consuming species. J. Wildl. Dis. 25 (4), 540–547.

Brochier, B., Blancou, J., Thomas, I., Languet, B., Artois, M., Kieny, M.P., et al., 1989b. Use of recombinant vaccinia-rabies glycoprotein virus for oral vaccination of wildlife against rabies: innocuity to several non-target bait consuming species. J. Wildl. Dis. 25 (4), 540–547.

Brun, A., Albina, E., Barret, T., Chapman, D.A.G., Czub, M., Dixon, L.K., et al., 2008. Antigen delivery systems for veterinary vaccine development. Vaccine 26 (51), 6508–6528.

Bush, M., Montali, R.J., Reid, F.L., Sumner, J.W., Yager, P.A., Phillips, L.G., 1985. Antibody response in zoo animals to a killed virus rabies vaccine. In: En Proc Am Assoc Zoo Vet Annu Meet, p. 30.

Castillo-Neyra, R., Brown, J., Borrini, K., Arevalo, C., Levy, M.Z., Buttenheim, A., et al., 2017. Barriers to dog rabies vaccination during an urban rabies outbreak: qualitative findings from Arequipa, Peru. PLoS Negl. Trop. Dis. 11 (3), e0005460.

Cliquet, F., Barrat, J., Guiot, A.L., Caël, N., Boutrand, S., Maki, J., et al., 2008. Efficacy and bait acceptance of vaccinia vectored rabies glycoprotein vaccine in captive foxes (*Vulpes vulpes*), raccoon dogs (*Nyctereutes procyonoides*) and dogs (*Canis familiaris*). Vaccine 26 (36), 4627–4638.

Day, M.J., 2006. Vaccine side effects: fact and fiction. Vet. Microbiol. 117 (1), 51–58.

De Martini, J.C., Bickle, H.M.S., Brodie, J., He, B.X., Esposito, J.J., 1993. Raccoon poxvirus rabies virus glycoprotein recombinant vaccine in sheep. Arch. Virol. 133, 211–222.

Diallo, A., Minet, C., Berhe, G., Le Goff, C., Black, D.N., Fleming, M., et al., 2002. Goat immune response to capripox vaccine expressing the hemagglutinin protein of peste des petits ruminants. Ann. N. Y. Acad. Sci. 969, 88–91.

Escobar, L.E., Peterson, T., Favi, M., Yung, V., Medina-Vogel, G., 2015. Bat-borne rabies in Latin America. Rev. Inst. Med. Trop. Sao. Paulo 57 (1), 63–72.

Esposito, J., Knight, C., Shaddock, J., Novembre, F., Baer, G., 1988. Successful oral rabies vaccination of raccoons with raccoon poxvirus recombinants expressing rabies virus glycoprotein. Virology 165 (1), 313–316.

Fehlner-Gardiner, C., Rudd, R., Donovan, D., Slate, D., Kempf, L., Badcock, J., 2012. Comparing ONRAB® and RABORAL V-RG® oral rabies vaccine field performance in raccoons and striped skunks, New Brunswick, Canada, and Maine, USA. J. Wildl. Dis. 48 (1), 157–167.

Fekadu, M., Shaddock, J.H., Sumner, J.W., Sanderlin, D.W., Knight, J.C., Esposito, J.J., et al., 1991. Oral vaccination of skunks with raccoon poxvirus recombinants expressing the rabies glycoprotein or the nucleoprotein. J. Wildl. Dis. 27 (4), 681–684.

Fishbein, D.B., Frontini, M.G., Dobbins, J.G., Flores Collins, E., Quiroz Huerta, G., Gamez Rodriguez, J.J., et al., 1992. Prevention of canine rabies in rural Mexico: an epidemiologic study of vaccination campaigns. Am. J. Trop. Med. Hyg. 47 (3), 317–327.

Fry, T., Van Dalen, K., Hurley, J., Nash, P., 2012. Mucosal adjuvants to improve wildlife rabies vaccination. J. Wildl. Dis. 48 (4), 1042–1046.

Fry, T.L., Vandalen, K.K., Duncan, C., Vercauteren, K., 2013. The safety of ONRAB® in select non-target wildlife. Vaccine 31 (37), 3839–3842.

Gaskell, R.M., Dawson, S., Radford, A.D., 2006. Duration of immunity (DOI)—the regulatory issues. Vet. Microbiol. 117 (1), 80–85.

Ge, J., Wang, X., Tao, L., Wen, Z., Feng, N., Yang, S., et al., 2011. Newcastle disease virus-vectored rabies vaccine is safe, highly immunogenic, and provides long-lasting protection in dogs and cats. J. Virol. 85 (16), 8241–8252.

Grosenbaugh, D.A., Maki, J.L., Rupprecht, C.E., Wall, D.K., 2007. Rabies challenge of captive striped skunks (*Mephitis mephitis*) following oral administration of a live vaccinia-vectored rabies vaccine. J. Wildl. Dis. 43 (1), 124–128.

Gupta, P.K., Dahiya, S.S., Kumar, P., Rai, A., Patel, C.L., Sonwane, A.A., et al., 2009. Sindbis virus replicon-based DNA vaccine encoding rabies virus glycoprotein elicits specific humoral and cellular immune response in dogs. Acta Virol. 53, 83–88.

Hampson, K., Coudeville, L., Lembo, T., Sambo, M., Kieffer, A., Attlan, M., et al., 2015. Estimating the global burden of endemic canine rabies. PLoS Negl. Trop. Dis. 9 (4), 1–20.

Henderson, H., Jackson, F., Bean, K., Panasuk, B., Niezgoda, M., Slate, D., et al., 2009. Oral immunization of raccoons and skunks with a canine adenovirus recombinant rabies vaccine. Vaccine 27 (51), 7194–7197.

Hu, R., Zhang, S., Fooks, A.R., Yuan, H., Liu, Y., Li, H., et al., 2006. Prevention of rabies virus infection in dogs by a recombinant canine adenovirus type-2 encoding the rabies virus glycoprotein. Microbes Infect. 8 (4), 1090–1097.

Hu, R.L., Liu, Y., Zhang, S.F., Zhang, F., Fooks, A.R., 2007. Experimental immunization of cats with a recombinant rabies-canine adenovirus vaccine elicits a long-lasting neutralizing antibody response against rabies. Vaccine 25 (29), 5301–5307.

Jackman, J., Rowan, A., 2007. Free-roaming dogs in developing countries: the benefits of capture, neuter, and return programs. In: Salem, D.J., Rowan, A.N. (Eds.), The state of the animals 2007. Humane Society Press, Washington, DC, pp. 55–78.

Jas, D., Coupier, C., Edlund Toulemonde, C., Guigal, P.-M., Poulet, H., 2012. Three-year duration of immunity in cats vaccinated with a canarypox-vectored recombinant rabies virus vaccine. Vaccine 30, 6991–6996.

Kieny, M.P., Lathe, R., Drillien, R., Spehner, D., Skory, S., Schmitt, D., et al., 1984. Expression of rabies virus glycoprotein from a recombinant vaccinia virus. Nature 312 (5990), 163–166.

King, A., Adams, M., Carstens, E., Lefkowitz, E., 2012. Virus taxonomy. Ninth Report of the International Committee on Taxonomy of Viruses. Academic Press.

Knowles, M.K., Nadin-Davis, S.A., Sheen, M., Rosatte, R., Mueller, R., Beresford, A., 2009. Safety studies on an adenovirus recombinant vaccine for rabies (AdRG1.3-ONRAB) in target and non-target species. Vaccine 27 (47), 6619–6626.

Liu, Y., Zhang, S., Ma, G., Zhang, F., Hu, R., 2008. Efficacy and safety of a live canine adenovirus-vectored rabies virus vaccine in swine. Vaccine 26 (42), 5368–5372.

Ma, W., Wei, J., Wei, Y., Guo, H., Jin, Y., Xue, Y., et al., 2014. Immunogenicity of the capsid precursor and a nine-amino-acid site-directed mutant of the 3C protease of foot-and-mouth disease virus coexpressed by a recombinant goatpox virus. Arch. Virol. 159 (7), 1715–1722.

Marrow, J.C., Padilla, L.R., Hayek, L.-A.C., Bush, M., Murray, S., 2014. Comparison of antibody response to a non-adjuvanted, live canarypox-vectored recombinant rabies vaccine and a killed, adjuvanted rabies vaccine in Eld's deer (*Rucervus eldi thamin*). J. Zoo Wildl. Med. 45 (2), 315–320.

Martins, M., Joshi, L.R., Rodrigues, F.S., Anziliero, D., Frandoloso, R., Kutish, G.F., et al., 2017. Immunogenicity of ORFV-based vectors expressing the rabies virus glycoprotein in livestock species. Virology 511, 229–239.

Matter, H.C., Fico, R., 1992. Accessibility of Dogs to Oral and Parenteral Vaccination Against Rabies in Tunisia and Turkey. Geneva, World Health Organization, (WHO/Rabies/93.206).

Mayen, F., 2003. Haematophagous bats in Brazil, their role in rabies transmission, impact on public health, livestock industry and alternatives to an indiscriminate reduction of bat population. J. Vet. Med. B. Infect. Dis. Vet. Public Health 50 (10), 469–472.

Müller, T.F., Schröder, R., Wysocki, P., Mettenleiter, T.C., Freuling, C.M., 2015. Spatio-temporal use of oral rabies vaccines in fox rabies elimination programmes in Europe. PLoS Negl. Trop. Dis. 9 (8), e0003953.

Muthiani, Y., Traoré, A., Mauti, S., Zinsstag, J., Hattendorf, J., 2015. Low coverage of central point vaccination against dog rabies in Bamako, Mali. Prev. Vet. Med. 120 (2), 203–209.

Osorio, J.E., Frank, R.S., Moss, K., Taraska, T., Powell, T., Stinchcomb, D.T., 2003. Raccoon poxvirus as a mucosal vaccine vector for domestic cats. J. Drug Target. 11 (8–10), 463–470.

Prevec, L., Campbell, J.B., Christie, B.S., Belbeck, L., Graham, F.L., 1990. A recombinant human adenovirus vaccine against rabies. J. Infect. Dis. 161 (1), 27–30.

Root, J.J., McLean, R.G., Slate, D., MacCarthy, K.A., Osorio, J.E., 2008. Potential effect of prior raccoonpox virus infection in raccoons on vaccinia-based rabies immunization. BMC Immunol. 9, 57.

Rupprecht, C.E., Hanlon, C.A., Blanton, J., Manangan, J., Morrill, P., Murphy, S., et al., 2005. Oral vaccination of dogs with recombinant rabies virus vaccines. Virus Res. 111 (1), 101–105.

Schultz, R.D., 2006. Duration of immunity for canine and feline vaccines: a review. Vet. Microbiol. 117 (1), 75–79.

Stading, B.R., Osorio, J.E., Velasco-Villa, A., Smotherman, M., Kingstad-Bakke, B., Rocke, T.E., 2016. Infectivity of attenuated poxvirus vaccine vectors and immunogenicity of a raccoonpox vectored rabies vaccine in the Brazilian Free-tailed bat (*Tadarida brasiliensis*). Vaccine 34 (44), 5352–5358.

Taylor, J., Weinberg, R., Languet, B., Desmettre, P., Paoletti, E., 1988. Recombinant fowlpox virus inducing protective immunity in non-avian species. Vaccine 6 (6), 497–503.

Taylor, J., Trimarchi, C., Weinberg, R., Languet, B., Guillemin, F., Desmettre, P., et al., 1991. Efficacy studies on a canarypox-rabies recombinant virus. Vaccine 9 (3), 190–193.

Taylor, J., Meignier, B., Tartaglia, J., Languet, B., VanderHoeven, J., Franchini, G., et al., 1995. Biological and immunogenic properties of a canarypox-rabies recombinant, ALVAC-RG (vCP65) in non-avian species. Vaccine 13 (6), 539–549.

Tims, T., Briggs, D.J., Davis, R.D., Moore, S.M., Xiang, Z., Ertl, H.C., et al., 2000. Adult dogs receiving a rabies booster dose with a recombinant adenovirus expressing rabies virus glycoprotein develop high titers of neutralizing antibodies. Vaccine 18 (25), 2804–2807.

Velasco-Villa, A., Escobar, L.E., Sanchez, A., Shi, M., Streicker, D.G., Gallardo-Romero, N. F., et al., 2017. Successful strategies implemented towards the elimination of canine rabies in the Western Hemisphere. Antiviral. Res. 143, 1–12.

Vos, A., Neubert, A., Pommerening, E., Müller, T., Döhner, L., Neubert, L., et al., 2001. Immunogenicity of an E1-deleted recombinant human adenovirus against rabies by different routes of administration. J. Gen. Virol. 82 (Pt 9), 2191–2197.

Walker, S.M., 2001. Conservation progress in Latin America. Bats Mag. 19 (1), 1–2.

Wallace, R.M., Undurraga, E.A., Blanton, J.D., Cleaton, J., Franka, R., 2017. Elimination of dog-mediated human rabies deaths by 2030: needs assessment and alternatives for progress based on dog vaccination. Front. Vet. Sci. 4, 9.

Wang, X., Feng, N., Ge, J., Shuai, L., Peng, L., Gao, Y., et al., 2012. Recombinant canine distemper virus serves as bivalent live vaccine against rabies and canine distemper. Vaccine 30 (34), 5067–5072.

Weyer, J., Rupprecht, C.E., Mansc, J., Viljoen, G.J., Nel, L.H., 2007. Generation and evaluation of a recombinant modified vaccinia virus Ankara vaccine for rabies. Vaccine 25, 4213–4222.

Wiktor, T.J., Macfarlan, R.I., Reagan, K.J., Dietzschold, B., Curtis, P.J., Wunner, W.H., et al., 1984. Protection from rabies by a vaccinia virus recombinant containing the rabies virus glycoprotein gene. Proc. Natl. Acad. Sci. U.S.A. 81 (22), 7194–7198.

World Health Organization, 1988. Report of WHO Consultation on Dog Ecology Studies Related to Rabies Control.

World Health Organization, 2007. Oral vaccination of dogs against rabies. In: Guidance for Research on Oral Rabies Vaccines and Field Application of Oral Vaccination of Dogs Against Rabies.

World Health Organization, 2013. WHO Expert Consultation on Rabies, Second Report. In: En WHO Technical Report Series 982.

Wright, N., Jackson, F.R., Niezgoda, M., Ellison, J.A., Rupprecht, C.E., Nel, L.H., 2013. High prevalence of antibodies against canine adenovirus (CAV) type 2 in domestic dog populations in South Africa precludes the use of CAV-based recombinant rabies vaccines. Vaccine 31 (38), 4177–4182.

Wunner, W.H., Larson, J.K., Dietzschold, B., Smith, C.L., 1988. The molecular biology of rabies viruses. Rev. Infect. Dis. 10 (Suppl 4), S771–S784.

Yang, D.-K., Kim, H.-H., Lee, K.-W., Song, J.-Y., 2013. The present and future of rabies vaccine in animals. Clin. Exp. Vaccine Res. 2 (1), 19.

Yarosh, O.K., Wandeler, A.I., Graham, F.L., Campbell, J.B., Prevec, L., 1996. Human adenovirus type 5 vectors expressing rabies glycoprotein. Vaccine 14 (13), 1257–1264.

Yuan, Z., Zhang, S., Liu, Y., Zhang, F., Fooks, A.R., Li, Q., et al., 2008. A recombinant pseudorabies virus expressing rabies virus glycoprotein: safety and immunogenicity in dogs. Vaccine 26 (10), 1314–1321.

Zhang, S., Liu, Y., Fooks, A.R., Zhang, F., Hu, R., 2008. Oral vaccination of dogs (*Canis familiaris*) with baits containing the recombinant rabies-canine adenovirus type-2 vaccine confers long-lasting immunity against rabies. Vaccine 26 (3), 345–350.

Zhao, J., Liu, Y., Zhang, S., Fang, L., Zhang, F., Hu, R., 2014. Experimental oral immunization of ferret badgers (*Melogale moschata*) with a recombinant canine adenovirus vaccine CAV-2-E3Δ-RGP and an attenuated rabies virus SRV9. J. Wildl. Dis. 50 (2), 374–377.

CHAPTER

13

Epidemiology and Ecology of Emerging Viruses in Two Freshwater Lakes of the Northern Hemisphere

Mahi M. Mohiuddin and Herb E. Schellhorn

Department of Biology, McMaster University, Hamilton, ON, Canada

INTRODUCTION

Viruses are the most abundant microorganisms on the earth (Bergh et al., 1989) and play important roles in global biogeochemical and ecological cycles (Roux et al., 2016). Because of their ubiquity and environmental complexity, the role of viruses in aquatic environments has been the focus of many investigations in recent years (Brum et al., 2015; Dell'Anno et al., 2015; Sweet and Bythell, 2017). However, the majority of such investigations have focused mostly on marine environments, while freshwater habitats have remained largely unexplored. Studies performed in freshwater habitats indicate that freshwater viral communities, although exhibit less diversity compared to marine viruses, likely contain diverse and novel viruses (Djikeng et al., 2009; Hewson et al., 2012; Roux et al., 2012).

The majority of the freshwater viruses identified are bacteriophages, human and animal viruses, and, to a lesser extent, some plant viruses (Djikeng et al., 2009; Fancello et al., 2013; Tseng et al., 2013). Identification of viruses from the environment poses a challenge, because most of the viruses are not culturable through conventional techniques. Unlike 16S rRNA of prokaryotes and 18S rRNA of

Emerging and Reemerging Viral Pathogens
DOI: https://doi.org/10.1016/B978-0-12-814966-9.00013-5

eukaryotes, viruses lack universal conserved sequences. This lack of a universal phylogenetic marker limits the direct detection of viruses from the environment (Rohwer and Edwards, 2002). Metagenomic approach, an alternative to culture-dependent techniques, does not rely on the presence of any shared gene between viruses, and since its first application for the identification of viruses at two marine locations (Breitbart et al., 2002), metagenomic approach has been the primary technique used to identify viruses. This approach relies on the extraction of DNA from mixed microbial and viral communities and has expanded our knowledge of viral and bacterial diversity in many environments including soil, freshwater sources, marine sediments, and the human gut (Qin et al., 2010; Reyes et al., 2010; Rodriguez-Brito et al., 2010).

The Great Lakes make up the largest freshwater reservoir in North America and the second largest in the world (Waples et al., 2008). These lakes are of long-standing interest because they provide essential services including drinking water, agricultural and industrial use, recreational activities, and routes for transportation and, therefore, are closely monitored by public health and municipal authorities. Failure to properly monitor the water quality and water-dependent agricultural activities may lead to serious economic and human health consequences as suggested in Walkerton tragedy (Brown and Hussain, 2003) and the *Escherichia coli* outbreak in Germany (Bielaszewska et al., 2011).

The Great Lakes basin harbors many beaches, approximately 800 of which are monitored by local municipal agencies because of their frequent closure due to high *E. coli* counts, yet high recreational usage during the summer months (Ulrich et al., 2009). Examination of freshwater beaches of the Great Lakes, particularly of the lower Great Lakes region, provides evidence for the presence of pathogenic viruses (Fong et al., 2007; Mohiuddin and Schellhorn, 2015; Fong et al., 2005). Despite this evidence, very little is known about the viral communities inhabiting this region. Therefore a comprehensive knowledge of bacterial and viral community composition, structure, and ecology is required to assess the presumed risk that these microorganisms pose to public health. As a result, we have compiled the information available from the lower Great Lakes region, specifically Lake Ontario and Lake Erie, with the aim of providing a better understanding of viral communities inhabiting this area and demonstrating the diverse nature of these freshwater microorganisms.

VIRAL COMMUNITY STRUCTURE

Viruses, the most abundant microorganisms in aquatic environments, infect cellular organisms, and viral infections kill approximately

10%–20% of cellular organisms each day (Evans and Brussaard, 2012; Suttle, 2007). Viruses have also been associated with a number of waterborne disease outbreaks in the Great Lakes region (Fong et al., 2007; Hlavsa et al., 2011). In addition, viruses have low infectious dose (Atmar et al., 2014; Yezli and Otter, 2011) and can persist in the environment for months (Kotwal and Cannon, 2014; Prevost et al., 2016). Because of their low infectious dose, persistence in water and association with recreational water-related disease outbreaks, viruses pose a serious threat to public health. Therefore continuous research on viral epidemiology and ecology is required for better understanding the dynamics of viruses in aquatic environments and associated viral risk. However, unlike bacterial communities, characteristics of viral communities as well as their impact on the environment did not receive much attention in Lake Ontario and Lake Erie. The majority of the studies performed in this region have focused primarily on the identification of either coliphages (somatic and F+-specific) (Fong et al., 2007), the viral indicator of fecal contamination, or select pathogenic viruses (Payment and Locas, 2011; Wu et al., 2011). More recently, shotgun metagenomic sequencing approach was used to identify a wide range of viruses from both the Lake Ontario and Lake Erie (Mohiuddin and Schellhorn, 2015).

Viral Pathogens

While viruses are abundant in aquatic environments, not all viruses are pathogenic to humans. Viruses that cause gastrointestinal infections are frequently found in fecal-contaminated water. Although viruses cannot replicate outside their host, they can persist in the environment for a long period without significant loss of infectivity, and their high rate and ease of transmissibility and capacity to cause infection at a very low concentration pose a significant threat to public health. Viruses that are mainly responsible for waterborne diseases include enteroviruses, adenoviruses, and caliciviruses (noroviruses) (Abbaszadegan et al., 1999; Rutjes et al., 2006; Xagoraraki et al., 2007). These viruses are transmitted via fecal–oral route and present a great risk of infection to both humans and animals. Both human and animal feces excrete a large number of pathogenic viruses, and these viruses are transported to the environment through malfunctioning sewage treatment plants, leakage of sewage systems, untreated wastewater [combined sewer overflow (CSO)], river water, and ground water.

The majority of the viruses identified in the lower Great Lakes region, particularly in Lake Ontario and Lake Erie, are bacteriophages. These viruses (phages) kill bacterial cells and play important roles in global biogeochemical and ecological cycles (Fuhrman, 1999; Suttle, 2007). In addition to bacteriophages, viruses including phycodnaviruses, cyanophages, noroviruses, and other human enteroviruses have also been identified in this region (Edge et al., 2013; Greer et al., 2009; Short et al., 2011; Short and Short, 2009). More recently, using shotgun metagenomic sequencing, sequences originating from an array of human and animal viruses including adenoviruses, poxviruses, and herpesviruses have been identified in Lake Ontario and Lake Erie (Mohiuddin and Schellhorn, 2015). The presence of these viruses in both Lake Ontario and Lake Erie suggests that both these lakes serve as a natural reservoir for many pathogenic viruses and therefore requires constant monitoring. A list of potential pathogenic viruses identified in the lower Great Lake area is included in Table 13.1.

Adenoviruses

Adenoviruses are mainly responsible for mild infections including gastroenteritis, respiratory diseases (tonsillitis, pharyngitis, otitis media, and bronchiolitis/bronchitis) involving upper or lower respiratory tract, conjunctivitis, encephalitis, and urinary tract diseases. Although to date 52 different serotypes and 7 different subgroups (A through G) have been described, the majority of the waterborne gastroenteritis diseases is caused by serotype 40 and 41 under subgroup F (van Heerden et al., 2005). Adenoviruses are transmitted through the fecal–oral route, and the viral load in the feces of infected individuals is very high (10^6 viral particles/g of fecal matter) (Jiang, 2006). Transmission occurs through direct contact with contaminated objects, particularly recreational waters, drinking water, and with contaminated food. Using shotgun metagenomic sequencing approach, sequences originating from both human and animal adenoviruses were identified in Lake Ontario and Lake Erie (Mohiuddin and Schellhorn, 2015).

Noroviruses

Noroviruses, previously known as Norwalk viruses, are the primary cause of nonbacterial gastroenteritis. Norovirus infection occurs throughout the year, though the incidence of infection increases during the winter (Mounts et al., 2000). Contaminated water is the primary source of infection (Maunula et al., 2005), and these viruses are transmitted mainly through the fecal–oral route. Infection can occur through the intake of contaminated water, consumption of contaminated food, or contact with contaminated environmental surfaces and infected individuals. Norovirus infection occurs in people of all ages, and this infection is characterized by the onset of vomiting or diarrhea or both. Symptoms may also include nausea, mild fever, abdominal pain, muscle aches, and mild fever. Norovirus infections are self-limiting, and symptoms develop within 24–48 hours of infection and typically last for 2–3 days

TABLE 13.1 List of Pathogenic Viruses Identified in the Lower Great Lakes Region

Potential pathogenic viruses	Virus family	Primary host	Identification method[a]	Location	References[b]
Adenoviruses	Adenoviridae	Human and other vertebrates	Shotgun metagenomic sequencing, immunoperoxidase assay	Lake Ontario, Lake Erie	Edge et al. (2013), Mohiuddin and Schellhorn (2015)
Noroviruses	Caliciviridae	Human	RT-PCR, immunoperoxidase assay	Lake Ontario	Edge et al. (2013), Greer et al. (2009)
Iridoviruses	Iridoviridae	Insects, amphibians, fish, invertebrates	Shotgun metagenomic sequencing	Lake Ontario, Lake Erie	Mohiuddin and Schellhorn (2015)
Cyanophages	Myoviridae, Podoviridae, Siphoviridae	Bacteria	Shotgun metagenomic sequencing, qPCR	Lake Ontario, Lake Erie	Matteson et al. (2011), Mohiuddin and Schellhorn (2015)
Phycodnaviruses	Phycodnaviridae	Algae, human (rare)	PCR, shotgun metagenomic sequencing, qPCR	Lake Ontario, Lake Erie	Mirza et al. (2015), Mohiuddin and Schellhorn (2015), Short and Short (2009)
VHSv	Rhabdoviridae	Fish	PCR, qPCR	Lake Ontario, Lake Erie	Cornwell et al. (2012), Thompson et al. (2011)
Giant viruses	Mimiviridae, Marseilleviridae	Ameba, human (still unclear)	Shotgun metagenomic sequencing	Lake Ontario, Lake Erie	Mohiuddin and Schellhorn (2015)
Poxviruses	Poxviridae	Human and other vertebrates, arthropods	Shotgun metagenomic sequencing	Lake Ontario, Lake Erie	Mohiuddin and Schellhorn (2015)

[a] *RT-PCR, Reverse transcriptase-polymerase chain reaction; qPCR, quantitative polymerase chain reaction.*
[b] *Edge et al. (2013) used immunoperoxidase method to identify enteroviruses that include both adenoviruses and noroviruses. However, the types of viruses present were not specified in the study.*
VHSv, Viral hemorrhagic septicemia viruses.

(Graham et al., 1994). However, infected individuals excrete a large number of viral particles ranging from 10^7 to 10^8 copies per gram of stool (Chan et al., 2006), and shedding can continue for more than 1 month for patients with acute gastroenteritis (Murata et al., 2007). Noroviruses are present in watersheds of the Lake Ontario region (Greer et al., 2009).

Other Human Enteroviruses

Association between recreational water-related outbreaks and other human enteroviruses, including coxsackieviruses and echoviruses, has also been identified (Sinclair et al., 2009). These viruses are present in recreational beaches of Lake Michigan and in many other freshwater beaches across the globe (Aslan et al., 2011; Sinclair et al., 2009; Xagoraraki et al., 2007). While human enteric viruses were identified at the offshore intakes of Lake Ontario, the types of viruses identified were not specified (Edge et al., 2013). In a separate experiment, using coliphages as an indicator, the presence of human enteroviruses was also confirmed in Lake Erie (Fong et al., 2007).

Cyanophages

Cyanophages, although not directly pathogenic to humans, are present in Lake Ontario and Lake Erie (Matteson et al., 2011; Mohiuddin and Schellhorn, 2015). The majority of the cyanophages identified are *Synechococcus* phages, followed by *Prochlorococcus* phages. Cyanophages prey on cyanobacteria, which are responsible for harmful algal blooms (HABs) (Cheung et al., 2013). Therefore monitoring their abundance in freshwater lakes may provide important insight into the cyanobacteria–cyanophage relationship. This, in turn, may provide useful information for controlling HABs in the lower Great Lakes watershed.

Phycodnaviruses

Phycodnaviruses are also present in both Lake Ontario and Lake Erie (Mirza et al., 2015; Short and Short, 2009; Mohiuddin and Schellhorn, 2015). Although phycodnaviruses are generally considered to exclusively infect algal species, recent findings suggest that chlorovirus Acanthocystis turfacea virus 1 belonging to the Phycodnaviridae family can infect humans and is associated with diminished cognitive function (Yolken et al., 2014).

Viral Hemorrhagic Septicemia Virus

Viral hemorrhagic septicemia virus (VHSv), one of the most serious fish pathogen that kills over 80 species of marine and freshwater finfish

(Faisal et al., 2012), has also been identified from the Great Lakes region including Lake Ontario and Lake Erie (Cornwell et al., 2012; Thompson et al., 2011). VHSv is widely abundant in the Laurentian Great Lakes basin, and understanding the genetics, epidemiology, and ecology of this virus may provide useful information to the surveillance programs employed in this region.

Giant Viruses

Using shotgun metagenomic sequencing approach, sequences originating from giant viruses (also known as nucleocytoplasmic large DNA viruses) belonging to the families of Marseilleviridae and Mimiviridae have also been identified in the lower Great Lakes region (Mohiuddin and Schellhorn, 2015). Compared to other viruses, giant viruses possess extremely large genomes and often carry genes that are common in bacteria and eukaryotes (Aherfi et al., 2014; Arslan et al., 2011). While the giant viruses mostly infect ameba (Aherfi et al., 2016; Philippe et al., 2013), more recently, these viruses have been identified in patients with unexplained pneumonia (Colson et al., 2016) and lymph node adenitis (Popgeorgiev et al., 2013).

Other Viruses

Similar to other aquatic environments, the majority of the viruses identified in the lower Great Lakes region are bacteriophages belonging to the families of Myoviridae, Podoviridae, and Siphoviridae (Mohiuddin and Schellhorn, 2015). In addition to bacteriophages, viruses infecting animals and insects have also been identified in Lake Ontario and Lake Erie. These include iridoviruses, poxviruses, alloherpesviruses, and baculoviruses (Table 13.1) (Mohiuddin and Schellhorn, 2015).

SOURCES OF PATHOGENIC VIRUSES

Studies investigating potential pathogens in recreational beaches have indicated that pathogenic viral species can be introduced into aquatic environments through several sources. These include point-source (wastewater) (Aslan et al., 2011; Tamaki et al., 2012) and nonpoint-source pollution (Bae and Wuertz, 2012), direct fecal discharge from humans and other animals (Wright et al., 2009; Ahmed et al., 2010). Multiple microbial source-tracking methods are used to investigate the source of fecal indicator organisms as well as some pathogens in the Great Lakes area (Fong et al., 2007).

Malfunctioning wastewater treatment facilities and leaking septic tanks were associated with the presence of noroviruses in Lake Erie (Fong et al., 2007). In addition, runoffs from agricultural farms, wastewater effluents, and CSO were responsible for adenovirus and other enteric virus occurrence in rivers (Hundesa et al., 2006). Ballast water of commercial vessels has also been well established as another vehicle for and/or source of pathogens in aquatic environments including the Great Lakes (Drake et al., 2007; Grigorovich et al., 2003; Great Lakes Environmental Research Laboratory, 2010). VHSv, a fish pathogen responsible for killing many finfish in North America, is a recent example of pathogenic microorganism introduced into the Great Lakes through ballast water (Bain et al., 2010; Elsayed et al., 2006).

CURRENT APPROACHES FOR THE IDENTIFICATION OF PATHOGENS, THEIR LIMITATIONS AND SCOPE

Pathogenic bacteria and viruses are not directly measured in recreational water since there is a wide range of possible agents, and detection methods vary from pathogen to pathogen. In addition, simultaneous identification of a large number of pathogens would be time consuming and expensive. Therefore fecal indicator microorganisms including *E. coli*, *Enterococcus* spp., *Bacteroides* spp., and F-specific RNA coliphages are used as proxy for pathogen occurrence in freshwater environments, and identification of fecal indicators is the primary choice of method for pathogen identification in the lower Great Lakes area. Using culture-based approaches, *E. coli*, *Enterococcus* spp., and coliphages were identified in both Lake Ontario and Lake Erie (Edge and Hill, 2007; Edge et al., 2013; Fong et al., 2007). While fecal indicators are useful for predicting pathogen occurrence, their concentration is rarely predictive of individual pathogens (Payment and Locas, 2011; Wu et al., 2011). In addition, indicator microorganisms are not always reliable because of the differences in properties between indicator microorganisms and target pathogens. For example, indicator microorganisms are inactivated effectively through the water treatment processes (such as UV irradiation and chlorination treatment), whereas human viral and protozoan parasites are not inactivated efficiently through these processes indicating that some pathogens are more persistent in the environment than indicator organisms (Omarova et al., 2018; Qiu et al., 2015). To circumvent the limitations of these culture-based methods, molecular techniques, such as polymerase chain reactions (PCRs) and quantitative PCR (qPCR), have been developed in recent years and are being used regularly to detect individual pathogens. More recently, because of the rapid development and cost reduction of sequencing techniques, metagenomic studies are being used regularly to identify microorganisms from the environment.

Traditional fecal indicators including *E. coli*, *Enterococcus* spp., and coliphages are commonly found in the intestine of birds, humans, and other animals. Therefore the presence of these indicators does not provide information regarding the source of fecal contamination and emphasizes the need for developing alternative indicators that can distinguish human fecal contamination (or wastewater contamination) from fecal contamination by birds or other animals. One such indicator organism is *Bacteroidales* and use of human-specific *Bacteroidales* makers, such as HF183, can identify human fecal contamination in aquatic environments (Seurinck et al., 2005). *Bacteroidales* DNA markers have also been used to identify the source of fecal contamination in Lake Ontario (Edge et al., 2010, 2013).

Another approach to identify the source of fecal contamination as well as the type of pathogen present is the sequence-based approach. Both shotgun metagenomic sequencing and amplicon sequencing are used to identify the characteristics of bacterial and viral populations inhabiting the Great Lakes area (Bouzat et al., 2013; Hotto et al., 2007; Mohiuddin et al., 2017; Rinta-Kanto and Wilhelm, 2006; Mohiuddin and Schellhorn, 2015). However, due to the absence of any signature sequence in viruses (Rohwer and Edwards, 2002), amplicon sequencing, which relies on the presence of signature sequences, cannot be applied for the detection of viruses.

The majority of the metagenomic studies use relative abundance of bacterial and viral taxa as an estimate of microbial community composition. Relative abundance is defined as the relative amounts of different taxa present within an environment. Relative abundance estimates are useful for identifying the characteristics of microbial and viral communities, but they do not provide any information regarding the true abundance of pathogens. An alternative parameter to consider is absolute abundance that quantifies the absolute quantities of taxa present in a sample and can be useful for identifying the true abundance of pathogens. However, factors including the sequencing technology used and the primer pairs used to target the hypervariable region of the 16S rRNA gene may introduce biases in the measurement of microbial community abundances (Tessler et al., 2017; Tremblay et al., 2015), and therefore sequencing data alone cannot be used to estimate absolute abundance (Nayfach and Pollard, 2016). Sequence-based approaches, if complemented with culture-based and/or qPCR-based approaches, may provide an accurate estimation of the true abundance of pathogens.

While coliphages are used as a viral indicator of fecal contamination and are routinely used in the lower Great Lake area for monitoring water quality (Fong et al., 2007), reverse transcriptase-PCRs (Greer et al., 2009), and immunoperoxidase methods (Edge et al., 2013) can also be used to identify noroviruses and cultivable human enteric viruses, respectively. More recently, metagenomic approaches are being used to identify viruses in both Lake Ontario and Lake Erie (Matteson et al., 2011; Short et al., 2011; Mohiuddin and Schellhorn, 2015). Compared to bacteria, the detection of viruses is difficult since the majority of the viruses are not culturable and lack universal genetic marker. Therefore targeted amplification and shotgun metagenomic sequencing remain the only choice of identification for these pathogens.

SUMMARY AND CONCLUSION

Here we have reviewed the diverse nature of viral communities of the lower Great Lakes region. We reviewed studies showing that freshwater environments serve as a reservoir of many viral pathogens, which pose potential threat to public health, may lead to the impairment of recreational opportunities, and could be harmful to the economy. Although wastewater contamination is primarily responsible for the presence of many of these pathogens in recreational water, ballast water and beachgoers also contribute to the contamination of these aquatic environments. Currently, there are many surveillance programs being employed in North America to monitor recreational water quality. However, the majority of them use source-tracking methods, which are unable to identify the types of pathogens present in water. While sequencing-based approaches provide information of total microbial and viral communities, routine use of such approaches is not feasible due to their high cost and the time required for analysis. However, information obtained from sequencing-based approaches can be used to complement these traditional monitoring programs. Developing sequence-based identification methods for pathogens and using adequate quantification standards will improve the sensitivity of pathogen detection and therefore will augment the existing monitoring programs.

Acknowledgment

Research in the Schellhorn Laboratory is supported by an NSERC Discovery grant (RGPIN-2015-06187).

References

Abbaszadegan, M., Stewart, P., LeChevallier, M., 1999. A strategy for detection of viruses in groundwater by PCR. Appl. Environ. Microbiol. 65, 444–449.

Aherfi, S., La Scola, B., Pagnier, I., Raoult, D., Colson, P., 2014. The expanding family Marseilleviridae. Virology 466–467, 27–37.

Aherfi, S., Colson, P., La Scola, B., Raoult, D., 2016. Giant viruses of amoebas: an update. Front. Microbiol. 7, 349.

Ahmed, W., Goonetilleke, A., Gardner, T., 2010. Human and bovine adenoviruses for the detection of source-specific fecal pollution in coastal waters in Australia. Water Res. 44, 4662–4673.

Arslan, D., Legendre, M., Seltzer, V., Abergel, C., Claverie, J.M., 2011. Distant mimivirus relative with a larger genome highlights the fundamental features of Megaviridae. Proc. Natl. Acad. Sci. U.S.A. 108, 17486–17491.

Aslan, A., Xagoraraki, I., Simmons, F.J., Rose, J.B., Dorevitch, S., 2011. Occurrence of adenovirus and other enteric viruses in limited-contact freshwater recreational areas and bathing waters. J. Appl. Microbiol. 111, 1250–1261.

Atmar, R.L., Opekun, A.R., Gilger, M.A., Estes, M.K., Crawford, S.E., Neill, F.H., et al., 2014. Determination of the 50% human infectious dose for Norwalk virus. J. Infect. Dis. 209, 1016–1022.

Bae, S., Wuertz, S., 2012. Survival of host-associated *Bacteroidales* cells and their relationship with *Enterococcus* spp., *Campylobacter jejuni, Salmonella enterica* serovar *Typhimurium,* and adenovirus in freshwater microcosms as measured by propidium monoazide-quantitative PCR. Appl. Environ. Microbiol. 78, 922–932.

Bain, M.B., Cornwell, E.R., Hope, K.M., Eckerlin, G.E., Casey, R.N., Groocock, G.H., et al., 2010. Distribution of an invasive aquatic pathogen (viral hemorrhagic septicemia virus) in the Great Lakes and its relationship to shipping. PLoS One 5, e10156.

Bergh, O., Borsheim, K.Y., Bratbak, G., Heldal, M., 1989. High abundance of viruses found in aquatic environments. Nature 340, 467–468.

Bielaszewska, M., Mellmann, A., Zhang, W., Kock, R., Fruth, A., Bauwens, A., et al., 2011. Characterisation of the *Escherichia coli* strain associated with an outbreak of haemolytic uraemic syndrome in Germany, 2011: a microbiological study. Lancet Infect. Dis. 11, 671–676.

Bouzat, J.L., Hoostal, M.J., Looft, T., 2013. Spatial patterns of bacterial community composition within Lake Erie sediments. J. Great Lakes Res. 39, 344–351.

Breitbart, M., Salamon, P., Andresen, B., Mahaffy, J.M., Segall, A.M., Mead, D., et al., 2002. Genomic analysis of uncultured marine viral communities. Proc. Natl. Acad. Sci. U.S.A. 99, 14250–14255.

Brown, R.S., Hussain, M., 2003. The Walkerton tragedy—issues for water quality monitoring. Analyst 128, 320–322.

Brum, J.R., Ignacio-Espinoza, J.C., Roux, S., Doulcier, G., Acinas, S.G., Alberti, A., et al., 2015. Ocean plankton. Patterns and ecological drivers of ocean viral communities. Science 348, 1261498.

Chan, M.C., Sung, J.J., Lam, R.K., Chan, P.K., Lee, N.L., Lai, R.W., et al., 2006. Fecal viral load and norovirus-associated gastroenteritis. Emerg. Infect. Dis. 12, 1278–1280.

Cheung, M.Y., Liang, S., Lee, J., 2013. Toxin-producing cyanobacteria in freshwater: a review of the problems, impact on drinking water safety, and efforts for protecting public health. J. Microbiol. 51, 1–10.

Colson, P., Aherfi, S., La Scola, B., Raoult, D., 2016. The role of giant viruses of amoebas in humans. Curr. Opin. Microbiol. 31, 199–208.

Cornwell, E.R., Eckerlin, G.E., Thompson, T.M., Batts, W.N., Getchell, R.G., Groocock, G. H., et al., 2012. Predictive factors and viral genetic diversity for viral hemorrhagic septicemia virus infection in Lake Ontario and the St. Lawrence River. J. Great Lakes Res. 38, 278–288.

Dell'Anno, A., Corinaldesi, C., Danovaro, R., 2015. Virus decomposition provides an important contribution to benthic deep-sea ecosystem functioning. Proc. Natl. Acad. Sci. U.S.A. 112, E2014–E2019.

Djikeng, A., Kuzmickas, R., Anderson, N.G., Spiro, D.J., 2009. Metagenomic analysis of RNA viruses in a fresh water lake. PLoS One 4, e7264.

Drake, L.A., Doblin, M.A., Dobbs, F.C., 2007. Potential microbial bioinvasions via ships' ballast water, sediment, and biofilm. Mar. Pollut. Bull. 55, 333–341.

Edge, T.A., Hill, S., 2007. Multiple lines of evidence to identify the sources of fecal pollution at a freshwater beach in Hamilton Harbour, Lake Ontario. Water Res. 41, 3585–3594.

Edge, T.A., Hill, S., Seto, P., Marsalek, J., 2010. Library-dependent and library-independent microbial source tracking to identify spatial variation in faecal contamination sources along a Lake Ontario beach (Ontario, Canada). Water Sci. Technol. 62, 719–727.

Edge, T.A., Khan, I.U.H., Bouchard, R., Guo, J., Hill, S., Locas, A., et al., 2013. Occurrence of waterborne pathogens and *Escherichia coli* at offshore drinking water intakes in Lake Ontario. Appl. Environ. Microbiol. 79, 5799–5813.

Elsayed, E., Faisal, M., Thomas, M., Whelan, G., Batts, W., Winton, J., 2006. Isolation of viral haemorrhagic septicaemia virus from muskellunge, *Esox masquinongy* (Mitchill), in Lake St Clair, Michigan, USA reveals a new sublineage of the North American genotype. J. Fish Dis. 29, 611–619.

Evans, C., Brussaard, C.P., 2012. Regional variation in lytic and lysogenic viral infection in the Southern Ocean and its contribution to biogeochemical cycling. Appl. Environ. Microbiol. 78, 6741–6748.

Faisal, M., Shavalier, M., Kim, R.K., Millard, E.V., Gunn, M.R., Winters, A.D., et al., 2012. Spread of the emerging viral hemorrhagic septicemia virus strain, genotype IVb, in Michigan, USA. Viruses 4, 734–760.

Fancello, L., Trape, S., Robert, C., Boyer, M., Popgeorgiev, N., Raoult, D., et al., 2013. Viruses in the desert: a metagenomic survey of viral communities in four perennial ponds of the Mauritanian Sahara. ISME J. 7, 359–369.

Fong, T.T., Griffin, D.W., Lipp, E.K., 2005. Molecular assays for targeting human and bovine enteric viruses in coastal waters and their application for library-independent source tracking. Appl. Environ. Microbiol. 71, 2070–2078.

Fong, T.T., Mansfield, L.S., Wilson, D.L., Schwab, D.J., Molloy, S.L., Rose, J.B., 2007. Massive microbiological groundwater contamination associated with a waterborne outbreak in Lake Erie, South Bass Island, Ohio. Environ. Health Perspect. 115, 856–864.

Fuhrman, J.A., 1999. Marine viruses and their biogeochemical and ecological effects. Nature 399, 541–548.

Graham, D.Y., Jiang, X., Tanaka, T., Opekun, A.R., Madore, H.P., Estes, M.K., 1994. Norwalk virus infection of volunteers: new insights based on improved assays. J. Infect. Dis. 170, 34–43.

Great Lakes Environmental Research Laboratory, 2010. Great Lakes Aquatic Nonindigenous Species Information System (GLANSIS). Great Lakes Environmental Research Laboratory, National Oceanic and Atmospheric Administration, Ann Arbor, MI.

Greer, A.L., Drews, S.J., Fisman, D.N., 2009. Why "Winter" vomiting disease? seasonality, hydrology, and norovirus epidemiology in Toronto, Canada. EcoHealth 6, 192–199.

Grigorovich, I.A., Colautti, R.I., Mills, E.L., Holeck, K., Ballert, A.G., MacIsaac, H.J., 2003. Ballast-mediated animal introductions in the Laurentian Great Lakes: retrospective and prospective analyses. Can. J. Fish. Aquat. Sci. 60, 740–756.

Hewson, I., Barbosa, J.G., Brown, J.M., Donelan, R.P., Eaglesham, J.B., Eggleston, E.M., et al., 2012. Temporal dynamics and decay of putatively allochthonous and autochthonous viral genotypes in contrasting freshwater lakes. Appl. Environ. Microbiol. 78, 6583–6591.

Hlavsa, M.C., Roberts, V.A., Anderson, A.R., Hill, V.R., Kahler, A.M., Orr, M., et al., 2011. Surveillance for waterborne disease outbreaks and other health events associated with recreational water—United States, 2007–2008. Morb. Mortal. Wkly Rep. Surveill. Sum. (Washington, DC: 2002) 60, 1–32.

Hotto, A.M., Satchwell, M.F., Boyer, G.L., 2007. Molecular characterization of potential microcystin-producing cyanobacteria in Lake Ontario embayments and nearshore waters. Appl. Environ. Microbiol. 73, 4570–4578.

Hundesa, A., Maluquer De Motes, C., Bofill-Mas, S., Albinana-Gimenez, N., Girones, R., 2006. Identification of human and animal adenoviruses and polyomaviruses for determination of sources of fecal contamination in the environment. Appl. Environ. Microbiol. 72, 7886–7893.

Jiang, S.C., 2006. Human adenoviruses in water: occurrence and health implications: a critical review. Environ. Sci. Technol. 40, 7132–7140.

Kotwal, G., Cannon, J.L., 2014. Environmental persistence and transfer of enteric viruses. Curr. Opin. Virol. 4, 37–43.

Matteson, A.R., Loar, S.N., Bourbonniere, R.A., Wilhelm, S.W., 2011. Molecular enumeration of an ecologically important cyanophage in a Laurentian Great Lake. Appl. Environ. Microbiol. 77, 6772–6779.

Maunula, L., Miettinen, I.T., von Bonsdorff, C.H., 2005. Norovirus outbreaks from drinking water. Emerg. Infect. Dis. 11, 1716–1721.

Mirza, S.F., Staniewski, M.A., Short, C.M., Long, A.M., Chaban, Y.V., Short, S.M., 2015. Isolation and characterization of a virus infecting the freshwater algae *Chrysochromulina parva*. Virology 486, 105–115.

Mohiuddin, M., Schellhorn, H.E., 2015. Spatial and temporal dynamics of virus occurrence in two freshwater lakes captured through metagenomic analysis. Front. Microbiol. 6, 960.

Mohiuddin, M.M., Salama, Y., Schellhorn, H.E., Golding, G.B., 2017. Shotgun metagenomic sequencing reveals freshwater beach sands as reservoir of bacterial pathogens. Water Res. 115, 360–369.

Mounts, A.W., Ando, T., Koopmans, M., Bresee, J.S., Noel, J., Glass, R.I., 2000. Cold weather seasonality of gastroenteritis associated with Norwalk-like viruses. J. Infect. Dis. 181 (Suppl 2), S284–S287.

Murata, T., Katsushima, N., Mizuta, K., Muraki, Y., Hongo, S., Matsuzaki, Y., 2007. Prolonged norovirus shedding in infants $<$ or $=$ 6 months of age with gastroenteritis. Pediatr. Infect. Dis. J. 26, 46–49.

Nayfach, S., Pollard, K.S., 2016. Toward accurate and quantitative comparative metagenomics. Cell 166, 1103–1116.

Omarova, A., Tussupova, K., Berndtsson, R., Kalishev, M., Sharapatova, K., 2018. Protozoan parasites in drinking water: a system approach for improved water, sanitation and hygiene in developing countries. Int. J. Environ. Res. Public Health 15.

Payment, P., Locas, A., 2011. Pathogens in water: value and limits of correlation with microbial indicators. Ground Water 49, 4–11.

Philippe, N., Legendre, M., Doutre, G., Coute, Y., Poirot, O., Lescot, M., et al., 2013. Pandoraviruses: amoeba viruses with genomes up to 2.5 Mb reaching that of parasitic eukaryotes. Science 341, 281–286.

Popgeorgiev, N., Michel, G., Lepidi, H., Raoult, D., Desnues, C., 2013. Marseillevirus adenitis in an 11-month-old child. J. Clin. Microbiol. 51, 4102–4105.

Prevost, B., Goulet, M., Lucas, F.S., Joyeux, M., Moulin, L., Wurtzer, S., 2016. Viral persistence in surface and drinking water: suitability of PCR pre-treatment with intercalating dyes. Water Res. 91, 68–76.

Qin, J., Li, R., Raes, J., Arumugam, M., Burgdorf, K.S., Manichanh, C., et al., 2010. A human gut microbial gene catalogue established by metagenomic sequencing. Nature 464, 59–65.

Qiu, Y., Lee, B.E., Neumann, N., Ashbolt, N., Craik, S., Maal-Bared, R., et al., 2015. Assessment of human virus removal during municipal wastewater treatment in Edmonton, Canada. J. Appl. Microbiol. 119, 1729–1739.

Reyes, A., Haynes, M., Hanson, N., Angly, F.E., Heath, A.C., Rohwer, F., et al., 2010. Viruses in the faecal microbiota of monozygotic twins and their mothers. Nature 466, 334–U381.

Rinta-Kanto, J.M., Wilhelm, S.W., 2006. Diversity of microcystin-producing cyanobacteria in spatially isolated regions of Lake Erie. Appl. Environ. Microbiol. 72, 5083–5085.

Rodriguez-Brito, B., Li, L., Wegley, L., Furlan, M., Angly, F., Breitbart, M., et al., 2010. Viral and microbial community dynamics in four aquatic environments. ISME J. 4, 739–751.

Rohwer, F., Edwards, R., 2002. The phage proteomic tree: a genome-based taxonomy for phage. J. Bacteriol. 184, 4529–4535.

Roux, S., Enault, F., Robin, A., Ravet, V., Personnic, S., Theil, S., et al., 2012. Assessing the diversity and specificity of two freshwater viral communities through metagenomics. PLoS One 7, e33641.

Roux, S., Brum, J.R., Dutilh, B.E., Sunagawa, S., Duhaime, M.B., Loy, A., et al., 2016. Ecogenomics and potential biogeochemical impacts of globally abundant ocean viruses. Nature 537, 689–693.

Rutjes, S.A., van den Berg, H.H.J.L., Lodder, W.J., de Roda Husman, A.M., 2006. Real-time detection of noroviruses in surface water by use of a broadly reactive nucleic acid sequence-based amplification assay. Appl. Environ. Microbiol. 72, 5349–5358.

Seurinck, S., Defoirdt, T., Verstraete, W., Siciliano, S.D., 2005. Detection and quantification of the human-specific HF183 Bacteroides 16S rRNA genetic marker with real-time PCR for assessment of human faecal pollution in freshwater. Environ. Microbiol. 7, 249–259.

Short, S.M., Short, C.M., 2009. Quantitative PCR reveals transient and persistent algal viruses in Lake Ontario, Canada. Environ. Microbiol. 11, 2639–2648.

Short, C.M., Rusanova, O., Short, S.M., 2011. Quantification of virus genes provides evidence for seed-bank populations of phycodnaviruses in Lake Ontario, Canada. ISME J. 5, 810–821.

Sinclair, R.G., Jones, E.L., Gerba, C.P., 2009. Viruses in recreational water-borne disease outbreaks: a review. J. Appl. Microbiol. 107, 1769–1780.

Suttle, C.A., 2007. Marine viruses—major players in the global ecosystem. Nat. Rev. Microbiol. 5, 801–812.

Sweet, M., Bythell, J., 2017. The role of viruses in coral health and disease. J. Invertebr. Pathol. 147, 136–144.

Tamaki, H., Zhang, R., Angly, F.E., Nakamura, S., Hong, P.-Y., Yasunaga, T., et al., 2012. Metagenomic analysis of DNA viruses in a wastewater treatment plant in tropical climate. Environ. Microbiol. 14, 441–452.

Tessler, M., Neumann, J.S., Afshinnekoo, E., Pineda, M., Hersch, R., Velho, L.F.M., et al., 2017. Large-scale differences in microbial biodiversity discovery between 16S amplicon and shotgun sequencing. Sci. Rep. 7, 6589.

Thompson, T.M., Batts, W.N., Faisal, M., Bowser, P., Casey, J.W., Phillips, K., et al., 2011. Emergence of viral hemorrhagic septicemia virus in the North American Great Lakes region is associated with low viral genetic diversity. Dis. Aquat. Organ. 96, 29–43.

Tremblay, J., Singh, K., Fern, A., Kirton, E.S., He, S., Woyke, T., et al., 2015. Primer and platform effects on 16S rRNA tag sequencing. Front. Microbiol. 6, 771.

Tseng, C.H., Chiang, P.W., Shiah, F.K., Chen, Y.L., Liou, J.R., Hsu, T.C., et al., 2013. Microbial and viral metagenomes of a subtropical freshwater reservoir subject to climatic disturbances. ISME J. 7, 2374–2386.

Ulrich, D., Whitman, R., Alley, D., 2009. Work group report on beaches and recreational water quality. In: Great Lakes Water Quality Agreement Priorities. International Joint Commission, Great Lakes Regional Office, Windsor, ON, Canada, p. 20.

van Heerden, J., Ehlers, M.M., Heim, A., Grabow, W.O., 2005. Prevalence, quantification and typing of adenoviruses detected in river and treated drinking water in South Africa. J. Appl. Microbiol. 99, 234–242.

Waples, J.T., Eadie, B., Klump, J.V., Squires, M., Cotner, A.J., Mckinley, G., 2008. The Laurential Great Lakes. In: Continental Margins: A Synthesis and Planning Workshop.

Wright, M.E., Solo-Gabriele, H.M., Elmir, S., Fleming, L.E., 2009. Microbial load from animal feces at a recreational beach. Mar. Pollut. Bull. 58, 1649–1656.

Wu, J., Long, S.C., Das, D., Dorner, S.M., 2011. Are microbial indicators and pathogens correlated? A statistical analysis of 40 years of research. J. Water Health 9, 265–278.

Xagoraraki, I., Kuo, D.H.W., Wong, K., Wong, M., Rose, J.B., 2007. Occurrence of human adenoviruses at two recreational beaches of the great lakes. Appl. Environ. Microbiol. 73, 7874–7881.

Yezli, S., Otter, J.A., 2011. Minimum infective dose of the major human respiratory and enteric viruses transmitted through food and the environment. Food Environ. Virol. 3, 1–30.

Yolken, R.H., Jones-Brando, L., Dunigan, D.D., Kannan, G., Dickerson, F., Severance, E., et al., 2014. Chlorovirus ATCV-1 is part of the human oropharyngeal virome and is associated with changes in cognitive functions in humans and mice. Proc. Natl Acad. Sci. U.S.A. 111, 16106–16111.

CHAPTER

14

Global Epidemiology and Genetic Variability of Rabies Viruses

Darkaoui Sami[1] *and Moulay Mustapha Ennaji*[2]

[1]Moroccan Food Safety Office (ONSSA), Rabat, Morocco [2]Laboratory of Virology, Microbiology, Quality, Biotechnologies/Eco-Toxicology and Biodiversity, Faculty of Sciences and Techniques, Mohammedia, University Hassan II of Casablanca, Casablanca, Morocco

INTRODUCTION

Rabies is one of the oldest and most virulent zoonoses. It has been described 2300 BCE (Bourhy et al., 2008). It is a zoonotic disease that causes about 74,000 deaths annually worldwide (WHO and GARC, 2014). Dogs are responsible for 99% of human infections and are the main vector and reservoir of rabies (WHO, 2013). Several viruses of the *Lyssavirus* genus that are spread worldwide causes rabies in nonflying mammals and chiroptera (CDC, 2011). The epidemiological situation of rabies is different and evolving in different continents:

- Most European countries eradicated canine rabies before the 1940s. Vulpine rabies began to be effectively controlled by oral vaccination in the late 1970s. Since then, it has been eliminated in the 1990s in certain countries. To date, this method has made Central Europe free from sylvatic rabies (Müller et al., 2014).

Emerging and Reemerging Viral Pathogens
DOI: https://doi.org/10.1016/B978-0-12-814966-9.00014-7

- The Americas record cases of rabies in nonflying wild mammals in North America with different reservoirs such as raccoon, skunk, coyote, and bats. In Latin America, dogs and vampire bats are the reservoir of rabies (WHO, 2013). Canine rabies was eradicated in North America in the beginning of the last century. The situation in Latin America has been improved significantly as a result of dog mass vaccination efforts since the 1980s (Cima, 2013).
- Australia is free from canine rabies against which it undertakes strict quarantine measures; however, this continent recorded three human cases due to flying fox bites in 1996, 1998, and 2013 (CDNA, 2013).
- The situation in Asia is very heterogeneous regarding rabies. Indeed, some countries are free (Japan, Malaysia, Singapore, Hong Kong, and Pacific Islands), and the rest of the continent is endemic (Kaare et al., 2009). Asia is the continent that records most cases of human rabies with about 30,000 deaths per year (WHO, 2013).
- Africa records about 24,000 human deaths each year (Talbi et al., 2009), and no African country is free of rabies (Nel, 2013). Canine rabies is responsible for the majority of human cases in Africa (Talbi et al., 2009).

ETIOLOGY AND CLASSIFICATION OF RABIES VIRUSES

Rabies is caused by viruses belonging to species of the genus *Lyssavirus* (order Mononegavirales, family Rhabdoviridae) with neurological symptoms similar to those of classical rabies (OIE, 2016). Lyssaviruses are divided into 14 species as registered by the International Committee of Viral Taxonomy up to 2018 (ICTV, 2018). Species are identified on the basis of demarcation criteria such as genetic distance (genogroup) and antigenic characteristics vis-à-vis monoclonal antibodies antinucleocapsid (serogroup). This demarcation is supported by geographic distribution and host species (see Table 14.1). Lyssaviruses have two phylogroups (WHO, 2013): Phylogroup 1 includes the following species: classical rabies virus (RABV), Aravan virus, Australian bat lyssavirus, Bokeloh beats lyssavirus, Duvenhage virus, European bat lyssavirus 1, European bat lyssavirus 2, Ikoma lyssavirus (IKOV), Irkut virus, Khujand virus. Phylogroup 2 includes the following species: Lagos beats virus, Mokola virus (MOKV), Shimoni beats viruses. West Caucasian bat viruses, IKOV, and Lleida bat lyssavirus can form independent phylogroups. Within each phylogroup, different viruses can have cross-seroneutralization and induce cross-protection. On the other hand, between the two phylogroups, there is neither serum neutralization nor

TABLE 14.1 Species of the *Lyssavirus* Genus, With Their Serotype, Genotype, Phylogroup, Host Species, and Geographical Distribution

Species name	Shortened name	Serotype	Genotype	Phylogroup	Host species (reservoir)	Geographical distribution
Classical rabies virus	RABV	1	1	1	Carnivorous mammals and bats	Worldwide except Australia, Antarctica, and some island nations
Aravan virus	ARAV	ND	ND	1	Insectivorous bats (*Myotis blythi*)	Central Asia
Australian bat lyssavirus	ABLV	1	7	1	Insectivorous and frugivorous bats (*Megachiroptera*/*Microchiroptera*)	Australia
Duvenhage virus	DUVV	4	4	1	Insectivorous bats	South of Africa
European bat lyssavirus 1	EBLV 1	5	5	1	Insectivorous bats (*Eptesicus serotinus*)	Europe
European bat lyssavirus 2	EBLV 2	5	6	1	Insectivorous bats (*Myotis daubentonii*, *Myotis dasycneme*)	Europe
Irkut virus	IRKV	ND	ND	1	Insectivorous bats (*Murina leucogaster*)	Eastern Siberia
Khujand virus	KHUV	ND	ND	1	Insectivorous bats (*Myotis mystacinus*)	Central Asia
Lagos bat virus	LBV	2	2	2	Frugivorous bats	Africa
Mokola virus	MOKV	3	3	2	Unknown	Sub-Saharan Africa
Shimoni bat virus	SHIBV	ND	ND	2	Insectivorous bats (*Hipposideros commersoni*)	East Africa: Kenya
Bokeloh bat lyssavirus	BBLV	ND	ND	ND	Insectivorous bats (*Myotis nattereri*)	Europe
Ikoma lyssavirus	IKOV	ND	ND	ND	isolated on *Civettictis civetta*	Africa (Tanzania)
West Caucasian Lleida bat virus	WCBV	ND	ND	ND	Insectivorous bats (*Miniopterus schreibersi*)	Caucasus region

ND, Not determined.
Adapted from FLI, 2014. Rabies – Bulletin – Europe: Classification [Institutionnel]. Retrieved November 21, 2016, from: <http://www.who-rabies-bulletin.org/about_rabies/classification.aspx>.

cross-protection. Conventional vaccines that protect lyssaviruses from phylogroup 1 may be less effective against those of phylogroup 2 (OIE, 2013). The RABV is distributed worldwide and is responsible for almost all cases of rabies in animals and humans (WHO, 2013). However, all lyssaviruses are likely to cause fatal encephalitis in humans and animals (WHO, 2013). Table 14.1 summarizes the *Lyssavirus* viral species as well as their serotype, genotype, and phylogroup. The host species and geographical distribution are also indicated. We can notice that all viral species have as their primary host the bats with the exception of "MOKV" for which the host species is not yet specified (FLI, 2014).

STRUCTURE OF THE RABIES VIRUS

The rabies virus (*Lyssavirus* prototype) is, like all Rhabdoviridae, an unsegmented single-stranded RNA-enveloped virus of negative polarity (Delmas et al., 2008; Johnson et al., 2010; Walker et al., 2011). The genome is about 12 kb in size (Bourhy et al., 2008).

The plane projection of the rabies virus (Fig. 14.1A) reveals surface spicules, an envelope and a nucleocapsid with helical symmetry. The thin fringe of spicules on the surface is 8 nm thick and does not cover the surface of the plane end of the virus particle. Projections are placed 4.9 nm apart (Asaye and Getachew, 2014).

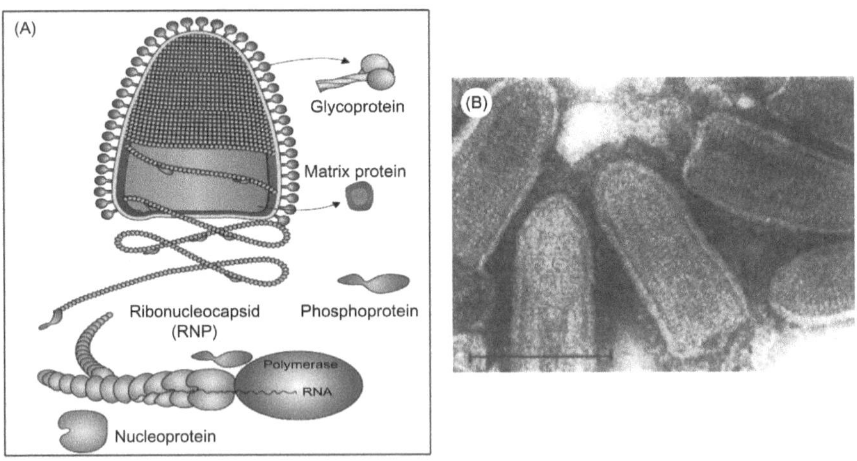

FIGURE 14.1 **Structure of the rabies virus:** (A) Diagram of the virus with its proteins, (B) photo of electron microscopy of viral particles showing the shape in revolver bullet. *Source: Adapted from Asaye, M., Getachew, N., 2014. Rabies virus proteins and their mechanism of pathogenecity. Indian J. Drugs 2, 89–95.*

The viral particle has a characteristic bullet-shape (Fig. 14.1B), helical symmetry, length of about 180 nm (130–300 nm), diameter ranging from 45 to 100 nm with one round and one flatted end or as bacilliform particle that appears hemiserphical at both ends when matured (Asaye and Getachew, 2014).

The rabies virus has a simple organization genome (Fig. 14.2) that codes for five structural proteins (Asaye and Getachew, 2014; Nadin-Davis and Real, 2011):

- Nucleocapsid protein (N) (400 aa)
- Phosphoprotein (P) (200 aa): cofactor of RNA polymerase
- Protein of the matrix (M) (200 aa)
- Glycoprotein of the envelope (G) (500 aa)
- RNA polymerase (L) (200 aa)

The parts of the genome encoding the five rabies proteins (N, P, M, G, and L) are separated by short noncoding intergene regions except for a long noncoding region between genes that code for G and L proteins (Nadin-Davis and Real, 2011).

The rabies virus replication cycle according to Rupprecht et al. (2002) is as follows (see Fig. 14.3):

1. Fixation: The spicules on the surface of the viral particle are fixed to the cellular receptors.
2. Penetration: The virion is endocyted by the cell, and the envelope fuses with the membrane of the endosome.
3. Eclipse: The nucleocapsid is released in the cytoplasm.
4. Transcription: The viral RNA is transcribed into messenger RNA.
5. Protein translation: The messenger RNAs are translated into proteins by the ribosomes of the host cell. Synthesis of the five structural proteins.
6. Secondary transcription: Protein G, a transmembrane protein, is synthesized in the endoplasmic reticulum and glycosylated by the Golgi apparatus.

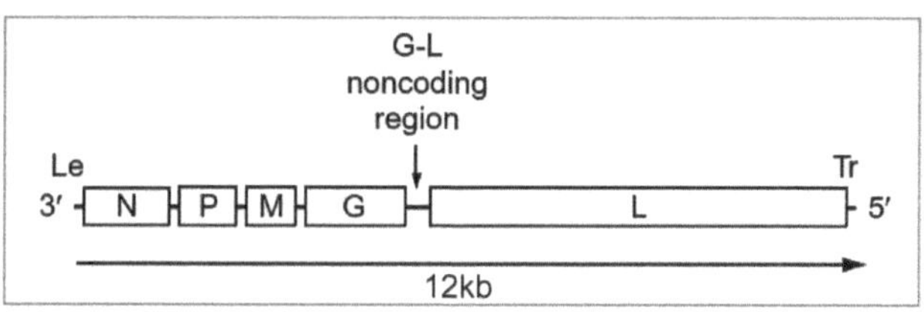

FIGURE 14.2 **Organization of the genome of the "Pasteur" strain.** *Source: Adapted from Nadin-Davis, S.A., Real, L.A., 2011. Molecular phylogenetics of the Lyssaviruses—insights from a coalescent approach. In: Advances in Virus Research, vol. 79, Elsevier, pp. 203–238.*

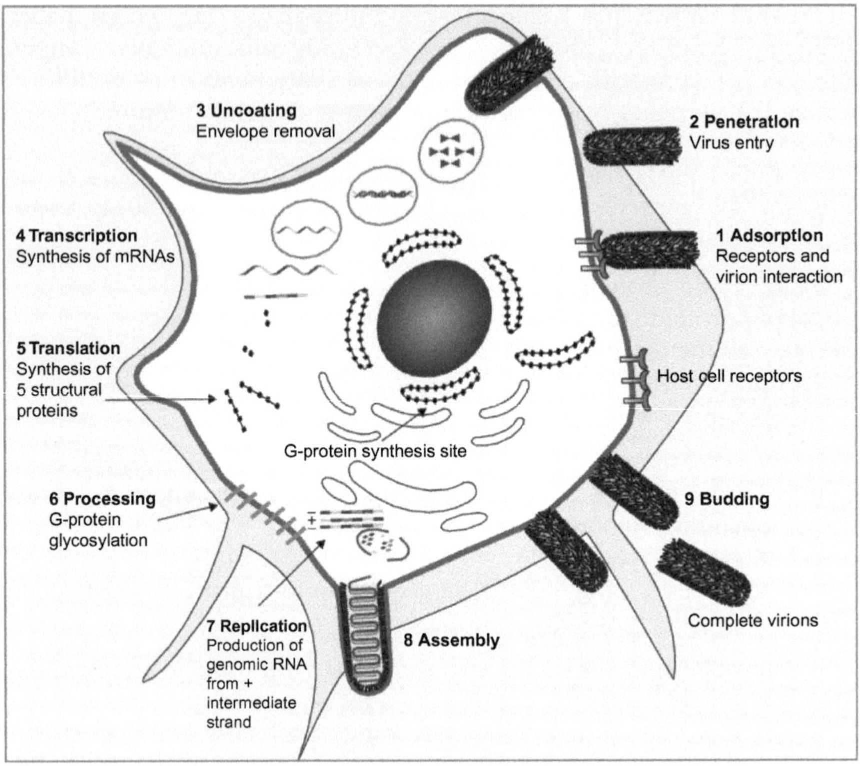

FIGURE 14.3 **Rabies virus replication cycle.** *Source: Adapted from Rupprecht, C.E., Hanlon, C.A., Hemachudha, T., 2002. Rabies re-examined. Lancet Infect. Dis. 2, 327–343.*

7. Replication: The vesicles carry the glycoprotein G toward the baso-lateral region of the cytoplasmic membrane with which they merge. The glycoprotein G is thus included in several membrane sites: the baso-lateral region of the cytoplasmic membrane but also the endoplasmic reticulum and the different compartments of the Golgi apparatus.
8. Assembly: The M protein is deposited on the inner surface of the cytoplasmic membrane. It interacts
   a. with the spicules of glycoproteins that it gathers,
   b. with the nucleocapsids which it ensures the condensation in the characteristic helicoidal form.

   Although the RNA + and RNA strands are covered by the N protein, only the viral nucleocapsids (strand −) bind to the protein M.
9. Release (budding): One can observe an external budding of the viral particles, starting from the cell membrane and an internal budding

from the other membranes (reticulum and Golgi); the virions will then gain the cytoplasmic membrane in vesicles.

PATHOGENESIS OF RABIES

The rabies development process can be summarized in four steps that can be described as follows:

Inoculation of the Virus to Tissues

Rabies virus infections occur more commonly through more or less severe and deep bites, especially in children and in the arms since the victims try to protect themselves. Such bites cause the virus to be deeply deposited in muscle tissue (Asaye and Getachew, 2014). In contrast, bats appear to transmit the rabies virus more effectively through shallow intradermal (ID) bites often without the victim noticing. Aerosol transmission has been observed under extreme conditions; however, this transmission mechanism does not appear to contribute to the level of natural transmission of the disease. It is also possible that the infection in these cases was done orally (Johnson et al., 2010).

The effectiveness of rabies virus transmission depends on the severity of the bite and the viral load of the infected saliva. Particularly, deep wounds present a significant risk due to the abundance of nicotinic acetylcholine receptors in the muscle cells. Virulent saliva deposition in the conjunctival, oral, and genital mucosa has been implicated as a pathway for rabies transmission. It has been estimated that the rabies virus migrates through axons at a rate of 3 mm per hour (Dietzschold et al., 2008; Finke and Conzelmann, 2005). However, the exact mechanism that facilitates intersynaptic dissemination is still unknown (Dietzschold et al., 2008).

Migration of the Virus From the Periphery to the Central Nervous System

The incubation period can range from 2 weeks to a few months and in rare cases for years as reported in a few human cases. Long incubation periods can be explained by the location of the virus at the neuromuscular junction of muscle cells or other cell types (Asaye and Getachew, 2014). It is also recognized that head and neck bites will lead to clinical rabies with shorter incubation periods than bites at the peripheral limbs (Asaye and Getachew, 2014).

After inoculation, the virus can persist at the site of inoculation and replicate for hours or even weeks, or follow a relatively fast centripetal pathway to the central nervous system (CNS) with replication and spread of the virus (Asaye and Getachew, 2014).

In the case of an inoculum with a high viral load, it is possible that the virus is taken simultaneously at the level of the muscle cells and at the level of the nerve endings of the site of inoculation. In this case the predominant route of infection is the nerve pathway with transmission of rabies virus through the axoplasmic retrograde flow to the CNS or spinal ganglia. This gives rise to relatively short incubation periods. When the inoculum has a low viral load, there is a high probability that the virus enters the nerve endings, or the muscle fibers (in some cases neither). This leads to an early CNS transit via nerve endings or muscle retention of virus during varying periods and its transmission to the peripheral nervous system through neuromuscular junctions (Asaye and Getachew, 2014). This may contribute to the explanation of the wide variability of rabies incubation periods (Charlton et al., 1997). It is estimated that the rabies virus migrates to the CNS at a rate of approximately 50–100 mm/day (Dietzschold et al., 2008; Finke and Conzelmann, 2005).

Key factors involved in the neuroinvasion of rabies virus are virus contamination, axonal transport, transsynaptic diffusion, and viral replication rate (Asaye and Getachew, 2014).

Invasion of the Central Nervous System

In general, examination of the brain suggests moderate congestion of the meningeal vessels. Microscopic examination demonstrates a minor perivascular cuff, limited tissue necrosis, and acidophilic intracytoplasmic neuronal inclusions (Asaye and Getachew, 2014).

Centrifugal Release From the Central Nervous System

After infection of the brain, the virus diffuses centrifugally to the peripheral nervous system and to several peripheral organs. In the final stage of the infection cycle, the rabies virus migrates to the salivary glands, and after local replication, it is excreted in the saliva and is ready to be passed on to the next host (Asaye and Getachew, 2014).

VIRULENCE FACTORS OF RABIES VIRUS

According to Dietzschold et al. (2008) and Franka et al. (2009), the major factors that determine the pathogenicity of the rabies virus are the dose of virus inoculated, the route of inoculation, the absorption of the virus, cell-to-cell diffusion, viral replication rate, G-glycoprotein expression, and host species and susceptibility to rabies virus. Indeed, the pathogenicity is directly correlated with the uptake of the virus and its diffusion but inversely correlated with the viral replication rate and the level of expression of the G protein. The success of rabies virus expression in a host and the success of its transmission to other organisms depends on the integrity of transport mechanisms over long distances; this is perhaps one of the reasons for the development of the mechanism that downregulates the transcription of the genome of the rabies virus (Dietzschold et al., 2008; Finke and Conzelmann, 2005).

The rabies virus has developed mechanisms of adjustment of gene expression from the maximal level to the optimal level and which are in harmony with sufficient viral replication and maintenance below the critical threshold that allows the preservation of the host cells (Finke and Conzelmann, 2005). Indeed, a submaximal expression of the viral genome can postpone apoptosis or any other response reaction of the host; this allows the spread of the virus. It should be noted that the onset of apoptosis is correlated with the expression levels of the G protein gene (Finke and Conzelmann, 2005).

The spicules of the glycoprotein on the surface of the viral particles are major determinants of rabies virus neuropathogenicity since these structures are responsible for the specific binding of the viral particle to cellular receptors (Finke and Conzelmann, 2005). Indeed, the pathogenicity of a particular strain is inversely correlated to the level of G protein expression and its ability to induce cell apoptosis (Dietzschold et al., 2008; Finke and Conzelmann, 2005); this is more relevant on the periphery (Finke and Conzelmann, 2005).

From a molecular point of view, the absence of R333 at the level of the gene that encodes the glycoprotein has a negative effect on pathogenicity (Badrane et al., 2001). In addition, the regulation of machinery for the synthesis of rabies virus RNA by M protein can be an important factor of pathogenicity for the virus. Indeed, the protein M attenuates the maximal expression of the basic machinery for synthesizing rabies virus RNA and allows the survival of the host cells. M protein is therefore an important factor in the pathogenicity of the virus and emphasizes the close relationship between viral functions at the molecular level and adaptations of host defense mechanisms (Finke and Conzelmann, 2005).

The regulation of viral replication appears to be one of the important mechanisms involved in the pathogenicity of rabies virus. In order to escape the immune response and preserve the integrity of the nervous network, the virulent strains of the rabies virus, unlike the attenuated strains, can regulate their rabies replication rate (Dietzschold et al., 2008; Finke and Conzelmann, 2005). A low level of replication benefits virulent strains of rabies virus by retaining the structure of the neurons used to reach the CNS. Another explanation for the low level of replication of virulent strains is that to escape the immune system, these strains maintain the level of expression of their antigen at minimum (Dietzschold et al., 2008).

HUMAN ANTIRABIES VACCINES

Nerve Tissue-Based Vaccines

Rabies vaccines produced in mammalian neural tissues (brain of adult animals such as sheep and goat; brain of suckling animals such as mouse, rat, and rabbit) have been in worldwide use for many years. They were used for postexposure prophylaxis. Nerve tissue-based vaccines are very reactogenic and may cause severe, even fatal, encephalitis and polyneuritis. Moreover, there is evidence for a lack of potency of these neural tissue vaccines, leading to inadequate protection in humans, making a strong argument for the discontinuation of their production and use (WHO, 2013, 2017).

Cell-Culture Vaccines and Embryonated Egg-Based Rabies Vaccines

For many years, safe and highly effective rabies vaccines, produced on various cell cultures ranging from primary cells (hamster kidney and chick embryo fibroblasts), diploid cells, to continuous cell lines such as Vero cells, including embryonated eggs, are on the market. In some rabies enzootic countries, vaccines prepared in cell culture are difficult to obtain and/or too expensive. In contrast, ID administration has emerged as an effective and economical alternative to conventional intramuscular regimens (WHO, 2013, 2017).

VETERINARY ANTIRABIES VACCINES

The only effective method of rabies control is vaccination. Vaccines used for rabies control can be classified according to their route of administration (WHO, 2013):

Injectable Vaccines

Injectable vaccines can be manufactured according to several methods presented below:

Live Modified Injectable Vaccines

Produced from modified rabies strains adapted to embryonated eggs (e.g., Flury strain) by successive passages on embryonated eggs of chickens or duck (Briggs, 2011). They can be produced on cell cultures [e.g., Street Alabama Dufferin (SAD)/Evelyn (Gaynor) Rokitniki Abelseth (ERA) strains]. These vaccines are no longer considered "safe" and should no longer be used in domestic animals because of their ability to induce rabies (WHO, 2013).

Inactivated (Monovalent or Multivalent) Injectable Vaccines

These vaccines are the most used in domestic animals. They are inactivated (killed) and are considered "safe" and not expensive. Their efficacy, safety, and purity should be evaluated by validated methods as recommended by international organizations before use (EDQM, 2013a; OIE, 2013). Inactivated vaccines can be produced on different cell culture systems: primary cells or continuous cell lines (Syrian hamster kidney cells, primary cell lines produced from embryonated chicken or duck eggs, continuous cell lines produced from Vero cells) (Briggs, 2011). Inactivated rabies vaccines, whether adjuvanted or not, may be used alone or in combination with other vaccines for other animal diseases (WHO, 2013).

Recombinant Vectorized (Monovalent or Multivalent) Injectable Vaccines

A canarypox vaccine expressing rabies glycoprotein is available for cats. It combines feline panleukopenia, feline calicivirosis, and feline herpesvirus (WHO, 2013).

Control of Inactivated Vaccines

It is important to market only good-quality inactivated rabies vaccines. WHO recommends that inactivated veterinary rabies vaccines have a titer of at least 1 IU/dose (WHO, 2013). This activity is measured historically by the National Institutes of Health (NIH) test, which is based on a virulent test of vaccinated mice (Meslin et al., 1996). This test has been criticized by many authors because of its high variability and the high use of laboratory animals (mice) (Hraenhart et al., 1989; Servat et al., 2015). A new test has recently been developed and included in edition 8.0 of the European Pharmacopoeia (EDQM, 2013a). It is based on the vaccination of mice and the determination of serum-neutralizing antirabies antibody titers induced by the vaccine to be tested. This new method does not allow to measure the activity of vaccines with a low titer close to the limit of 1 IU/dose and requires the use of animals of experience certainly in smaller number (Servat et al., 2015). Proposals for the use of enzyme-linked immunosorbent assay (ELISA) tests to titrate rabies glycoprotein G during veterinary vaccine manufacturing could be retained by international standardization organizations and thus lead to a reduction in the number of animals used; the presence of adjuvant at the finished product level remains a challenge for any alternative test (De Mattia et al., 2015).

Oral Vaccines

Since 1978, oral antirabies vaccination has been a great success in the fight against rabies in Europe. This result has been achieved through the use of about 10 oral vaccine strains. The combination of factors such as vaccine strain, vaccine bait, and distribution strategy is a key element of any oral vaccination strategy (Cliquet and Aubert, 2004).

The application of oral vaccination requires prior knowledge of all factors that may influence it such as the dynamics and density of the target species, the presence of nontarget species that may compete with the target species for the taking of vaccine baits, the behavioral factors of the target species (seasonality, preference, and feeding behavior) (Linhart, 1993). The stability of bait and virus strain in external environmental conditions is also a determining factor (Masson et al., 1996a,b).

Oral rabies vaccines are either derived from live attenuated rabies strains or recombinant strain. Fig. 14.4 traces the relationships between the different oral antirabies strains; all oral strains were developed from SAD strain (Cliquet, 2008).

Oral vaccine strains can be classified into three families (Cliquet, 2007):

Street Alabama Dufferin Vaccine Strains

The modified live vaccine strains are all derived from the attenuated strain SAD, which was isolated in the United States in 1935 in Alabama from a rabid dog. This strain underwent several multiplications on mouse brain cells and gave rise to the ERA strain, and was then adapted by successive passages on baby Hamster kidney fibroblasts (BHK cells) to give rise to the SAD Bern strain. The SAD B19 and SAD P5/88 vaccine strains (Impfstoffwerk Dessau-Tornau) were produced by successive passages on BHK clone cells of the SAD Bern strain (Cliquet and Aubert, 2004; Cliquet, 2007; Müller et al., 2015). These successive selections from the original strain may produce random and uncontrolled results and variants may remain pathogenic for nontarget species (Cliquet, 2007; Vuta et al., 2016).

Monoclonal Antibody Selected Vaccine Strains

The SAG1 and SAG2 (Street Alabama Gif) vaccine strains (Centre national de la recherche scientifique (CNRS) at Gif-sur-Yvette being the

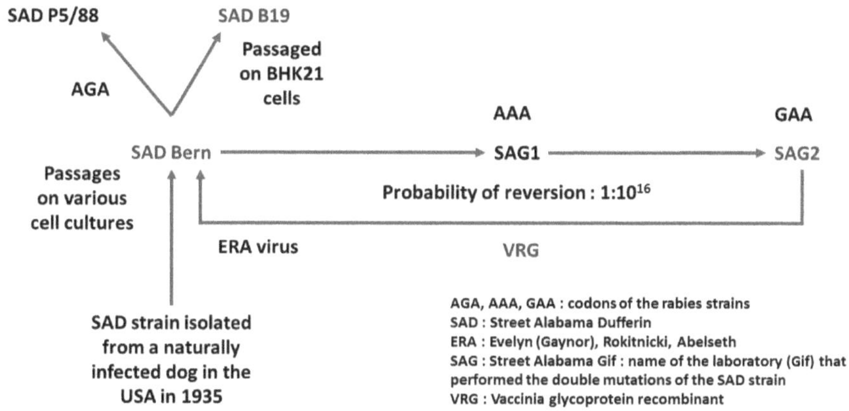

FIGURE 14.4 **Relationship between oral antirabies vaccine strains derived from the SAD strain.** *SAD*, Street Alabama Dufferin. *Source: Adapted from Cliquet, F., 2007. Oral vaccines used for rabies control programmes: types, storage, quality control and performance in different species. In: Meeting on Enhancing Rabies Eradication in the EU: International Co-Operation, Helsinki, Finland. Available from: <http://ec.europa.eu/food/animal/diseases/eradication/rabies_pres_19.pdf> (accessed 29.12.14.). Citeseer. Retrieved from: <http://citeseerx.ist.psu.edu/viewdoc/download?doi=10.1.1.577.8853&rep=rep1&type=pdf>.*

name of the laboratory where the isolation was performed) are selected from the SAD Bern strain after one or two successive mutations of the codon of the arginine 333, respectively. These mutations were performed using monoclonal antibodies specific for rabies glycoprotein at a portion of the genome whose integrity is necessary for the conservation of pathogenicity by the oral route (Cliquet, 2007; Müller et al., 2015).

Recombinant Live Vaccines

The vaccine strain vaccinia recombinant glycoprotein is a recombinant vaccinia virus (Copenhagen strain) encoding the glycoprotein gene of the ERA strain (Cliquet, 2008). The Copenhagen strain already attenuated was even more so following the replacement of the thymidine kinase gene by the complementary DNA of the rabies glycoprotein; this confers antirabies immunity to the strain (Cliquet, 2007; Müller et al., 2015).

Quality Criteria for Oral Rabies Vaccines

Several guides exist concerning the quality criteria of oral rabies vaccines before they are placed on the market. The most accurate were produced by WHO, the European Pharmacopoeia, the European Commission, and more recently by European Food Safety Authority (EFSA) (Commission EU, 2002; EDQM, 2013b; EFSA, 2015; OMS, 2007). Oral rabies vaccines have been more or less extensively tested on different animal species using different routes of inoculation (intracerebral, intramuscular, and oral) on puppies, carnivores, avian species, nonhuman primates, rodents, and immunodeficient mice (Bingham et al., 1997, 1999; Blancou et al., 1989; Cliquet et al., 2007; Masson et al., 1996a,b). Nonhuman primates have been added to the list since the discovery in 1992 that the SAD Bern strain is highly pathogenic to baboons by oral route (Bingham et al., 1992).

WHO (2013) guidelines indicate that vaccine strains should be genetically characterized preferably by complete genome sequencing. This control is an important factor, indeed, the study of (Geue et al., 2008) which focuses on rabies vaccine strains used orally for the control of rabies in wildlife (SAD B19, SAD P5/88, SAG2, SAD VA1, SAD Bern, ERA, and SAD 1-3670 Wistar) concluded that

- Most commercial strains currently available appear to be derived from SAD strain B19 rather than SAD Bern.
- A commercial vaccine did not contain the SAD strain that was mentioned by the producer.
- Two SAD vaccines did not have a pure vaccine strain but a mixture of several strains.

DNA Vaccines

DNA vaccine technology is based on the injection (intramuscularly or intradermally) of DNA plasmids that encode the G protein. The plasmid thus injected integrates the host cell and encodes the immunogenic protein. DNA vaccines can be adjuvanted to improve their efficacy (Ullas et al., 2014). The use of DNA vaccines may be an alternative to adjuvanted inactivated vaccines and live attenuated (oral) vaccines. This technology is adapted to fight against viruses that induce protein-based immunity (case of rabies G protein) and has the ability to induce immunity against several pathologies at the same time and could be used orally (Kumar et al., 2013; Redding and Weiner, 2009).

Field and laboratory trials compared a DNA vaccine and an adjuvanted inactivated vaccine produced in cell culture. They demonstrated that the DNA vaccine has the ability to induce a higher level of antibodies and for up to 4 years after a single injection, making it an excellent candidate for vaccination of domestic animals and wild animals (Bahloul et al., 2006; Redding and Weiner, 2009).

References

Asaye, M., Getachew, N., 2014. Rabies virus proteins and their mechanism of pathogenecity. Indian J. Drugs 2, 89–95.

Badrane, H., Bahloul, C., Perrin, P., Tordo, N., 2001. Evidence of two lyssavirus phylogroups with distinct pathogenicity and immunogenicity. J. Virol. 75, 3268–3276.

Bahloul, C., Taieb, D., Diouani, M., Ahmed, S., Chtourou, Y., Bchir, B., et al., 2006. Field trials of a very potent rabies DNA vaccine which induced long lasting virus neutralizing antibodies and protection in dogs in experimental conditions. Vaccine 24, 1063–1072.

Bingham, J., Foggin, C.M., Gerber, H., Hill, F., Kappeler, A., King, A.A., et al., 1992. Pathogenicity of SAD rabies vaccine given orally in chacma baboons (*Papio ursinus*). Vet. Rec. 55–56.

Bingham, J., Schumacher, C.L., Aubert, M.F.A., Hill, F.W.G., Aubert, A., 1997. Innocuity studies of SAG-2 oral rabies vaccine in various Zimbabwean wild non-target species. Vaccine 15, 937–943.

Bingham, J., Schumacher, C., Hill, F.W., Aubert, A., 1999. Efficacy of SAG-2 oral rabies vaccine in two species of jackal (*Canis adustus* and *Canis mesomelas*). Vaccine 17, 551–558.

Blancou, J., Artois, M., Brochier, B., Thomas, I., Pastoret, P.P., Desmettre, P., et al., 1989. Innocuité et efficacité d'un vaccin antirabique recombinant des virus de la vaccine et de la rage administré par voie orale au renard, au chien et au chat. Ann. Rech. Vet. 20, 195–204.

Bourhy, H., Reynes, J.-M., Dunham, E.J., Dacheux, L., Larrous, F., Huong, V.T.Q., et al., 2008. The origin and phylogeography of dog rabies virus. J. Gen. Virol. 89, 2673–2681.

Briggs, D.J., 2011. Immunological Basis for Immunization. Module 17: Rabies. World Health Organization, Geneva, Retrieved from: <http://apps.who.int/iris/bitstream/10665/44517/1/9789241501088_eng.pdf>.

CDC, 2011, February. Operationalizing "One Health": A Policy Perspective—Taking Stock and Shaping an Implementation Roadmap. Centers for Disease Control and Prevention. Retrieved from: <https://www.cdc.gov/onehealth/pdfs/atlanta/meeting-overview.pdf>.

CDNA, 2013, January 21. Rabies Virus and Other Lyssavirus (Including Australian Bat Lyssavirus) Exposures and Infections. Australian Government Department of Health. Retrieved from: <http://www.health.gov.au/internet/main/publishing.nsf/Content/ED62D139B56F7B80CA257BF0001B7422/$File/ABLV-rabies-song-gl-June2014.pdf>.

Charlton, K., Nadin-Davis, S., Casey, G., Wandeler, A., 1997. The long incubation period in rabies: delayed progression of infection in muscle at the site of exposure. Acta Neuropathol. 1, 73–77.

Cima, G., 2013. Eliminating dog-transmitted rabies: end of human deaths in Americas from dog-transmitted rabies targeted for 2015. J. Am. Vet. Med. Assoc. 243, 1662–1663.

Cliquet, F., 2007. Oral vaccines used for rabies control programmes: types, storage, quality control and performance in different species. In: Meeting on Enhancing Rabies Eradication in the EU: International Co-Operation, Helsinki, Finland. Available from: <http://ec.europa.eu/food/animal/diseases/eradication/rabies_pres_19.pdf> (accessed 29.12.14.). Citeseer. Retrieved from: <http://citeseerx.ist.psu.edu/viewdoc/download?doi = 10.1.1.577.8853&rep = rep1&type = pdf>.

Cliquet, F., 2008, December. Vaccines used for rabies control programmes: types, performance in different species, quality control, storage. In: Exposé presented at the Workshop on Rabies: Regional Cooperation Towards Eradicating the Oldest Known Zoonotic Disease in Europe, Anatlaya, Turkey. Retrieved from: <http://www.fve.org/news/presentations/taiex/2008/2008_4_12_vaccines_rabies_fcliquet.pdf>.

Cliquet, F., Aubert, M.F., 2004. Elimination of terrestrial rabies in Western European countries. Dev. Biol. (Basel) 185–204.

Cliquet, F., Gurbuxani, J.P., Pradhan, H.K., Pattnaik, B., Patil, S.S., Regnault, A., et al., 2007. The safety and efficacy of the oral rabies vaccine SAG2 in Indian stray dogs. Vaccine 25, 3409–3418.

Commission EU, 2002. The oral vaccination of foxes against rabies. In: Report of the Scientific Committee on Animal Health and Animal Welfare. European Commission, p. 55. Retrieved from: <http://foxrabiesblueprint.org/IMG/pdf/pdf_116_eu_scientific_report_orv_en.pdf>.

Delmas, O., Holmes, E.C., Talbi, C., Larrous, F., Dacheux, L., Bouchier, C., et al., 2008. Genomic diversity and evolution of the lyssaviruses. PLoS One 3, e2057.

De Mattia, F., Hendriksen, C., Buchheit, K.-H., Chapsal, J.-M., Halder, M., Lambrigts, D., et al., 2015. The vaccines consistency approach project: an EPAA initiative. Eur. Directorate Qual. Med. 2015, 30–56.

Dietzschold, B., Li, J., Faber, M., Schnell, M., 2008. Concepts in the pathogenesis of rabies. Fut. Virol. 3, 481–490.

EDQM, 2013a. Rabies vaccine (inactivated) for veterinary use, monograph 0451, European Pharmacopoeia, seventh ed. Council of Europe, EDQM, Strasbourg, pp. 1008–1010.

EDQM, 2013b. Vaccin vivant oral de la rage pour renards et chiens viverrins (monographie 0746), European Pharmacopoeia, eighth ed. Council of Europe, EDQM, Strasbourg.

EFSA, 2015. Update on oral vaccination of foxes and raccoon dogs against rabies: oral vaccination against rabies. EFSA J. 13, 4164.

Finke, S., Conzelmann, K.-K., 2005. Replication strategies of rabies virus. Virus Res. 111, 120–131.

FLI, 2014. Rabies – Bulletin – Europe: Classification [Institutionnel]. Retrieved November 21, 2016, from: <http://www.who-rabies-bulletin.org/about_rabies/classification.aspx>.

Franka, R., Wu, X., Jackson, F.R., Velasco-Villa, A., Palmer, D.P., Henderson, H., et al., 2009. Rabies virus pathogenesis in relationship to intervention with inactivated and attenuated rabies vaccines. Vaccine 27, 7149–7155.

Geue, L., Schares, S., Schnick, C., Kliemt, J., Beckert, A., Freuling, C., et al., 2008. Genetic characterisation of attenuated SAD rabies virus strains used for oral vaccination of wildlife. Vaccine 26, 3227–3235.

Hraenhart, O., Blancou, J., Aubert, M.F.A., 1989. Results of an inquiry on potency testing of rabies vaccine by the NIH test: suggestions for further improvement. Rev. Sci. Tech. 8, 917–919.

ICTV, 2018, February 25. International Committee on Taxonomy of Viruses (ICTV) [Official]. Retrieved February 25, 2018, from: <https://talk.ictvonline.org/>.

Johnson, N., Cunningham, A.F., Fooks, A.R., 2010. The immune response to rabies virus infection and vaccination. Vaccine 28, 3896–3901.

Kaare, M., Lembo, T., Hampson, K., Ernest, E., Estes, A., Mentzel, C., et al., 2009. Rabies control in rural Africa: evaluating strategies for effective domestic dog vaccination. Vaccine 27, 152–160.

Kumar, U., Kumar, S., Varghese, S., Chamoli, R., Barthwal, P., 2013. DNA vaccine: a modern biotechnological approach towards human welfare and clinical trials. Int. J. Res. Biomed. Biotechnol. 3, 17–20.

Linhart, S.B., 1993. Some factors affecting the oral rabies vaccination of free-ranging carnivores. Rev. Sci. Tech. 12, 109–113.

Masson, E., Aubert, M.F.A., Barrat, J., Vuillaume, P., 1996a. Comparison of the efficacy of the antirabies vaccines used for foxes in France. Vet. Res. 255–266.

Masson, E., Cliquet, F., Aubert, M., Barrat, J., Aubert, A., Artois, M., et al., 1996b. Safety study of the SAG2 rabies virus mutant in several non-target species with a view to its future use for the immunization of foxes in Europe. Vaccine 14, 1506–1510.

Meslin, F.-X., Kaplan, M.M., Koprowski, H., World Health Organization, et al., 1996. Laboratory Techniques in Rabies, fourth ed. World Health Organization, Geneva, Switzerland, Retrieved from: <http://libdoc.who.int/publications/1996/9241544791_eng.pdf>.

Müller, T., Freuling, C.M., Wysocki, P., Roumiantzeff, M., Freney, J., Mettenleiter, T.C., et al., 2014. Terrestrial rabies control in the European Union: historical achievements and challenges ahead. Vet. J. (London, England: 1997). Available from: https://doi.org/10.1016/j.tvjl.2014.10.026.

Müller, T.F., Schröder, R., Wysocki, P., Mettenleiter, T.C., Freuling, C.M., 2015. Spatio-temporal use of oral rabies vaccines in fox rabies elimination programmes in Europe. PLoS Negl. Trop. Dis. 9, e0003953.

Nadin-Davis, S.A., Real, L.A., 2011. Molecular phylogenetics of the lyssaviruses—insights from a coalescent approach, Advances in Virus Research, vol. 79. Elsevier, pp. 203–238.

Nel, L.H., 2013. Discrepancies in data reporting for rabies, Africa. Emerg. Infect. Dis. 19, 529–533.

OIE, 2013, May. Rabies, Chapter 2.1.13. Retrieved December 6, 2014, from: <http://www.oie.int/fr/normes-internationales/manuel-terrestre/acces-en-ligne/>.

OIE, 2016. OIE Fiche d'information générale – RAGE.pdf. Retrieved from: <http://www.oie.int/fileadmin/Home/fr/Media_Center/docs/pdf/Disease_cards/RABIES-FR.pdf>.

OMS, 2007. Oral Vaccination of Dogs Against Rabies. WHO, Geneva, Retrieved from: <http://www.who.int/rabies/resources/guidelines%20for%20oral%20vaccination%20of%20dogs%20against%20rabies_with%20cover.pdf>.

Redding, L., Weiner, D.B., 2009. DNA vaccines in veterinary use. Expert Rev. Vaccines 8, 1251–1276.

Rupprecht, C.E., Hanlon, C.A., Hemachudha, T., 2002. Rabies re-examined. Lancet Infect. Dis. 2, 327–343.

Servat, A., Kempff, S., Brogat, V., Litaize, E., Schereffer, J.-L., Cliquet, F., 2015. A step forward in the quality control testing of inactivated rabies vaccines—extensive evaluation of European vaccines by using alternative methods to the in vivo potency tests. Altern. Lab. Anim. 43, 19–27.

Talbi, C., Holmes, E.C., Benedictis, P., de, Faye, O., Nakouné, E., Gamatié, D., et al., 2009. Evolutionary history and dynamics of dog rabies virus in western and central Africa. J. Gen. Virol. 90, 783–791.

Ullas, P.T., Desai, A., Madhusudana, S.N., 2014. Immunogenicity and efficacy of a plasmid DNA rabies vaccine incorporating Myd88 as a genetic adjuvant. Clin. Exp. Vaccine Res. 3, 202.

Vuta, V., Picard-Meyer, E., Robardet, E., Barboi, G., Motiu, R., Barbuceanu, F., et al., 2016. Vaccine-induced rabies case in a cow (*Bos taurus*): molecular characterisation of vaccine strain in brain tissue. Vaccine 34, 5021–5025.

Walker, P.J., Dietzgen, R.G., Joubert, D.A., Blasdell, K.R., 2011. Rhabdovirus accessory genes. Virus Res. 162, 110–125.

WHO, 2013. WHO expert consultation on rabies. Second report. World Health Organization Technical Report Series. Back Cover, pp. 1–139.

WHO (Ed.), 2017. WHO Expert Committee on Biological Standardization: Sixty-Seventh Report. WHO, Geneva, Retrieved from: <http://apps.who.int/iris/bitstream/10665/255657/1/9789241210133-eng.pdf?ua = 1>.

WHO, GARC, 2014. Rabies, Epidemiology and Burden of Disease. WHO, Morocco, Retrieved from: <http://www.who.int/rabies/epidemiology/Rabies_CP_Morocco_09_2014.pdf?ua = 1>.

CHAPTER

15

Reemerging Virus: Case of Norovirus

Yassine Amraouza[1], *Moulay Mustapha Ennaji*[2] *and Jamal Hafid*[1]

[1]Laboratory of Foods, Environment and Health, Faculty of Sciences and Techniques-Gueliz, Cadi Ayyad University, Marrakech, Morocco
[2]Laboratory of Virology, Microbiology, Quality, Biotechnologies/Eco-Toxicology and Biodiversity, Faculty of Sciences and Techniques, Mohammedia, University Hassan II of Casablanca, Casablanca, Morocco

INTRODUCTION

Norovirus (NoV) is a nonenveloped virus classified under the family Caliciviridae. It is a 27 nm particle, 7.6 kb genome length encoding for eight proteins: six nonstructural and two structural one VP1 and VP2. The diversity of genomic sequence coding for VP1 is the basis of 7 genogroups diversity (GI–GVII) and 40 genotypes (Vinjé, 2015). A recent study of 40 Chinese bat species virome led to some conclusion about an eighth genogroup (Wu et al., 2016).

Historically, NoVs were known as a winter vomiting disease long time before the discovery of the pathogen viral particle. J. Zahorsky was the first to publish a paper describing the main two particularities of the NoV infection: symptoms shown by patients are vomiting, acute gastroenteritis with diarrhea, and minor fever; seasonality preponderance of the disease in winters (Zahorsk, 1929). In 1969 two waves of epidemic winter vomiting disease happened. The first wave concerned students

Emerging and Reemerging Viral Pathogens
DOI: https://doi.org/10.1016/B978-0-12-814966-9.00015-9

and teachers of a school in Norwalk city (Ohio, United States), and the second wave implicated families of the infected persons in the first stage. This outbreak episode was the reference to the appellation of Norwalk virus for NoV, or the Norwalk-like virus for unknown virus causing same symptoms as NoV (Chan et al., 2017b). The highlighting of the NoV dates from 1972 according to two researches studies of Dolin et al. (1972) and Kapikian et al. (1972). Before that, some researchers proved the viral nature of the contamination (Gordon et al., 1947; Kojima et al., 1948; Reimann et al., 1945).

EPIDEMIOLOGY

NoVs are well known as the first cause of severe sporadic nonbacterial gastroenteritis outbreaks affecting all human age-groups (Ahmed et al., 2014; Zheng et al., 2006). Also it became the first cause leading to young children gastroenteritis since the introduction of rotavirus vaccination (Bucardo et al., 2008; Payne et al., 2013). GI and GII are the infective genogroups for human, including 9 and 22 genotypes, respectively, for each genogroup (Iturriza-Gómara and Lopman, 2014). Still GIV appear to be also infective but in rare cases (Ao et al., 2014). In a group of Ecuadorian children a NoV belonging to a novel GII genotype, proposed to be a GII.23, was reported (Lopman et al., 2015). Another new genotype GII.24, which seems to be detected in a stool sample from Nicaragua, was sequenced, and its complete genome was released in 2016 in GenBank under number KU306738 (Chan et al., 2017b).

As humans are the one natural host of NoV, high concentration of NoV could be detected from feces of an infected individual even after healing for a long period (Atmar et al., 2008), about 28 days according to Glass et al. (2009), which enhance risks of spreading the virus through some main vector: fecal–oral route through the ingestion of contaminated food and water or close contact with infected persons (Hall, 2012). In addition, conventional wastewater treatments are only effective against bacterial pathogens, but not for viral elimination (Corsi et al., 2014). The detection of both NoV genotypes I and II was reported in sewage (Mans et al., 2013), rivers and lakes (Corsi et al., 2014; Lenaker et al., 2017), seawater (Love et al., 2014), and in drinking water also (Xue et al., 2014).

About 12 to 48 hours after infection, persons can develop symptoms such as vomiting, nonbloody diarrhea, nausea, abdominal cramps, and low-grade fever. Infected immuno-competent individuals show symptoms no longer than 48h and the disease is self-limiting in most cases, counter to young children that require at increased risk to be hospitalized. NoV disease is more and more recognized as a cause of chronic gastroenteritis for immunocompromised patients (Bok and Green, 2012). More often, NoV outbreaks emerge in close spaces where people have to share common facilities such as in homes, schools, hospitals, and hotels (Dancer, 2014; Iturriza-Gómara and Lopman, 2014).

According to the World Health Organization, among just foodborne in 2010, 125 million cases of illness and 35000 death cases were estimated caused by NoV infection (Foodborne Disease Burden Epidemiology Reference Group, 2015). NoV occurrence in worldwide foodborne illnesses alone represents 23% [Centers for Disease Control and Prevention (CDC), 2013].

Water and foods are important vehicles of NoV. Bivalves are particularly concerned since they accumulate infectious particles if present in their surrounding water and lead to outbreaks in community (Loury et al., 2015). In Chinese marketed oysters 20.71% were positive for NoV; 21.48% for GI; 62.96% for GII; and 15.56% for both genogroups (Tao et al., 2018).

NOROVIRUS IN MOROCCO

In Morocco a study related to the presence of NoV in shellfish was conducted from October 2006 to August 2010 on cockles (*Acanthocardia tuberculatum*) and smooth clams (*Callista chione*) in the Mediterranean Sea coasts, and oysters (*Crassostrea gigas*) in the Atlantic coasts. Results show a great contamination by NoV in seacoasts more than oceans ones. NoV was detected in 41.4% of clams and 71% of cockles form the Mediterranean Sea, and in just 3% of oysters from the Atlantic. Both GI and GII were equally represented (Benabbes et al., 2013).

About the occurrence of the NoV in the Moroccan population, two studies interested in children less than 5 years were published. From January to December 2011, 335 samples of children's stools were tested for NoV. As a result, 22.2% of patients were positive for GI and 77.8% for GII, among this last GII.4 was predominant with a prevalence of 8.06%. GII.3, GII.13, GII.16, and GII.17 were also detected (El Qazoui et al., 2014). From March 2011 to March 2012, 122 children were included in a study where 0.82% of patients were positive for NoV (Benmessaoud et al., 2015).

NOROVIRUS IN SPAIN

A study was conducted during 4 years between 2008 and 2011, while population gastroenteritis were observed in the region of Valencia. NoV was found as origin of 42 (76.3%) out of the 55 outbreaks and 26 (7.8%)

out of 332 sporadic cases of severe gastroenteritis. Genogroup GII was predominant with genotype GII.4. The study also concerned about the assessment of the serological status for the GII.4 of the Valencian people. A total of 434 serum samples was analyzed and for which 429 (98.8%) were positive for antibodies to the P domain specific for the GII.4 variants (Carmona-Vicente et al., 2015).

NOROVIRUS IN INDIA

A prevalence study among 287 individuals (111 children and 176 adults) using stool samples was conducted from July 2013 to February 2015. Results obtained by applying reverse transcription polymerase chain reaction (RT-PCR) showed that four cases of NoV (one GI and three GII) were detected just among children patients and no traces in adults. This result indicates the low circulation of NoV in the region unlike majority of studies from India (Jain et al., 2016).

NOROVIRUS IN INDONESIA

This study proposes to assess the potential of the presence of NoV in an asymptomatic population in Indonesia. A total of 18 individuals (20–42 years old) was followed since July 2015 to June 2016 during which 512 stool specimens were sampled. About 14 (2.7%) stool samples were found positive to NoV GII during the study period, with 7 individuals having at least 1 positive stool sample. Those 14 positive samples were distributed as 10 GII.2, 2 GII.7, 1 GII.4 Sydney 2012 and 1 GII.1. This study suggests that the excretion of NoV from healthy individuals is one of the sources of NoV outbreak (Utsumi et al., 2017).

NOROVIRUS IN AUSTRALIA

In Victoria, 222 gastroenteritis outbreaks that occurred during the period January–September 2017 were tested for NoV. The results showed that 160 of outbreaks were NoV positive by RT-PCR. From the 160 outbreaks, 118 accrued during a peak period between May and August 2017. According to the authors, this prevalence peak was primarily due to the emergence of a GII.P4_NewOrleans_2009/GII.4_Sydney_2012 recombinant that had genetically changed sufficiently to escape herd immunity, and it seem to be specific to Australia (Bruggink et al., 2018).

NOROVIRUS IN ARGENTINA

The first report in South America checked the circulation of the new emergent strain GII.17 in Argentina. Ten stool samples from individuals with gastroenteritis symptoms were collected from San Martín de los Andes. This city had seen an increased number of acute gastroenteritis during the period from December 2014 and January 2015. RT-PCR showed that out of the 10th stool samples, 4 were positive for NoV, and the partial sequencing of 3 of them revealed that 2 samples belong to genotype GII.4 and 1 sample to genotype GII.17. GII.17 genotype was discovered in stool sample of a little girl (3 years old) with no traveling abroad or domestic outbreak in concern. This study talks about the emerging GII.17 cluster C strains and the GII.17 Argentinean strain (Degiuseppe et al., 2017).

NOROVIRUS AND REEMERGING STRAINS

Epidemiological studies show that GII.4 has been the dominant circulating genotype from last 28 years (Vinjé, 2015) and involved in more than 90% of outbreaks (Chen and Chiu, 2012). Then during winter 2014–15, an emerging variant of NoV GII.17, Kawasaki 308 like, was reported in China and Japan (De Graaf et al., 2015). GII.17 caused 66% of hospitalized NoV cases in Hong Kong during winter 2014–15, while GII.4 caused just 19% cases. In addition, it seems that this new variant is more infective for the older people (Chan et al., 2015). The Kawasaki strain was reported in a surveillance for enteric viruses conducted in raw sewage in Kansai area, central part of Japan from July 2015 to June 2016. GII.17 was the most frequently detected strain (50%) during the study and was the same as that one implicated in a previous outbreak in central Japan in 2015 (Thongprachum et al., 2018). A foodborne outbreak occurred during a lunch celebration in July 2016, Brazil. All stool samples tested were positive NoVs, and phylogenetic analysis revealed that strains were genetically close to GII.17 Kawasaki_2014. These findings indicated the circulation of NoV GII.17 Kawasaki_2014 in the Brazilian population (da Silva Ribeiro de Andrade et al., 2018). In Italy, NoV GII positive sample of sewage were subjected to genotypic characterization and revealed 14 sewage sample of 110 positive for NoV GII.17 Kawasaki (Suffredini et al., 2017). Like the Italian study, South African found a prevalence of 35% of the NoV GII.17 Kawasaki-2014 strain in sewage (Mabasa et al., 2018). This variant was also detected sporadically in Romania and the United States (Dinu et al., 2016; Parra and Green, 2015). Globally, GII.17 Kawasaki viruses were found in 13 countries across 4 continents: Canada, China, Germany, Hungary, Italy, Japan, the Netherlands, New Zealand, Russia, Slovenia, South Korea, Thailand, and the United States (Chan et al., 2017a).

GII.17 Kawasaki 308 variant is distinct from the other GII.17 strains and even from the cocirculating Kawasaki 323 by the insertion of two characteristic amino acid in VP1, the most surface-exposed antigenic region protein (Chan et al., 2015).

NOROVIRUS VACCINE

Genetic engineering technology was one of the ways used to develop a vaccine against NoV strains. It results in the formation of VLP (virus-like particle) structurally similar to NoV but lacks any viral genetic material. VLPs were obtained by the expression of the VP1 capsid structural protein in recombinant system (Jiang et al., 1992). The application of this technology leads to a significant innate, humoral, and cellular immune response (Fang et al., 2013). The VLP technique is in a clinical stage of development, and it has been used as an intranasal monovalent (GI.1) and bivalent (GI.1 and GII.4) vaccine, or as an intramuscular bivalent (GI.1 and GII.4) vaccine (Lucero et al., 2017).

P particles (P for protruding), another technique consisting of the expression of P2 domain, which is located on the surface of NoV capsid and the main subcomponent responsible of the VP1 variability and so of the NoV strains diversity(Lochridge et al., 2005). It has also a key role in binding of the NoV to the histobloodgroup antigens and in its recognition by antibodies (Tan et al., 2011). This technology will allow a stand-alone NoV vaccine, a dual, or a combined vaccine with other viruses (RotaV, HEV, AstroV, influenza) but remain in a preclinical stage of development (Lucero et al., 2017).

References

Ahmed, S.M., Hall, A.J., Robinson, A.E., Verhoef, L., Premkumar, P., Parashar, U.D., et al., 2014. Global prevalence of norovirus in cases of gastroenteritis: a systematic review and meta-analysis. Lancet Infect. Dis. 14 (8), 725–730. Available from: https://doi.org/10.1016/S1473-3099(14)70767-4.

Ao, Y., Yu, J., Li, L., Jin, M., Duan, Z., 2014. Detection of human norovirus GIV.1 in China: a case report. J. Clin. Virol. 61 (2), 298–301. Available from: https://doi.org/10.1016/j.jcv.2014.08.002.

Atmar, R.L., Opekun, A.R., Gilger, M.A., Estes, M.K., Crawford, S.E., Neill, F.H., et al., 2008. Norwalk virus shedding after experimental human infection. Emerg. Infect. Dis. 14 (10), 1553–1557. Available from: https://doi.org/10.3201/eid1410.080117.

Benabbes, L., Ollivier, J., Schaeffer, J., Parnaudeau, S., Rhaissi, H., Nourlil, J., et al., 2013. Norovirus and other human enteric viruses in Moroccan shellfish. Food Environ. Virol. 5 (1), 35–40. Available from: https://doi.org/10.1007/s12560-012-9095-8.

Benmessaoud, R., Jroundi, I., Nezha, M., Moraleda, C., Tligui, H., Seffar, M., et al., 2015. Aetiology, epidemiology and clinical characteristics of acute moderate-to-severe diarrhoea in children under 5 years of age hospitalized in a referral paediatric hospital in Rabat, Morocco. J. Med. Microbiol. 64 (Pt_1), 84–92. Available from: https://doi.org/10.1099/jmm.0.079830-0.

Bok, K., Green, K.Y., 2012. Norovirus gastroenteritis in immunocompromised patients. N. Engl. J. Med. 367 (22), 2126–2132. Available from: https://doi.org/10.1056/NEJMra1207742.

Bruggink, L., Witlox, K., Huang, B., Warrilow, D., Marshall, J., 2018. GII.4 norovirus recombinant causes gastroenteritis epidemic in Eastern Australia, winter 2017. J. Med. Virol. 90 (6), 1168–1171. Available from: https://doi.org/10.1002/jmv.25063.

Bucardo, F., Nordgren, J., Carlsson, B., Paniagua, M., Lindgren, P.-E., Espinoza, F., et al., 2008. Pediatric norovirus diarrhea in Nicaragua. J. Clin. Microbiol. 46 (8), 2573–2580. Available from: https://doi.org/10.1128/JCM.00505-08.

Carmona-Vicente, N., Fernández-Jiménez, M., Ribes, J.M., Téllez-Castillo, C.J., Khodayar-Pardo, P., Rodríguez-Diaz, J., et al., 2015. Norovirus infections and seroprevalence of genotype GII.4-specific antibodies in a Spanish population. J. Med. Virol. 87 (4), 675–682. Available from: https://doi.org/10.1002/jmv.24112.

Centers for Disease Control and Prevention (CDC), 2013. Surveillance for foodborne disease outbreaks—United States, 2009–2010. MMWR. Morb. Mortal. Wkly. Rep. 62 (3), 41–47. Available from: https://doi.org/10.1016/j.annemergmed.2009.11.004.

Chan, M.C.W., Lee, N., Hung, T.-N., Kwok, K., Cheung, K., Tin, E.K.Y., et al., 2015. Rapid emergence and predominance of a broadly recognizing and fast-evolving norovirus GII.17 variant in late 2014. Nat. Commun. 6 (1), 10061. Available from: https://doi.org/10.1038/ncomms10061.

Chan, M.C.W., Hu, Y., Chen, H., Podkolzin, A.T., Zaytseva, E.V., Komano, J., et al., 2017a. Global spread of norovirus GII.17 Kawasaki 308, 2014–2016. Emerg. Infect. Dis. 23 (8), 1350–1354. Available from: https://doi.org/10.3201/eid2308.161138.

Chan, P.K.S., Guan, H., Chan, M.C.W., 2017b. The Norovirus: Features, Detection and Prevention of Foodborne Disease. Academic Press.

Chen, S.-Y., Chiu, C.-H., 2012. Worldwide molecular epidemiology of norovirus infection. Paediatr. Int. Child Health 32 (3), 128–131. Available from: https://doi.org/10.1179/2046905512Y.0000000031.

Corsi, S.R., Borchardt, M.A., Spencer, S.K., Hughes, P.E., Baldwin, A.K., 2014. Human and bovine viruses in the Milwaukee River watershed: hydrologically relevant representation and relations with environmental variables. Sci. Total Environ. 490, 849–860. Available from: https://doi.org/10.1016/j.scitotenv.2014.05.072.

Dancer, S.J., 2014. Controlling hospital-acquired infection: focus on the role of the environment and new technologies for decontamination. Clin. Microbiol. Rev. 27 (4), 665–690. Available from: https://doi.org/10.1128/CMR.00020-14.

da Silva Ribeiro de Andrade, J., Fumian, T.M., Leite, J.P.G., de Assis, M.R., Fialho, A.M., Mouta, S., et al., 2018. Norovirus GII.17 associated with a foodborne acute gastroenteritis outbreak in Brazil, 2016. Food Environ. Virol. 10 (2), 212–216. Available from: https://doi.org/10.1007/s12560-017-9326-0.

De Graaf, M., Van Beek, J., Vennema, H., Podkolzin, A.T., Hewitt, J., Bucardo, F., et al., 2015. Emergence of a novel GII.17 norovirus – end of the GII.4 era? Euro Surveill. 20 (26), 1–8. Available from: https://doi.org/10.2807/1560-7917.ES2015.20.26.21178.

Degiuseppe, J.I., Gomes, K.A., Hadad, M.F., Parra, G.I., Stupka, J.A., 2017. Detection of novel GII.17 norovirus in Argentina, 2015. Infect. Genet. Evol. 47, 121–124. Available from: https://doi.org/10.1016/j.meegid.2016.11.026.

Dinu, S., Nagy, M., Negru, D.G., Popovici, E.D., Zota, L., Oprişan, G., 2016. Molecular identification of emergent GII.P17-GII.17 norovirus genotype, Romania, 2015. Euro Surveill. 21 (7), 30141. Available from: https://doi.org/10.2807/1560-7917.ES.2016.21.7.30141.

Dolin, R., Blacklow, N.R., DuPont, H., Buscho, R.F., Wyatt, R.G., Kasel, J.A., et al., 1972. Biological properties of Norwalk agent of acute infectious nonbacterial gastroenteritis. Exp. Biol. Med. 140 (2), 578–583. Available from: https://doi.org/10.3181/00379727-140-36508.

El Qazoui, M., Oumzil, H., Baassi, L., El Omari, N., Sadki, K., Amzazi, S., et al., 2014. Rotavirus and Norovirus infections among acute gastroenteritis children in Morocco. BMC Infect. Dis. 14 (1), 300. Available from: https://doi.org/10.1186/1471-2334-14-300.

Fang, H., Tan, M., Xia, M., Wang, L., Jiang, X., 2013. Norovirus P particle efficiently elicits innate, humoral and cellular immunity. PLoS One 8 (4), e63269. Available from: https://doi.org/10.1371/journal.pone.0063269.

Foodborne Disease Burden Epidemiology Reference Group, on behalf of the WHO, 2015. WHO estimates of the global burden of foodborne diseases. <https://extranet.who.int/sree/Reports?op = vs&path = /WHO_HQ_Reports/G36/PROD/EXT/Foodborne DiseaseBurden> (accessed 27.05.18).

Glass, R.I., Parashar, U.D., Estes, M.K., 2009. Norovirus gastroenteritis. N. Engl. J. Med. 361 (18), 1776–1785. Available from: https://doi.org/10.1056/NEJMra0804575.

Gordon, I., Ingraham, H.S., Korns, R.F., 1947. Transmission of epidemic gastroenteritis to human volunteers by oral administration of fecal fitrates. J. Exp. Med. 86 (5), 409–422. Available from: https://doi.org/10.1084/JEM.86.5.409.

Hall, A.J., 2012. Noroviruses: the perfect human pathogens? J. Infect. Dis. 205 (11), 1622–1624. Available from: https://doi.org/10.1093/infdis/jis251.

Iturriza-Gómara, M., Lopman, B., 2014. Norovirus in healthcare settings. Curr. Opin. Infect. Dis. 27 (5), 437–443. Available from: https://doi.org/10.1097/QCO.0000000000000094.

Jain, S., Thakur, N., Grover, N., Vashistt, J., Changotra, H., 2016. Prevalence of rotavirus, norovirus and enterovirus in diarrheal diseases in Himachal Pradesh, India. Virusdisease 27 (1), 77–83. Available from: https://doi.org/10.1007/s13337-016-0303-2.

Jiang, X., Wang, M., Graham, D.Y., Estes, M.K., 1992. Expression, self-assembly, and antigenicity of the Norwalk virus capsid protein. J. Virol. 66 (11), 6527–6532. Available from: https://doi.org/10.1021/la8017419.

Kapikian, A.Z., Wyatt, R.G., Dolin, R., Thornhill, T.S., Kalica, A.R., Chanock, R.M., 1972. Visualization by immune electron microscopy of a 27-nm particle associated with acute infectious nonbacterial gastroenteritis. J. Virol. 10 (5), 1075–1081. Available from: http://www.ncbi.nlm.nih.gov/pubmed/4117963. Accessed 1 June 2018.

Kojima, S., Fukumi, H., Kusama, H., Yamamoto, S., Suzuki, S., Uchida, T., et al., 1948. Studies on the causative agent of the infectious diarrhoea. Records of the experiments on human volunteers. Jpn. Med. J. 1 (6), 467–476. Available from: https://doi.org/10.7883/yoken1948.1.467.

Lenaker, P.L., Corsi, S.R., Borchardt, M.A., Spencer, S.K., Baldwin, A.K., Lutz, M.A., 2017. Hydrologic, land cover, and seasonal patterns of waterborne pathogens in Great Lakes tributaries. Water Res. 113, 11–21. Available from: https://doi.org/10.1016/j.watres.2017.01.060.

Lochridge, V.P., Jutila, K.L., Graff, J.W., Hardy, M.E., 2005. Epitopes in the P2 domain of norovirus VP1 recognized by monoclonal antibodies that block cell interactions. J. Gen. Virol. 86 (10), 2799–2806. Available from: https://doi.org/10.1099/vir.0.81134-0.

Lopman, B.A., Trivedi, T., Vicuña, Y., Costantini, V., Collins, N., Gregoricus, N., et al., 2015. Norovirus infection and disease in an Ecuadorian birth cohort: Association of certain norovirus genotypes with host FUT2 secretor status. J. Infect. Dis. 211 (11), 1813–1821. Available from: https://doi.org/10.1093/infdis/jiu672.

Loury, P., Le Guyader, F.S., Le Saux, J.C., Ambert-Balay, K., Parrot, P., Hubert, B., 2015. A norovirus oyster-related outbreak in a nursing home in France, January 2012. Epidemiol. Infect. 143 (12), 2486–2493. Available from: https://doi.org/10.1017/S0950268814003628.

Love, D.C., Rodriguez, R.A., Gibbons, C.D., Griffith, J.F., Yu, Q., Stewart, J.R., et al., 2014. Human viruses and viral indicators in marine water at two recreational beaches in Southern California, USA. J. Water Health 12 (1), 136. Available from: https://doi.org/10.2166/wh.2013.078.

Lucero, Y., Vidal, R., O'Ryan, G.M., 2017. Norovirus vaccines under development. Vaccine . Available from: https://doi.org/10.1016/j.vaccine.2017.06.043.

Mabasa, V.V., Meno, K.D., Taylor, M.B., Mans, J., 2018. Environmental surveillance for noroviruses in selected South African wastewaters 2015–2016: emergence of the novel GII.17. Food Environ. Virol. 10 (1), 16–28. Available from: https://doi.org/10.1007/s12560-017-9316-2.

Mans, J., Netshikweta, R., Magwalivha, M., Van Zyl, W.B., Taylor, M.B., 2013. Diverse norovirus genotypes identified in sewage-polluted river water in South Africa. Epidemiol. Infect. 141 (2), 303–313. Available from: https://doi.org/10.1017/S0950268812000490.

Parra, G.I., Green, K.Y., 2015. Genome of emerging norovirus GII.17, United States, 2014. Emerg. Infect. Dis. 21 (8), 1477–1479. Available from: https://doi.org/10.3201/eid2108.150652.

Payne, D.C., Vinjé, J., Szilagyi, P.G., Edwards, K.M., Staat, M.A., Weinberg, G.A., et al., 2013. Norovirus and medically attended gastroenteritis in U.S. children. N. Engl. J. Med. 368 (12), 1121–1130. Available from: https://doi.org/10.1056/NEJMsa1206589.

Reimann, H.A., Price, A.H., Hodges, J.H., 1945. The cause of epidemic diarrhea, nausea and vomiting. (Viral dysentery?). Exp. Biol. Med. 59 (1), 8–9. Available from: https://doi.org/10.3181/00379727-59-14959P.

Suffredini, E., Iaconelli, M., Equestre, M., Valdazo-González, B., Ciccaglione, A.R., Marcantonio, C., et al., 2017. Genetic diversity among genogroup II noroviruses and progressive emergence of GII.17 in wastewaters in Italy (2011–2016) revealed by next-generation and Sanger sequencing. Food Environ. Virol. 1–10. Available from: http://dx.doi.org/10.1007/s12560-017-9328-y.

Tan, M., Huang, P., Xia, M., Fang, P.-A., Zhong, W., McNeal, M., et al., 2011. Norovirus P particle, a novel platform for vaccine development and antibody production. J. Virol. 85 (2), 753–764. Available from: https://doi.org/10.1128/JVI.01835-10.

Tao, J., Chunhui, H., Fanning, S., Nan, L., Jiahui, W., Hongyuan, Z., et al., 2018. Norovirus contamination in retail oysters from Beijing and Qingdao, China. Food Control 86, 415–419. Available from: https://doi.org/10.1016/j.foodcont.2017.11.043.

Thongprachum, A., Fujimoto, T., Takanashi, S., Saito, H., Okitsu, S., Shimizu, H., et al., 2018. Detection of nineteen enteric viruses in raw sewage in Japan. Infect. Genet. Evol. 63, 17–23. Available from: https://doi.org/10.1016/j.meegid.2018.05.006.

Utsumi, T., Lusida, M.I., Dinana, Z., Wahyuni, R.M., Yamani, L.N., Juniastuti, et al., 2017. Occurrence of norovirus infection in an asymptomatic population in Indonesia. Infect. Genet. Evol. 55, 1–7. Available from: https://doi.org/10.1016/j.meegid.2017.08.020.

Vinjé, J., 2015. Advances in laboratory methods for detection and typing of norovirus. J. Clin. Microbiol. 53 (2), 373–381. Available from: https://doi.org/10.1128/JCM.01535-14.

Wu, Z., Yang, L., Ren, X., He, G., Zhang, J., Yang, J., et al., 2016. Deciphering the bat virome catalog to better understand the ecological diversity of bat viruses and the bat origin of emerging infectious diseases. ISME J. 10 (3), 609–620. Available from: https://doi.org/10.1038/ismej.2015.138.

Xue, C., Fu, Y., Zhu, W., Fei, Y., Zhu, L., Zhang, H., et al., 2014. An outbreak of acute norovirus gastroenteritis in a boarding school in Shanghai: a retrospective cohort study. BMC Public Health 14 (1), 1092. Available from: https://doi.org/10.1186/1471-2458-14-1092.

Zahorsk, J., 1929. Hyperemesis hiemis or the winter vomiting disease. Arch Pediatr. 46, 391–395.

Zheng, D.-P., Ando, T., Fankhauser, R.L., Beard, R.S., Glass, R.I., Monroe, S.S., 2006. Norovirus classification and proposed strain nomenclature. Virology 346 (2), 312–323. Available from: https://doi.org/10.1016/j.virol.2005.11.015.

CHAPTER

16

Risk Assessment and Biosecurity Considerations in Control of Emergent Plant Viruses

Amal Souiri[1], *Khadija Khataby*[1], *Yassine Kasmi*[1], *Mustapha Zemzami*[2], *Saaid Amzazi*[3] *and Moulay Mustapha Ennaji*[1]

[1]Laboratory of Virology, Microbiology, Quality, Biotechnologies/Eco-Toxicology and Biodiversity, Faculty of Sciences and Techniques, Mohammedia, University Hassan II of Casablanca, Mohammedia, Morocco [2]Laboratory of Sanitary Control, Control Unit of Plants, Domaines Agricoles Maâmora, Salé, Morocco [3]Laboratory of Biochemistry and Immunology, Faculty of Sciences, Agdal, University of Mohammed V, Rabat, Morocco

ABBREVIATIONS

PepMV Pepino mosaic virus
PRA pest risk analysis

INTRODUCTION

Viral diseases are an important limiting factor in many crop production systems. Because antiviral products are not available, control strategies rely on hygienic measures to prevent viral diseases, or on eradication of diseased crops to control such diseases and later genetic resistance. International trade in planting material, seeds and plants, has increased the risk of introduction of plant pathogens and pests to

Emerging and Reemerging Viral Pathogens
DOI: https://doi.org/10.1016/B978-0-12-814966-9.00016-0

horticulture, agriculture, forestry, and public and private gardens. The cost of new plant virus incursions can be high and result in the loss of trade and/or production for short or extended periods.

Plant biosecurity is defined as "a set of measures designed to protect crops from emergency plant pests at national, regional and individual farm level" (Rodoni, 2009; McKirdy et al., 2012).

At the policy level in many countries, risk analyses are applied as decision-making tools that include the identification and assessment of the risk or hazard, the consequences of establishment, and appropriate risk management strategies.

An initial biosecurity investigation might escalate to a full investigation or incursion response when a new or regulated virus, not yet established, is detected within the country's environment.

Risk assessment and analysis are the first steps in assessing the economic, social, environmental, or cultural impacts of the new incursion. Risk assessment is part of plant epidemiology. Certain principles of disease risk assessment can be found in disease epidemic predicting in the early development stage of epidemiology. In the 1970s, threat analysis was used in regulatory plant pathology to determine quarantine agents. Now, different terms such as threat analysis, risk analysis, or risk assessment are used for studies to determine the epidemic potential of exotic, new, and emerging diseases.

If the impacts are high, quarantine agencies will assess the various options available. If the likely impacts are low, normally, no further official action is taken. It is therefore important to understand the existing and potential plant virome in our ecosystems. Viruses are part of the wider ecosystem, and some interactions may result in disease while other interactions may be beneficial to the host plant (MacDiarmid et al., 2013).

Some of the more serious plant virus disease epidemics are the result of the introduction of either the host plant or the insect vector into a new region that exposes the crop plant to an endemic or "native" virus, resulting in disease. For example, introduction of cassava into sub-Saharan Africa has seen the emergence of aggressive strains of Cassava mosaic virus that have resulted in many farmers abandoning cassava cultivation and consequently destabilizing food security in east Africa (Legg and Fauquet, 2004). The introduction of the biotype B whitefly, *Bemisia tabaci*, into Brazil in the early 1990s resulted in the horizontal transfer of previously unrecorded indigenous begomoviruses from native onto cultivated plants (Rojas et al., 2005).

The quality of a risk assessment's results is entirely dependent upon the quality of the information collected while conducting the risk assessment. In other words, a risk assessment requires the collection and evaluation of accurate information. Without plant–virus interaction information, biosecurity risks cannot be defined (Wren et al., 2006; Malmstrom et al., 2011).

In addition, new risks are arising from the implementation of more environmentally friendly methods of biodegradable waste disposal, such as composting, because certain plant pathogens and pests can survive waste treatment processes, sometimes through inadequate methods of time and temperature exposure.

This chapter reviews recent developments in risk assessment of emerging new viral diseases with special reference to *Pepino mosaic virus* (PepMV) infecting tomato crops worldwide. Biosecurity issues related to the potential environmental risks such as plant wastes and the development of resistant varieties are also discussed. The bias potential of risk communication are defined. Lack of predictability of dispersal potential contributes to uncertainty of risk assessment for airborne diseases. The misuse potential of risk information in agricultural research is discussed as well.

PEPINO MOSAIC VIRUS: REVIEW

PepMV is an emerging pathogen infecting tomatoes and other solanaceous crops such as eggplant, tobacco, and potato through artificial inoculation studies. This viral disease became a limiting factor for tomato production worldwide causing significant economic losses (Soler-Aleixandre et al., 2005).

It was originally identified in Peru on pepino (melon-pear, *Solanum muricatum*) in 1974 (Jones et al., 1980). Later, it started infecting tomato (*Solanum lycopersicum*) crops in the Netherlands and United Kingdom (Wright and Mumford, 1999; Van der Vlugt et al., 2000). Since then, PepMV achieved rapidly a worldwide distribution, despite the attempts to control its spread. It was reported in many other European countries as well as in the United States and Chile (Cotillon et al., 2002; French et al., 2001; Hanssen et al., 2008; Hasiów et al., 2008; Mumford and Metcalfe, 2001; Pagán et al., 2006; Pospieszny et al., 2003; Van der Vlugt et al., 2000; Ling, 2007; Maroon-Lango et al., 2005). Besides, the virus was detected in China (Zhang et al., 2003), Middle East (Fakhro et al., 2010), and South Africa (Carmichael et al., 2011). In 2009, it was included in the European Plant Protection Organization (EPPO) alert list (EPPO, 2014).

PepMV belongs to genus *Potexvirus*, family Alphaflexiviridae with nonenveloped, flexuous rod-shaped virions of ≈ 580 nm in length (Adams et al., 2004; Martelli et al., 2007).

The genome structure of PepMV is a single-stranded RNA genome of approximately 6400 nt long having both a 5′ methylated guanine cap and 3′ polyadenylated tail. It comprises five open reading frames (ORFs), flanked by 5′ and 3′ untranslated regions (Aguilar et al., 2002). ORF1 encodes the putative RNA-dependent RNA polymerase (RdRp) containing methyl transferase, helicase, and polymerase domains. ORFs 2–4 overlap to form the triple gene block element, with roles indicated in viral movement, RNA silencing suppression, and symptom development (Hasiów-Jaroszewska et al., 2011; Morozov and Solovyev, 2003). ORF5 encodes the coat protein (CP), involved in movement and encapsidation of the virion (Morozov and Solovyev, 2003).

The virus causes a wide range of symptoms that reduce the economic value of crops, of which the typical fruit marbling and leaf or stem necrosis (Roggero et al., 2001; Spence et al., 2006; Hanssen et al., 2009). Many factors, such as genotype (Hasiów-Jaroszewska et al., 2011), climate (Spence et al., 2006), and cultivar may affect the development and intensity of the symptoms.

PepMV is known to infect a relatively broad host range of plants, most are in the family Solanaceae, but several solanaceous hosts may not support systemic infection (Jones et al., 1980; Salomone and Roggero, 2002; Córdoba et al., 2004).

The virus is transmitted mechanically from plant to plant without the involvement of an obvious vector. However, bumblebees, the soil-borne fungus *Olpidium virulentus* and whiteflies can operate as vectors for PepMV. Also, seed transmission may play a role in long distance spread, and water was confirmed to be the source of PepMV infection (Mehle et al., 2014).

Currently, four main PepMV genotypes are recognized, the Peruvian (LP) which includes the original pepino isolate (SM.74), the European (EU), the American (US1 and US2) and Chilean (CH2) (Maroon-Lango et al., 2005; Pagán et al., 2006; Ling, 2007; Moreno-Pérez et al., 2014), with an intergenotype RNA sequence identity of at least 80%. Hence, PepMV isolates display considerable genomic differences, possibly associated with geographic origins, specific host, and high mutation rates of RNA viruses, which has been suggested to increase the adaptation ability to new environments (Domingo, 2000). However, no correlation has been observed between different PepMV genotype and the severity of symptom development on infected tomato plants (Pagán et al., 2006; Hanssen et al., 2009). In contrast, single nucleotide variations have been associated with symptoms aggressiveness of a genotype (Hasiów-Jaroszewska et al., 2013).

Interestingly, a replacement of strains and/or coexistence have characterized the epidemic of PepMV (Gómez et al., 2009). In Europe, EU isolates have been replaced by CH2 strain, the American US1 genotype has been found in Canary Islands (Alfaro-Fernández et al., 2008) and in North America the EU strain became predominant (French et al., 2008) and then replaced by CH2 (Ling et al., 2013). Thus mixed infection and recombination between PepMV strains play a role in the virus evolutionary dynamics (Gómez et al., 2009).

Prevention of PepMV disease through strict hygiene measures is currently the main control strategy in tomato production. Cross-protection can be effective, but only if the conditions are well-defined and the effectiveness depends strongly on the PepMV genotype (Hanssen and Thomma, 2010).

PEST RISK ASSESSMENT OF BIOLOGICAL AND ECONOMIC IMPACT

Definitions

Risk assessment involves determining the potential epidemiological, environmental, and economic impact of emerging or new diseases. It is the process of identifying significant risks to the environment, estimating the level of risk, and determining those risks that require measures to reduce the level of risk [USEPA (United States Environmental Protection Agency), 1998]. A hazard is anything, including a situation or state that may cause harm, without considering its probability or the consequences. Harm is a negative outcome of effect of an action or event; in other words, an adverse effect.

A consequence is the result of an undesired event, such as harm to health, life, or the environment. A risk expresses a combination of the probability of an undesired event, and the scope of the consequences. Risk is used to compare various events in terms of having the highest or lowest risk. Risks can be classified qualitatively or quantitatively, if they are to be ranked. A common quantitative expression for risk is the consequence (expressed in a particular unit, e.g., number of deaths or financial loss) multiplied by the probability. Such an expression of risk is also called the expected loss.

Risk can be defined as

$$\text{Risk} = f(\text{Hazard, Exposure}).$$

Exposure refers to the contact or occurrence of a potential hazard with an environmental entity of value.

A study of risk assessment is a macroscale, long-term disease prediction that encompasses assessment of establishment potential, entry potential (when the range of a disease expands beyond a political border), and epidemic potential (epidemic frequency and epidemic severity or extent of the disease), and the potential losses in a region or country once an epidemic occurs (Yang, 2006).

While the potential impact of such activities may vary significantly, the methodology for risk assessment is fundamentally the same. First, potential risks or hazards are identified, and then the likelihood and consequence of these hazards are characterized. Thus the product of risk assessment is an estimate of the likelihood and magnitude of the harms or adverse effects that may result from an activity. Good risk assessment provides relevant and useful information to the decision-maker in a clear and comprehensible form. The rationale used in risk characterization must be unambiguous, and any uncertainties are explained.

Pest Risk Analysis

Pest risk analysis (PRA) is a form of risk analysis conducted by regulatory plant health authorities to identify the appropriate phytosanitary measures required to protect plant resources against new or emerging pests and regulated pests of plants or plant products. Specifically PRA is a term used within the International Plant Protection Convention (IPPC) (Article 2.1) and is defined within the glossary of phytosanitary terms (FAO, 2015) as "the process of evaluating biological or other scientific and economic evidence to determine whether an organism is a pest, whether it should be regulated, and the strength of any phytosanitary measures to be taken against it."

In a phytosanitary context, the term plant pest, or simply pest, refers to any species, strain, or biotype of plant, animal, or pathogenic agent injurious to plants or plant products and includes plant pathogenic bacteria, fungi, fungus-like organisms, viruses, and virus-like organisms, as well as insects, mites, nematodes, and weeds.

A PRA consists of three stages (FAO, 2004).

- Initiating the process
- Pest risk assessment
- Pest risk management

Stage 1: Initiation

The initiation involves identifying the pest(s) following a new detection during inspection, an interception on an imported commodity, a discovery of an established infestation, or an outbreak of a new pest within a PRA area or a new pest is reported in scientific literature.

This stage includes also pathways identification, for example, a mechanism potentially facilitating the entry or spread of a pest is identified. Mechanisms comprise new trade pathways, usually of plants or plant products but could also include articles used in the transport and distribution of traded goods such as pests carried as contaminants of passenger baggage and natural spread.

Finally, review of existing phytosanitary policy, for example, new information that impacts on an earlier phytosanitary decision can cause a review of policy as can disputes over phytosanitary measures.

Stage 2: Risk Assessment

This stage begins with the categorization of individual pests to determine whether the criteria for a quarantine pest are satisfied. Risk assessment continues with an evaluation of the probability of pest entry, establishment, and spread, and of their potential economic consequences (including environmental consequences).

Step 1: Categorization

The purpose of pest categorization is to determine whether a pest identified during the initiation stage satisfies the criteria of being a quarantine pest.

The categorization of a pest as a quarantine pest includes the following primary elements: (1) identity of the pest, (2) presence or absence in the PRA area, (3) regulatory status, (4) potential for establishment and spread in PRA area, (5) potential for economic consequences (including environmental consequences) in the PRA area.

The categorization step provides an opportunity to eliminate a pest from analysis at an early stage in the PRA process, thus avoiding unnecessary in-depth examination. Pest categorization can be done with relatively little information, provided that the information available is sufficient to carry out the categorization.

Step 2: Assessment of Pest Entry, Establishment, and Spread

Assessing the likelihood of pest entry involves assessment of each of the pathways a pest may be associated, from its origin to its establishment in the PRA area.

In a PRA initiated by a specific pathway, often an imported commodity or goods associated with an imported commodity, for example, packing materials, the probability of pest entry is evaluated for that specific pathway. For a PRA initiated for a specific pest, all probable pathways are evaluated for that individual pest.

Step 3: Assessment of Potential Consequences

In this step, the potential impacts that could be expected to result from a pest's introduction and spread is identified, described, and, as much as possible, quantified. Pest impacts can take many forms; they may be economic environmental or social impacts. Information on the species impacts in areas where it is already present, and particularly in areas where it has already spread to, together with information influencing the elements of risk in the PRA area, inform the assessment of potential consequences. Impacts reported from invaded areas are recognized as the best indicator of potential impacts in the PRA area. However, regarding environmental impacts, if the pest has not previously spread then the absence of any environmental impact in the area of pest origin should not be interpreted to mean that no environmental impact should be expected in the PRA area. This is because environmental impacts are difficult to predict, and a lack of impact in the origin is not a good predictor that there will be no impacts in regions where a pest is introduced (Kenis et al., 2012).

Recognizing that risk is a combination of likelihood and consequences, the results of steps 2 and 3 are combined to provide an overall estimation of pest risk.

Stage 3: Risk Management

Risk management stage involves identifying management options for reducing the risks identified at stage 2. These are evaluated for efficacy, feasibility, and impact in order to select those that are appropriate.

Quarantine and certification lists are regularly updated following new PRAs. They are based on six categories of information: (1) knowledge of the identity of the pest (and therefore being able to differentiate it from other viral agents); (2) data on its distribution and (3) host range; (4) information on the modes and efficiency of spread and on the identity of any vector(s); (5) suitability of the local agro-environmental conditions for the pest [and vector(s)]; and (6) the ability to cause a disease and impact the development, reproduction, or productivity of cultivated or wild host plants. Further, refinement of the risk assessment can be based on additional information such as the availability of efficient and easy-to-implement control methods.

The PRA provides the risk assessor with a risk assessment tool whose details will depend both on data availability concerning the six points listed above and on the needs of the risk manager, whose role is to consider the available scientific information (risk assessment) as well as other factors (economical, acceptable political risk, feasibility, and impact of measures) in reaching a decision on whether to regulate or address the pest in any specific way (Massart et al., 2017).

Biological Risk Related to the Viral Pathogen: Pepino Mosaic Virus

The amount of information needed to assess the risk posed by a new virus species to a certain commodity or region is huge. Scientists may indeed have to work for years to provide the answers needed to conduct a thorough PRA according to international phytosanitary standards (ISPM 2 and ISPM 11) (FAO, 2004, 2007).

This summary presents the main features of a PRA which has been conducted on PepMV as a key deliverable from the EU-funded PEPEIRA Project (Werkman and Sansford, 2010).

Stage 1: Initiation

- *Pest identification*:
  Virus/Family: Flexiviridae/Genus: *Potexvirus*/Species: PepMV.
- *Geographical distribution*:
  North America: Canada and the United States.
  Central and South America: Chile, Ecuador, Guatemala, and Peru
  Africa: Morocco
  Asia: China
  EU and EPPO region
- *Pathways*:
  Tomato fruit, seed of tomato, plants for planting of tomato, insect vectors (bumblebees)
- *Existing phytosanitary measures*:
  Yes, for tomato seeds and tomato plants. No, for tomato fruit

Stage 2: Risk Assessment of Pepino Mosaic Virus

- *Categorization*
  - PepMV could present a phytosanitary risk to the PRA area.
  - The pest does not qualify as a quarantine pest for the PRA area.
- *Likelihood of pest entry*
  - Tomato fruit: unlikely to very likely
  - Tomato seeds: unlikely to likely
  - Tomato plants: very unlikely to moderately
  - Bumblebees: very unlikely
- *Likelihood of establishment and spread*:
  - Plants or habitats at risk: tomato (high level), potato and eggplant (uncertain), weeds (risk not identified).
  - Climatic conditions affecting pest establishment of PepMV in the PRA area are considered completely similar.
  - Pest's biology aspects favoring the establishment: PepMV is very easily mechanically transmitted.
- *Potential geographical distribution*: the whole of the EU.

Stage 3: Risk Management of Pepino Mosaic Virus

The following figure lists the possible management measures and their feasibility depending on different pathways and origin. This is also summarized below with an individual assessment of whether the measures are realistic or impractical (Fig. 16.1).

Possible measure	Tomato fruit non-EU	Tomato fruit EU MS	Tomato seed non-EU	Tomato seed EU MS	Tomato plants non-EU	Tomato plants EU MS
Phytosanitary measures						
Visual inspection						
Specified testing						
Import under special licence/permit and post entry quarantine						
Specified treatment						
Removal of parts of plants from the consignment						
Specific handling/packing methods						
Import under special licence/permit and specified restrictions						
Specified treatment and/or period of treatment						
Consignment should be composed of specified cultivars						
Specified growing conditions						
Specified age of plant, growth stage, or time of year of harvest						
Certification scheme						
Pest freedom of crop or pest-free place of production or pest-free						

Legend:

Preexisting phytosanitary measures that have an impact on PepMV (including current emergency measures as well as those that are not specific to the pest).

Possible measure, realistic

Possible measure, not likely to be practical or reliable on its own

Measure ineffective

FIGURE 16.1 **Summary of pest risk management options for PepMV** (Werkman and Sansford, 2010). *PepMV*, Pepino mosaic virus.

Management of Pepino Mosaic Virus

Regulatory Framework

PRA is not strictly a scientific process but exists within an international regulatory framework that directs, defines, and disciplines the process.

PepMV was included early on the EPPO alert list. Currently, this virus is classified as a quarantine pest on tomato seeds for the European Union following Commission Decision 2004/200/EC of February 27, 2004 on measures to prevent the introduction and spread of virus. This decision defines the conditions for importation and circulation of tomato seeds in the European Union (Establishment of the European Plant Health Passport—PEP—on tomato seeds). Member states must ensure that all seed imported from third countries or put into circulation within the community has been obtained by an appropriate extraction method and meet other requirements for isolation of the area of production, or to the carrying out of official analyzes by credible laboratory tests.

The Moroccan decree of the Ministry of Agriculture, Rural Development and Water and Forests No. 832-02 of 30 Rabii II 1423 (June 12, 2002) amending and supplementing the decree of the Ministry of Agriculture and Forestry No. 467-84 of 15 Jumada II 1404 (March 19, 1984) regulates the importation of plants or parts of plants susceptible to infestation by certain pests of animal or plant pests.

The decree stipulates in Article 2 that the importation of plants or parts of plants must be with phytosanitary certificates, the period of which does not exceed 14 days before the date of dispatch of the consignment, and which conform to the model adopted by the IPPC of Rome 1951 (revised and amended in 1979 and 1997) and certifying that the shipment is free from pests and diseases. Article 3 adds that consignments recognized by the Official Plant Protection Officer as carriers of the pest species of animal or plant pests on the list annexed to this order are subject to quarantine treatment (fumigation, treatment with cold/heat/steam or others); in the event that this treatment appears ineffective, these items are immediately rejected or destroyed at the choice of the recipient and at his expense.

Hygiene Measures

As there is no specific treatment for the virus, prevention remains the watchword. The fight against the PepMV is necessary, especially its mandatory declaration, the destruction of contaminated plants and the implementation of prophylactic measures in tomato plants.

Here are some examples of preventative behaviors:

- raise awareness staff about the mode of transmission of the virus and its symptoms,

- limit access to crops including technical staff,
- ensure that all workers use tracksuits, shoes, gloves, and clean or disinfected equipment,
- regularly disinfect hands and tools while working in the crop,
- installation of a functional footbath at each entrance gate of production zone,
- eliminate any crop waste.

Periodic monitoring and careful inspection of the culture for symptoms are necessary to ensure disease detection and increase the chances of its eradication. All suspect plants should be diagnosed by appropriate means. If the disease is detected, the plants should be transported quickly in sealed plastic bags and then destroyed by incineration, in accordance with local regulations. The replacement or disinfection of contaminated equipment (pipes, pruning shears, crates, trailers, wheels, etc.), as well as the disinfection of irrigation facilities or irrigation water, if they are recycled, are necessary.

Therefore prevention through hygiene currently remains the most important strategy for the control of PepMV in commercial tomato production. However, as a result of the high infectivity of the virus, the prevention of infection through hygiene measures is a challenge, especially in dense tomato-growing areas (Hanssen and Thomma, 2010).

Cross-Protection

Cross-protection involves protecting a plant from infection with an aggressive viral strain in the inoculant beforehand with an attenuated viral strain isolated or mutagenized, with no effect on yield and with few symptoms (Zhou and Zhou, 2012).

Many tomato growers, especially in the Netherlands, have chosen to inoculate their crops with a mild PepMV isolate in an attempt to protect their crops from severe damage on natural infection by an aggressive isolate based on cross-protection (Spence et al., 2006). In addition to the cross-protection effect which is aimed for, many growers feel that an infection early in the growing season is less harmful than an infection that occurs later in the growing season. In support of this, glasshouse trials conducted in the United Kingdom from 2001 to 2003 showed that the time of infection had an impact on PepMV-associated damage, as inoculations in May were more damaging than inoculations in February (Spence et al., 2006). In addition, from a questionnaire conducted among Belgian tomato growers, it seems that early infections result in less damage than late infections (Hanssen et al., 2009).

Although cross-protection can be efficient, the enhanced symptom severity in the case of limited nucleotide sequence identity between protector and challenge isolate undermines the potential of cross-protection as a general PepMV control strategy. A management strategy based on cross-protection can only be used successfully in areas in which one single PepMV genotype is dominant, provided that continuous monitoring of the PepMV population is performed and that strict hygiene measures are undertaken.

Otherwise coinfection may result in the emergence of virus variants with new traits after recombination between the protective and challenge isolates. In addition, identifying effective attenuated viruses for each virus of economic importance might be very arduous (Hanssen and Thomma, 2010). Future strategies to combat PepMV epidemics in tomato production might also include transgenic approaches.

Recently in 2015, a "vaccine" protecting against the attack of PepMV, the PMV-01, has been authorized in Morocco by the National Office of Sanitary Safety of Food Products. It contains a benign PepMV strain of genotype CH2 (isolate 1906) (GenBank accession number: FJ457096) that has been carefully selected following a series of 10-year studies (Hanssen et al., 2009; Hoogenven, 2013).

Certification tests were launched in 2015–16 on 430 ha in order to understand the behavior of the product under Moroccan conditions and thus establish the cross-protection strategy. Satisfactory results were validated for these tests; these areas will be renewed in 2016–17 (Agriculture du maghreb, 2016).

Genetic Resistance

Plant viruses manifest a wide variety of pathogenicity that requires individual management strategies. The use of genetically resistant plants is one of the most effective, sustainable, and often used strategies for controlling virus infections in the field. For centuries, this has involved the selection of plants by breeders for their agronomic properties combined with the absence of symptoms of the disease. By the mid-20th century, however, plant breeding programs were capitalizing heavily on the knowledge associated with plant–virus interactions to develop resistant varieties exploitable in agriculture.

Conventional methods for the development of broad-spectrum resistant varieties used the cross-breeding method for resistance genes (Foxe, 1992). Cross-breeding is a long and difficult job: strong genes are rare, and when they are known, they do not always give lasting resistance. During the crossover processes between varieties or plant species, there is a transfer not only of the genes conferring the desired resistance, but also of the sometimes undesirable genes present on neighboring loci (Rommens et al., 2007).

The possibility offered by genetic engineering techniques to transfer quickly in a genotype a gene or a few genes while retaining the other characters, opened new perspectives explored in the fight against viruses (Nicaise, 2014).

The use of a resistant cultivar, against either two or more types of pathogen species, called broad-spectrum resistance, remains the best and cheapest control strategy in managing viral diseases. In addition, several in-depth reviews and books on the multiplication of RNA plant viruses, viral quasispecies, genetic diversity, and plant–virus interactions based on proteomic analysis have been published elsewhere (Mandahar, 2006; Domingo et al., 2012; Di Carli et al., 2012).

Phytosanitary Control

The objective of phytosanitary control is to prevent the introduction of new potentially dangerous quarantine organisms into the national territory and to limit their spread from one area to another, through verification at border posts, the state phytosanitary of imported plants and plant products, by the health surveillance of crops in the interior of the country, and by the preservation of the quality of products for export.

- On importation, phytosanitary controls are carried out at points of entry and aim to prevent the introduction of plant pests into the national territory;
- On the export side, phytosanitary controls aim to ensure compliance with the regulations of the country of destination with regard to the health of plants and plant products;
- Within the country, health surveillance is aimed at ensuring the health status of plants and plant products during their production, circulation, and planting;
- At plant nursery, the health control aims to apply the regulations in force to prevent the spread of pests from plant products intended for planting (ONSSA, 2016).

In fact, transport and in particular the sale of propagating material leads to the inadvertent introduction of pests. These introductions are usually of no consequence, but these organisms can sometimes establish themselves in the country and often lead to considerable damage to agriculture and to nature, threatening the natural balances by a strong reproduction.

Therefore it makes sense to act preventively and to prevent the introduction of these organisms, or at least to stem their spread. National regulations on health control of plants and plant products set out the measures to be taken to avoid as far as possible the spread of pests by establishing a list of organisms against which prevention and control measures are necessary. Given the high cost of preventive measures and their negative impact on trade, only the most important are taken into consideration.

The phytosanitary control system is composed of two elements: (1) a legal framework and technical support tools, (2) an official organization, in this case the plant protection services belonging to the National Office for Sanitary Safety of Food Products, with sufficient resources and infrastructure, to operate or supervise the control system (ONSSA, 2016).

The regulatory framework gives the plant protection services the legal authority to perform their plant health functions, to take the measures to which plants and plants The presentation of pests of a calamitous nature, or in the event of potential risks of introduction to the national territory of quarantine pests.

D'autre part, les services de la protection des végétaux sont chargés d'appliquer les obligations prises par notre pays dans le cadre de la Convention Internationale pour la Protection des Végétaux (CIPV) qui comprennent les responsabilités en matière de la délivrance de certificats phytosanitaires à l'export, la gestion de la surveillance de foyers et de la lutte contre des organismes nuisibles, la conduite d'inspection, les tests de laboratoire et, au besoin, la désinfection des envois de végétaux et produits végétaux, l'assurance de la sécurité phytosanitaire des envois depuis la certification jusqu'à l'exportation, l'établissement et la protection de zones exemptes d'organismes nuisibles, l'analyse du risque phytosanitaire, la formation et la valorisation des ressources humaines.

On the other hand the plant protection services are responsible for enforcing our country's obligations under the IPPC, which include responsibilities for delivering phytosanitary certificates for export, the monitoring of outbreaks and the control of pests, the conduct of inspection, the laboratory tests, and, if necessary, the disinfection of consignments of plants and plant products, the insurance of phytosanitary security of consignments from certification to export, establishment, and protection of pest free areas, PRA, training, and human resource development.

From these responsibilities arise functions in areas such as administration, audit and compliance audit, noncompliance measures, emergency actions, dispute resolution. Requirements are also prescribed for international and national liaison, documentation, and communication (ONSSA, 2016).

BIOLOGICAL RISK RELATED TO PLANT WASTE

Disposal of waste from crops, gardens, and from processing and handling of plant produce generates a risk of introduction or spread of plant pathogens and pests if this waste is not handled properly. New risks are now arising from the implementation of the European Union (EU) Landfill Directive (EU, 1999) for waste disposal to move away from landfill to more environmentally sensitive methods. All biodegradable waste, including that containing biological material (biowaste), will progressively be diverted toward composting, anaerobic digestion, or other modes of waste processing. An additional incentive for the move toward these methods for disposal of waste is provided by the introduction of constraints on burning plant waste (EU, 2006).

However, there is evidence that certain plant pathogens and pests can survive composting or other waste treatment processes, sometimes through inadequate methods or failures in the process (Sansford, 2003; Noble and Roberts, 2004). The most likely organisms to survive, and therefore those which have the potential to infect crop and noncrop plant species, are microorganisms with hardy resting spores, and heat-resistant viruses that are mechanically transmitted.

Safe management procedures are needed for disposal of crop and, in some instances, noncrop plants (plants in public gardens, etc.) and their associated wastes. Of particular concern are those plant residues produced after known introductions or outbreaks as the waste materials are infected with pathogens of quarantine or other regulatory importance or with other, heat-tolerant pathogens.

It is important that the disposal methods for biowaste-containing plant material are both environmentally sensitive and effective in preventing the introduction and further spread of plant pathogens.

Limited published information is available on particular lethal temperatures for different quarantine organisms, and there are no examples in which the full range of temperatures and required exposure times are known for any particular organism. Noble and Roberts (2004) work reported in showed *Tobacco mosaic virus* (TMV) and *Tobacco rattle virus* to be highly temperature tolerant, both surviving maximum compost temperatures of at least 64°C and composting durations of at least 6 days (Noble and Roberts, 2004).

The key parameters of phytosanitary risk assessment of composts as considered by Termorshuizen et al. (2005) are (1) the proportion of host biomass relative to the total quantity of biowaste, (2) the proportion of host infected with a pathogen, (3) the density of infected host material, (4) the proportion of propagules of a pathogen that survived the process, and (5) the threshold density of a pathogen in soil above which disease of the host is expected to develop.

The threshold density of soil-borne pathogens may depend on soil type, temperature, moisture, pH, cropping patterns, and time of year at which a pathogen is introduced, so this is not necessarily relevant when assessing the risks of using composted waste that may contain viable plant pathogens (Termorshuizen et al., 2005).

Methods have been developed for testing the phytosanitary safety of compost, based either on measuring the conditions within the compost or testing the effects of the composting process on indicator organisms. These methods are named "direct" and "indirect" methods.

Direct method uses indicator organisms for testing the sanitation of the composting process (EPPO, 2006, 2008).

This involves using known organisms such as *Pieris brassicae* and TMV with the composting process, followed by retrieval and testing for viability. TMV is known to be tolerant for temperature (Idelmann, 2005).

A significant disadvantage of using pathogens is that indicator organisms may themselves compromise the phytosanitary safety of the compost depending upon how they are introduced and retrieved from the material being composted.

On the other hand, indirect validation refers to monitoring the naturally occurring population of a microorganism (or organisms) or viable seeds in the compost (Christensen et al., 2002), and also to as the recording of temperatures at regular intervals throughout the composting process (EPPO, 2006, 2008).

When performing experimentation to determine the survival of organisms through composting treatments, it is important to ensure that (1) suitable levels and types of inoculum are used and (2) the detection techniques used are able to identify live organisms rather than the dead remnants of the inoculum used, to a high degree of sensitivity and specificity (Noble et al., 2009).

Noble et al. (2009) reviewed a selection of organisms on the basis of their importance to plant health quarantine legislation in the EU, and their likely survival through the composting process. They do provide a wide spectrum of examples illustrating the problems commonly encountered in assessing eradication in compost. The most important organisms are a quarantine fungi, oomycetes, and plasmodiophoromycetes such as *Guignardia citricarpa*, *Synchytrium endobioticum*, *Tamarindus indica*, and quarantine bacteria, viruses, and viroids such as *Clavibacter michiganensis*, PepMV, potato spindle tuber viroid (Noble et al., 2009).

BIOSAFETY ISSUES ASSOCIATED WITH VIRUS-RESISTANT TRANSGENIC PLANTS

About 20 years have passed since the first report showed that transgenic plants expressing the CP gene of TMV were resistant to that virus. Numerous reports have since proven that the concept of "pathogen-derived resistance" is an effective approach for controlling many plant viruses. Despite this proven technology, only a handful of virus-resistant transgenic crops have reached the stage of commercialization. These crops include squash, potato, and papaya, with papaya and squash still being produced commercially to any extent.

Although the environmental hazards associated with the release of such plants have been discussed widely, it has not been possible to reach generally acceptable conclusions about their safety. A case-by-case approach to the risk assessment of real examples is recommended.

The concerns on environmental risks that virus-resistant transgenic plants posed will be summarized in an attempt to develop a consensus on what environmental risks are important to consider. These points on application and environmental risks are presented in the hope that we plant pathologists would develop a better track record in the next 20 years toward the practical application of this extremely powerful approach for controlling plant viruses (Fuchs and Gonsalves, 2007) (Fig. 16.2).

AGROTERRORISM

The use of biological weapons is nothing new and has been practiced since centuries to reach war aims and terrorize enemies (Rogers et al., 1999). In most cases the targets were either warriors or their animals which they needed for transport or fight. In these cases, either pathogens of humans or animals were set free deliberately or otherwise biotoxins were delivered in form of poisoned food or drinking water. Rarely plant pathogens were used and if predominantly with the aim to cause shortage of food supply resulting in famine.

However, when the causal connection between microbes and diseases of humans, animals, and plants became elucidated by the end of the 19th century, this eventually led to the development of scientific research fields by their own, and the planned development of bioweapons started in several countries. In most cases, human and animal pathogens were weaponized, that is, mass propagation, development of ways to deliver them, protection of the own troops, and formulations to favor the spread, as well as the virulence under nonfavorable conditions after delivery. Due to the contagious nature of these pathogens also for the producer and deliverer severe security measurements had to be applied, especially during the mass propagation step. This is where probably the idea, to use plant pathogens as weapons, arose especially in a period where fungicides were more or less unknown, and the predictable production of crops for food supply from year to year was pretty uncertain. Also global production of agricultural crops was still unknown, and mostly the strategy of states even in Europe was self-supply. Only under such conditions, a scenario such as the "Irish

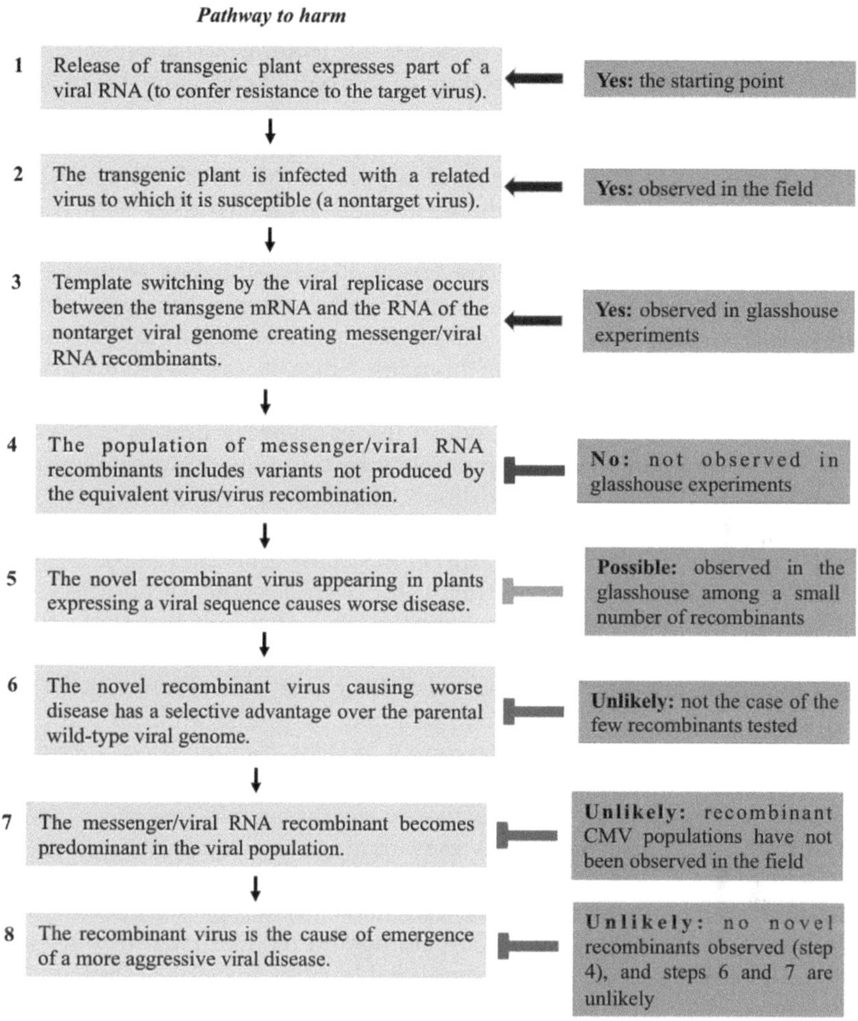

FIGURE 16.2 **Presentation of the risk hypothesis of recombination in transgenic plants expressing viral sequences will lead to emergence of novel viral disease and evaluation of the pathway to harm.** The present state of knowledge concerning CMV presented in the corresponding paragraphs in the text is summarized in the right column of boxes. The colored symbols between the columns indicate the likelihood of the step, shaded from most likely (green) to least likely (red) (Tepfer et al., 2015). *CMV*, Cucumber mosaic virus. *Note:* For interpretation of the references to color in this figure legend, the reader is referred to the web version of this chapter.

Famine" during 1845–50, caused by the potato pathogen *Phytophthora infestans*, was possible. This is the most famous example for the drastic impact a plant pathogen might have. Today, with the available resistant varieties, the possibilities to use highly effective plant protection chemicals and global production of a very efficient agroindustry, it appears highly unlikely that a scenario leading to another "Irish Famine" is in the realm of possibility. Plant pathogens against which no pesticides are available such as plant viruses offer an alternative.

Agroterrorism as a new type of terrorism does not try to directly attack humans, but rather their food supply at the most vulnerable site within the farm to fork chain, the farm itself. Agroterrorism, as it is discussed now, may use pathogens of animal or zoonotic diseases which are really threatening agents. The use of plant pathogens against crop plants is much easier. Access to these is rather simple, the mass propagation is possible, even under low level laboratory conditions, since no plant pathogen is infectious for humans. Since in our present agricultural conditions monocultures are characteristic for production, they are vulnerable for epidemic spread which under these conditions is predictable. The establishment of an epidemic infection would almost certainly cause significant damage. In addition, the delivery of plant pathogens under these condition appears to be easy and safe against immediate detection (Deen, 2000; Shawn Cupp et al., 2004). This makes the conviction of responsible terrorists difficult, if not impossible. In addition, the reaction of the population of an attacked country but also that of the world community may be much less influenced by the harmless looking loss of crop plants than by dying animals or even worse humans. So the damnation of the use of such weapons may be milder and the response of the attacked country to the terrorists, if they are identified, might be less aggressive (Adam, 2006).

Plant pathogens may be effective as terror agents in several ways (Scholthof, 2003). The pathogens may damage or destroy the plants on the field directly. They may also damage the yield by reduction of productivity or by produced toxins that make the crops useless as food (Madden and Wheelis, 2003). With storage crops, like most of our cash crops, the damage may not become visible prior to harvest, but develops later during storage such as potato tubers infected with PVYNTN (Weidemann, 1993).

CONCLUSION

The role of pathologists in presenting risk information has extended beyond the professional research domain and has become critical in influencing decision-making. The introduction of a new technology into a natural ecosystem, whether in the form of a physical product or a knowledge-based process, inevitably will have an effect on that ecosystem. It may even be argued that the purpose of introducing new technology is to engender change in the target system. In agriculture, ex ante and ex post evaluations of the impact of new technology on people and on the broader environment in which organisms exist has conducted for many decades.

Information supporting the pest risk analyses should be reviewed periodically by the pest risk analysts to ensure that any new information that becomes available does not invalidate the decision taken. The analysts should in particular be aware that new international trade may be initiated, host plants may newly be grown in the area which were not grown at the time the PRA was conducted, climate may change, new policy decisions may influence the result of a previous analysis.

References

Adam, G., 2006. Significant ways to spread plant virus diseases in agricultural ecosystems: is agroterrorism possible? In: Cooper, I., Kühne, T., Polishchuk, V.P. (Eds.), Virus Diseases and Crop Biosecurity. NATO Security through Science Series. Springer, Dordrecht.

Adams, M.J., Antoniw, J.F., Bar-Joseph, M., Brunt, A.A., Candresse, T., Foster, G.D., et al., 2004. The new plant virus family Flexiviridae and assessment of molecular criteria for species demarcation. Arch. Virol. 149, 1045–1060.

Agriculture du maghreb, 2016. PMV®-01 la solution efficace durable pour une culture de tomate sans pépino. *Agriculture du maghreb*, n° 96, pp. 24–25.

Aguilar, J.M., Hernández-Gallardo, M.D., Cenis, J.L., Lacasa, A., Aranda, M.A., 2002. Complete sequence of the Pepino mosaic virus RNA genome. Arch. Virol. 147 (10), 2009–2015.

Alfaro-Fernández, A., Cebrián, M.C., Córdoba-Selles, C., Herrera-Vasquez, J.A., Jordá, C., 2008. First report of the US1 strain of Pepino Mosaic Virus in tomato in the Canary Islands, Spain. Plant Dis. 92, 1590–1591. Available from: https://doi.org/10.1094/PDIS-92-11-1590B.

Carmichael, D.J., Rey, M.E.C., Naidoo, S., Cook, G., van Heerden, S.W., 2011. First report of Pepino mosaic virus infecting tomato in South Africa. Plant Dis. 95, 767.

Christensen, K.K., Carlsbeck, M., Kron, E., 2002. Strategies for testing the sanitary quality of composting. J. Appl. Microbiol. 92, 1143–1158.

Córdoba, M.C., Martínez-Priego, L., Jordá, C., 2004. New natural hosts of Pepino mosaic virus in Spain. Plant Dis. 88, 906.

Cotillon, A.C., Girard, M., Ducouret, S., 2002. Complete nucleotide sequence of the genomic RNA of a French isolate of Pepino mosaic virus (PepMV). Arch. Virol. 147, 2231–2238.

Deen, W.A., 2000. Trends in American agriculture: their implications for biological warfare against crop and animal resources. Ann. N.Y. Acad. Sci. 164–167.

Di Carli, M., Benvenuto, E., Donini, M., 2012. Recent insights into plant-virus interactions through proteomic analysis. J. Proteome. Res. 11 (10), 4765–4780. Available from: https://doi.org/10.1021/pr300494e.

Domingo, E., 2000. Viruses at the edge of adaptation. Virology 270, 251–253.

Domingo, E., Sheldon, J., Perales, C., 2012. Viral quasispecies evolution. Microbiol. Mol. Biol. Rev. 76 (2), 159–216. Available from: https://doi.org/10.1128/MMBR.05023-11.

EPPO, 2006. Phytosanitary procedures. Guidelines for the management of plant health risks of biowaste of plant origin. PM3/66 (1). EPPO Bull. 36, 353–358.

EPPO, 2008. Guidelines for the management of plant health risks of biowaste of plant origin. PM3/66 (2). EPPO Bull. 38, 4–9.

EPPO, 2014. *Pepino mosaic virus*. European and Mediterranean Plant Protection Organization. <http://www.eppo.int/QUARANTINE/Alert_List/viruses/PEPMV0.htm> (Panel review date 2014-03).

Fakhro, A., Von Bargen, S., Bandte, M., Büttner, C., 2010. Pepino mosaic virus, a first report of a virus infecting tomato in Syria. Phytopathol. Mediterr. 49, 99–101.

ISPM No. 11. Pest risk analysis for quarantine pests including analysis of environmental risks and living modified organisms. In: FAO International Standards for Phytosanitary Measures 1 to 29. 2007 ed. Food and Agriculture Organization of the United Nations, Rome.

ISPM No. 2. Framework for pest risk analysis. In: FAO International Standards for Phytosanitary Measures 1 to 29. 2007 Ed. Food and Agriculture Organization of the United Nations, Rome.

FAO, 2015. ISPM No 5. Glossary of Phytosanitary Terms. FAO, Rome, 34 pp.

Foxe, M.J., 1992. Breeding for viral resistance: conventional methods. Neth. J. Plant Pathol. 98 (2), 13–20.

French, C.J., Bouthillier, M., Bernardy, M., Ferguson, G., Sabourin, M., Johnson, R.C., et al., 2001. First report of Pepino mosaic virus in Canada and the United States. Plant Dis. 85, 1121.

French, C.J., Dubeau, C., Bunckle, A., Ferguson, G., Haesevoets, R., Bouthillier, M., et al., 2008. Overview of Pepino Mosaic Virus research. Can. J. Plant Pathol. 30, 373–374.

Fuchs, M., Gonsalves, D., 2007. Safety of virus-resistant transgenic plants two decades after their introduction: lessons from realistic field risk assessment studies. Annu. Rev. Phytopathol. 45, 173–202.

Gómez, P., Sempere, R.N., Elena, S.F., Aranda, M.A., 2009. Mixed infections of Pepino Mosaic Virus strains modulate the evolutionary dynamics of this emergent virus. J. Virol. 83, 12378–12387. Available from: https://doi.org/10.1128/JVI.01486-09.

Hanssen, I.M., Thomma, B.P., 2010. Pepino mosaic virus: a successful pathogen that rapidly evolved from emerging to endemic in tomato crops. Mol. Plant Pathol. 11 (2), 179–189.

Hanssen, I.M., Paeleman, A., Wittemans, L., Goen, K., Lievens, B., Bragard, C., et al., 2008. Genetic characterization of Pepino Mosaic Virus isolates from Belgian greenhouse tomatoes reveals genetic recombination. Eur. J. Plant Pathol. 121, 131–146.

Hanssen, I.M., Paeleman, A., Vandewoestijne, E., Van Bergen, L., Bragard, C., Lievens, B., et al., 2009. *Pepino mosaic virus* isolates and differential symptomatology in tomato. Plant Pathol. 58, 450–460. Available from: https://doi.org/10.1111/j.1365-3059.2008.02018.x.

Hasiów, B., Borodynko, N., Pospieszny, H., 2008. Complete genomic RNA sequence of the Polish Pepino mosaic virus isolate belonging to the US2 strain. Virus Genes 36, 209–214.

Hasiów-Jaroszewska, B., Borodynko, N., Jackowia, K.P., Figlerowicz, M., Pospieszny, H., 2011. Single mutation converts mild pathotype of the Pepino mosaic virus into necrotic one. Virus Res. 159, 57–61.

Hasiów-Jaroszewska, B., Paeleman, A., Ortega-Parra, N., Borodynko, N., Minicka, J., Czerwoniec, A., et al., 2013. Ratio of mutated versus wild-type coat protein sequences in Pepino mosaic virus determines the nature and severity of yellowing symptoms on tomato plants. Mol. Plant Pathol. 14 (9), 923–933. Available from: https://doi.org/10.1111/mpp.12059. Epub 2013 July 15.

Hoogenven, J.P., 2013. Besluit van de Staatssecretaris van Economische Zaken van 7 oktober 2013, nr 13168344, Wet ewasbeschermingsmiddelen en biociden ter bestrijding van Pepinomozaïekvirus in de productieteelt van tomaat (Tijdelijke vrijstelling voor het gewasbeschermingsmiddel PMV®-01 ter bescherming van de onbelichte teelt van tomaat, toe te passen in de productieteelt). Staatscourant, 29627.

Idelmann, M., 2005. Hygienisierung von Kompost; Möglichkeiten zum Nachweis einer erfolgreichen Abtötung von Pathogenen und Unkrautsamen (Ph.D. thesis). University of Kassel, Germany, 126 pp.

Jones, R.A.C., Koenig, R., Lesemann, D.E., 1980. Pepino mosaic virus, a new potexvirus from pepino (*Solanum muricatum*). Ann. Appl. Biol. 94, 61–68.

Kenis, M., Bacher, S., Baker, R.H.A., Branquart, E., Brunel, S., Holt, J., et al., 2012. New protocols to assess the environmental impact of pests in the EPPO decision-support scheme for pest risk analysis. EPPO Bull. 42 (1), 21–27.

Legg, J.P., Fauquet, C.M., 2004. Cassava mosaic geminiviruses in Africa. Plant Mol. Biol. 56, 585–599.

Ling, K.-S., 2007. Molecular characterization of two Pepino Mosaic Virus variants from imported tomato seed reveals high levels of sequence identity between Chilean and US isolates. Virus Genes 34, 1–8.

Ling, K.-S., Li, R.G., Bledsoe, M., 2013. Pepino Mosaic Virus genotype shift in North America and development of a loop-mediated isothermal amplification for rapid genotype identification. Virol. J. 10, 117. Available from: https://doi.org/10.1186/1743-422X-10-117.

MacDiarmid, R., Rodoni, B., Melcher, U., Ochoa-Corona, F., Roossinck, M., 2013. Biosecurity implications of new technology and discovery in plant virus research. PLoS Pathog. 9 (8), e1003337. Available from: https://doi.org/10.1371/journal.ppat.1003337.

Madden, L.V., Wheelis, M., 2003. The threat of plant pathogens as weapons against U.S. crops. Annu. Rev. Phytopathol. 41, 155–176.

Malmstrom, C.M., Melcher, U., Bosque-Pérez, N.A., 2011. The expanding field of plant virus ecology: historical foundations, knowledge gaps, and research directions. Virus Res. 9, 84–94.

Mandahar, C.L., 2006. Multiplication of RNA Plant Viruses. Springer, Netherlands, p. 327. Available from: https://doi.org/10.1007/1-4020-4725-8.

Maroon-Lango, C.J., Guaragna, M.A., Jordan, R.L., Hammond, J., Bandla, M., Marquardt, S.K., 2005. Two unique isolates of Pepino Mosaic Virus from a limited source of pooled tomato tissue are distinct from a third (European-like) US isolate. Arch. Virol. 150, 1187–1201.

Martelli, G.P., Adams, M.J., Kreuze, J.F., Dolja, V.V., 2007. Family Flexiviridae: a case study in virion and genome plasticity. Annu. Rev. Phytopathol. 45, 73–100.

Massart, S., Candresse, T., Gil, J., Lacomme, C., Predajna, L., Ravnikar, M., et al., 2017. A framework for the evaluation of biosecurity, commercial, regulatory, and scientific impacts of plant viruses and viroids identified by NGS technologies. Front. Microbiol. 8, 45. Available from: https://doi.org/10.3389/fmicb.2017.00045.

McKirdy, S., Rodoni, B., Moran, J., Sharma, S., 2012. Microbial threat – a growing concern for plant biosecurity. Microbiol. Aust. 33, 11–13.

Mehle, N., Gutiérrez-Aguirre, I., Prezelj, N., Delic, D., Vidic, U., Ravnikar, M., 2014. Survival and transmission of potato virus Y, Pepino mosaic virus, and potato spindle tuber viroid in water. Appl. Environ. Microbiol. 80 (4), 1455–1462. Available from: https://doi.org/10.1128/AEM.03349-13.

Moreno-Pérez, M.G., Pagán, I., Aragón-Caballero, L., Cáceres, F., Fraile, A., García-Arenal, F., 2014. Ecological and genetic determinants of *Pepino Mosaic Virus* emergence. J. Virol. 88 (6), 3359–3368. Available from: https://doi.org/10.1128/JVI.02980-13.

Morozov, S.Y., Solovyev, A.G., 2003. Triple gene block: modular design of a multifunctional machine for plant virus movement. J. Gen. Virol. 84, 1351–1366.

Mumford, R.A., Metcalfe, E.J., 2001. The partial sequencing of the genomic RNA of a UK isolate of *Pepino mosaic virus* and the comparison of the coat protein sequence with other isolates from Europe and Peru. Arch. Virol. 146, 2455–2460.

Nicaise, V., 2014. Crop immunity against viruses: outcomes and future challenges. Front. Plant Sci. 21 (5), 660. Available from: https://doi.org/10.3389/fpls.2014.00660.

Noble, R., Roberts, S.J., 2004. Eradication of plant pathogens and nematodes during composting: a review. Plant Pathol. 53, 548–568.

Noble, R., Elpinstone, J.G., Sansford, C.E., Budge, G.E., Henry, C.M., 2009. Management of plant health risks associated with processing of plant-based wastes. Bioresour. Technol. 100, 3431–3446.
ONSSA, 2016. Contrôles phytosanitaires, 2013-2016. Disponible sur: <http://www.onssa.gov.ma/fr/sante-vegetale/protection-des-vegetaux/protection-du-patrimoine-vegetale/controles-phytosanitaires> (consulté le 21.12.16.). (web site)
Pagán, I., Córdoba-Selles, M.C., Martínez-Priego, L., Fraile, A., Malpica, J.M., Jordá, C., et al., 2006. Genetic structure of the population of Pepino Mosaic Virus infecting tomato crops in Spain. Phytopathology 96, 274–279.
Pospieszny, H., Borodynko, N., Palczewska, M., 2003. First record of Pepino mosaic virus in Poland. J. Plant Dis. Protect. 100, 97.
Rodoni, B., 2009. The role of plant biosecurity in preventing and controlling emerging plant virus disease epidemics. Virus Res. 141, 150–157.
Rogers, P., Whitby, S., Dando, M., 1999. Erntevernichtende Bio-Waffen, Spektr. Wiss. October, 72–77.
Roggero, P., Masenga, V., Lenzi, R., Coghe, F., Ena, S., Winter, S., 2001. First report of Pepino mosaic virus in tomato in Italy. Plant Pathol. 50, 798–800.
Rojas, M.R., Hagen, C., Lucas, W.J., Gilbertson, R.L., 2005. Exploiting chinks in the plant's armour: evolution and emergence of geminiviruses. Annu. Rev. Phytopathol. 4, 361–394.
Rommens, C.M., Haring, M.A., Swords, K., Davies, H.V., Belknap, W.R., 2007. The intragenic approach as a new extension to traditional plant breeding. Trends Plant. Sci. 12, 397–403. Available from: https://doi.org/10.1016/j.tplants.2007.08.001.
Salomone, A., Roggero, P., 2002. Host range, seed transmission and detection by ELISA and lateral flow of an Italian isolate of Pepino mosaic virus. J. Plant Pathol. 84, 65–68.
Scholthof, K.B., 2003. One foot in the furrow: linkages between agriculture, plant pathology, and public health. Annu. Rev. Public Health 24, 153–174.
Shawn Cupp, O., Walker, D.E., Hillison, J., 2004. Agroterrorism in the U.S.: key security challenge for the 21st century. Biosecur. Bioterror. 2, 97–105.
Soler-Aleixandre, S., Lopez, C., Diez, M., de Castro, A., Nuez, F., 2005. Association of Pepino mosaic virus with tomato collapse. J. Phytopathol. 153, 464–469.
Spence, N.J., Basham, J., Mumford, R.A., Hayman, G., Edmondson, R., Jones, D.R., 2006. Effect of Pepino mosaic virus on the yield and quality of glasshouse-grown tomatoes in the UK. Plant Pathol. 55, 595–606.
Tepfer, M., Jacquemond, M., García-Arenal, F., 2015. A critical evaluation of whether recombination in virus-resistant transgenic plants will lead to the emergence of novel viral diseases. New Phytol. 207, 536–541.
Termorshuizen, A.J., van Rijn, E., Bolk, W.J., 2005. Phytosanitary risk assessment of composts. Compost. Sci. Util. 13, 108–115.
USEPA (United States Environmental Protection Agency), 1998. Guidelines for ecological risk assessment. In: EPA/630/R-95/002F. Risk Assessment Forum, Washington, DC.
Van der Vlugt, R.A.A., Stijger, C.C.M.M., Naaldwijk, A.A., Verhoeven, J.Th.J., Lesemann, D.E., 2000. First report of Pepino mosaic virus on tomato. Plant Dis. 84, 103.
Weidemann, H.-L., 1993. Necrotic ring symptoms on potato tubers caused by a new potato virus Y race. Kartoffelbau (Germany) 44, 308–309.
Werkman, A.W., Sansford, C., 2010. Pest risk analysis for Pepino mosaic virus for the EU. In: Deliverable Report 4.3. EU Sixth Framework Project PEPEIRA. <http://www.pepeira.com>.
Wren, J.D., Roossinck, M.J., Nelson, R.S., Scheets, K., Palmer, M.W., et al., 2006. Plant virus biodiversity and ecology. PLoS Biol. 4, e80. Available from: https://doi.org/10.1371/journal.pbio.0040080.

Wright, D., Mumford, R., 1999. Pepino mosaic Potexvirus (PepMV): first records in tomato in the United Kingdom. Plant Disease Notice. Central Science Laboratory, New York, No. 89.

Yang, X., 2006. Framework development in plant disease risk assessment and its application. Plant Disease Epidemiology: Facing Challenges of the 21st Century. Springer, Dordrecht.

Zhang, Y., Shen, Z.J., Zhong, J., Lu, X.L., Cheng, G., Li, R.D., 2003. Preliminary characterization of Pepino mosaic virus Shanghai isolate (PepMV-Sh) and its detection with ELISA. Acta Agricult. Shanghai 19, 90–92.

Zhou, C., Zhou, Y., 2012. Strategies for viral cross-protection in plants. Methods Mol. Biol. 894, 69–81. Available from: https://doi.org/10.1007/978-1-61779-882-5.

Further Reading

Gal-On, A., Shiboleth, Y.M., 2006. Cross protection. In: Loebenstein, G., Carr, J.P. (Eds.), Natural Resistance Mechanisms of Plants to Viruses. Kluwer Academic Publishers, Dordrecht, the Netherlands, pp. 261–268.

Golemboski, D.B., Lomonossoff, G.P., Zaitlin, M., 1990. Plants transformed with Tobacco mosaic virus nonstructural gene sequence are resistant to the virus. Proc. Natl. Acad. Sci. U.S.A. 87, 6311–6315.

Lapidot, M., Gafny, R., Ding, B., Wolf, S., Lucas, W.J., Beachy, R.N., 1993. A dysfunctional movement protein of Tobacco mosaic virus that partially modifies the plasmodesmata and limits virus spread in transgenic plants. Plant J. 4, 959–970.

Ling, K., Scott, J.W., 2007. Sources of resistance to Pepino mosaic virus in tomato accessions. Plant Dis. 91, 749–753.

Prins, M., Laimer, M., Noris, E., Schubert, J., Wasseneger, M., Tepfer, M., 2008. Strategies for antiviral resistance in transgenic plants. Mol. Plant Pathol. 9, 73–83.

Ratcliff, F.G., MacFarlane, S., Baulcombe, D.C., 1999. Gene silencing without DNA: RNA-mediated cross-protection between viruses. Plant Cell 11, 1207–1215.

Termorshuizen, A.J., 2006. Management of soil health in horticulture using compost. In: Final Report EU Project QLK5-CT-01442. Wageningen-UR, Wageningen, the Netherlands.

Wang, H.L., Gonsalves, D., Provvidenti, R., Lecoq, H.L., 1991. Effectiveness of cross protection by a mild strain of zucchini yellow mosaic virus in cucumber, melon, and squash. Plant Dis. 75, 203–207.

CHAPTER

17

Human Immunodeficiency Virus as Emergent Viral Infection With the Presence of the Immune Adaptive Response: Viral Dynamics

Karam Allali[1] *and Moulay Mustapha Ennaji*[2]

[1]Laboratory of Mathematics and Applications, University Hassan II of Casablanca, FST-Mohammadia, Casablanca, Morocco [2]Laboratory of Virology, Microbiology, Quality, Biotechnologies/Eco-Toxicology and Biodiversity, Faculty of Sciences and Techniques, Mohammedia, University Hassan II of Casablanca, Casablanca, Morocco

INTRODUCTION

Human immunodeficiency virus (HIV) was first recognized in the year 1983 as an emergent infectious disease (Morse, 1996, 2001; Satcher, 1995). The last stage of HIV pathogen is the acquired immunodeficiency syndrome (Blattner et al., 1988). In this end stage of the infection, the HIV viruses have attacked and killed the vast majority of $CD4^+$ cells reducing their amount to an account less than 200 cells per μL. Once arrived to this disastrous situation, the immune system fails to play its principal role that is to protect the whole body. It will be worth noting that HIV can be transmitted easily through any bodily fluid contact which makes this serious infection highly epidemic, especially when no

Emerging and Reemerging Viral Pathogens
DOI: https://doi.org/10.1016/B978-0-12-814966-9.00017-2

protection measure is taken. With more than 40 million people infected by HIV registered in the early years of this 21st century and more than 3 million deaths annually (Chen and Narasimhan, 2003), HIV is considered a major public health issue. For those reasons, many mathematical models have been developed during these two last decades. Indeed, the first model illustrating the interaction between the uninfected cells, the infected cells, and the free HIV viruses was tackled in Nowak and Bangham (1996); its mathematical form consists of three nonlinear differential equations that are given as follows:

$$\begin{cases} \dot{x} = \lambda - dx - \beta xv, \\ \dot{y} = \beta xv - ay, \\ \dot{v} = aNy - uv. \end{cases} \tag{17.1}$$

In this model the variables x, y, and v denote the concentration of uninfected $CD4^+$ cells, infected $CD4^+$ cells, and free HIV virus, respectively. Susceptible host cells $CD4^+$ T cells are produced at a rate λ, die at a rate dx, and become infected by virus at a rate βxv. Infected cells die at a rate ay. Free virus is produced by infected cells at a rate aNy and decays at a rate uv.

It is well known that the cellular immune response can play an essential role during the HIV viral infection. This cellular immune response is represented by the cytotoxic T-lymphocyte (CTL) cells, which is responsible to reduce the infection by reducing the account of the infected cells. Indeed, some patients are considered to be elite controllers, due to the CTL effect, since they are infected by HIV; however, they maintain a sufficient $CD4^+$ amount for several years in such a manner that they remain asymptomatic or have very delayed progression of the disease for the time of being infected (Betts et al., 2006; Saez-Cirion et al., 2007). Consequently, as an improvement of the first model, some mathematical models have taken this effect into consideration. The simplest form of them is as follows (Nowak and Bangham, 1996; De Boer and Perelson, 1998):

$$\begin{cases} \dot{x} = \lambda - dx - \beta xv, \\ \dot{y} = \beta xv - ay - pyz, \\ \dot{v} = aNy - uv, \\ \dot{z} = cyz - bz, \end{cases} \tag{17.2}$$

where the new component to the problem is z which stands for the CTL concentration. The CTLs expand in response to viral antigen derived from infected cells at a rate cyz and decay in the absence of antigenic stimulation at a rate bZ. Infected cells are killed by the CTL response at a rate pyz.

In addition to cellular immune response, one can consider another essential immune response which fights the free viruses. This immune response is represented by the antibodies that are produced by the B cells and are responsible to attack and kill the free viruses. Both of the two cited immune responses, the cellular and humoral ones, constitute the main components of what we can call the adaptive immune response (Burnet, 1959). Taking the adaptive immune response under consideration, the mathematical model can be reformulated as follows:

$$\begin{cases} \dot{x} = \lambda - dx - \beta xv, \\ \dot{y} = \beta xv - ay - pyz, \\ \dot{v} = aNy - uv - qvw, \\ \dot{z} = cyz - bz, \\ \dot{w} = gvw - hw, \end{cases} \tag{17.3}$$

where the new component is w which stands for the antibodies concentration. The free viruses are neutralized by antibodies at a rate qvw. Antibodies develop in response to free virus at a rate gvw and decay at a rate hw. For the three models the initial conditions $x(0) = x_0$, $y(0) = y_0$, $v(0) = v_0$, $z(0) = z_0$, and $w(0) = w_0$ will be chosen in order to complete each problem.

The main purpose of this chapter will be to highlight the role of the adaptive immune response in reducing the patient viral load by taking into consideration models (1)–(3). Despite the fact that many models have been developed taking into account new parameters, new variables, or new functions [see for instance Allali et al. (2017); Zhu and Zou (2009); Hattaf et al. (2015), and the references therein], we will be restricted to the above mentioned models because they are the simplest models to describe HIV disease with or without the adaptive immunity. This work will be concerned to whether this immunity can reduce viral replication in order to improve the life quality of the patient.

This present chapter is organized as follows. In the next section, we will study model (1), while Section 17.3 is devoted to an investigation of model (2). The effect of the adaptive immune response in controlling HIV infection is illustrated in Section 17.4. We conclude in the last section.

MATHEMATICAL ANALYSIS OF THE BASIC MODEL

In this section, we are interested in studying mathematical model (1). The schematics of the viral dynamics of such a model are illustrated in Fig. 17.1. The first subsection will be devoted to the analytical study, while the second will be devoted to some numerical simulations.

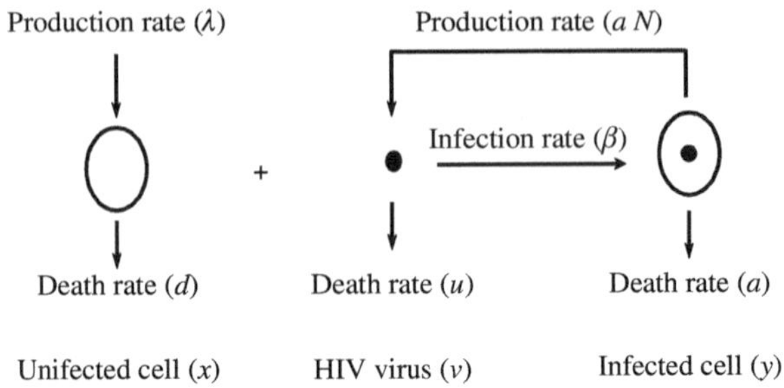

FIGURE 17.1 Viral dynamics schematic of the first model (1).

Disease-Free and Endemic Equilibria

It is straightforward that system (1) has the following reproduction number (Korobeinikov, 2004):

$$R_0 = \frac{\beta\lambda N}{du}.$$

Biologically, this number stands for the average number of secondary infections generated by one infected cell when all cells are susceptible. In addition, the same problem has the two following steady states:

$$E_0^1 = \left(\frac{\lambda}{d}; 0; 0\right)$$

and

$$E_1^1 = \left(\frac{u}{\beta N}; \frac{\lambda\beta N - du}{\beta a N}; \frac{\lambda\beta N - du}{u\beta}\right),$$

by means of R_0, this endemic steady state can be rewritten as follows:

$$E_1^1 = \left(\frac{\lambda}{dR_0}; \frac{\lambda}{a}\left(\frac{R_0 - 1}{R_0}\right); \frac{\lambda N}{u}\left(\frac{R_0 - 1}{R_0}\right)\right).$$

The first steady state represents the infection-free equilibrium, corresponding to the maximal level of healthy $CD4^+$ T cells. For this case the disease cannot invade the cell population. On the other hand, the second steady state represents the endemic equilibrium; this point exists when the basic reproduction number exceeds the unity, and the disease invasion will be always possible. The existence and the global stability of these two steady states were studied in Korobeinikov (2004).

TABLE 17.1 Parameters, Their Symbols and Default Values Used in the Suggested Human Immunodeficiency Virus Models.

Parameters	Meaning	Value	References
λ	Source rate of $CD4^+$ T cells	1–10 cells/μL/day	Culshaw et al. (2004)
d	Decay rate of healthy cells	0.007–0.1 day^{-1}	Culshaw et al. (2004)
β	Rate $CD4^+$ T cells become	0.00025–0.5 μL/virion/day	Culshaw et al. (2004)
a	Infected death rate infected $CD4^+$	0.2–0.3 day^{-1}	Culshaw et al. (2004)
	T cells, not by CTL killing		
u	Clearance rate of virus	2.06–3.81 day^{-1}	Perelson et al. (1996)
N	Number of virions produced by infected $CD4^+$ T cells	6.25–23599.9 virion/cell	Ciupe et al. (2006); Wang et al. (2013)
p	Clearance rate of infection	0.0001–4.048 $\times 10^{-4}$ mL virion/day	Ciupe et al. (2006); Pawelek et al. (2012)
c	Activation rate of CTL cells	0.0051–3.912 day^{-1}	Ciupe et al. (2006)
b	Death rate of CTL cells	0.004–8.087 day^{-1}	Ciupe et al. (2006)
g	Activation rate of B cells	No data	–
h	Death rate of B cells	No data	–
q	Neutralization rate of virions by antibodies	No data	–

CTL, Cytotoxic T-lymphocyte.

Convergence Toward the Steady States

In what follows, we will use Table 17.1 in order to perform all the numerical simulations of this present chapter. This table will give us the biological intervals of each used parameter.

Fig. 17.2 shows the evolution of the infection during the first days of observation for $\lambda = 2$, $d = 0.05$, $\beta = 0.00025$, $N = 120$, $a = 0.25$, and $u = 2.5$. For these parameters the basic reproduction number is below the unity, $R_0 = 0.48 < 1$, which means that the disease-free equilibrium E_0^1 is locally

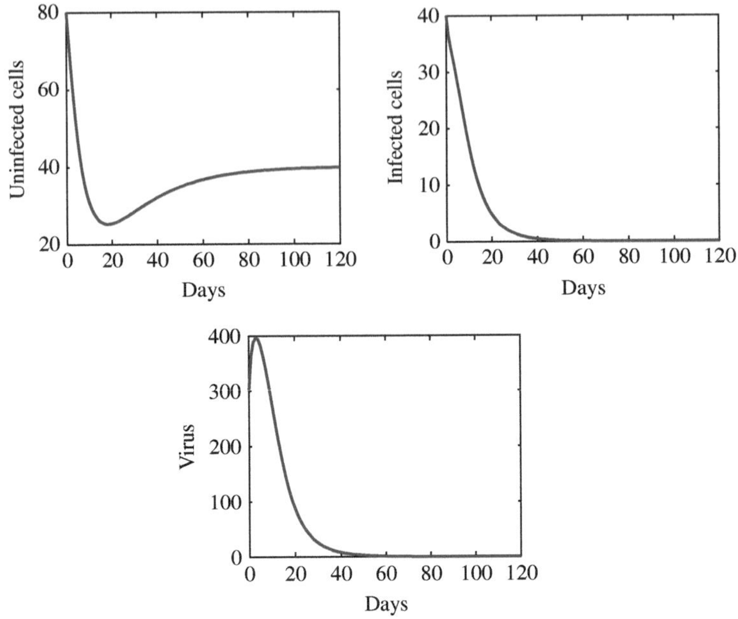

FIGURE 17.2 Behavior of the infection during the time for $\lambda = 2$, $d = 0.05$, $\beta = 0.00025$, $N = 120$, $a = 0.25$, and $u = 2.5$.

stable. From this figure, we observe that all the curves converge toward the equilibrium, $E_0^1 = (40, 0, 0)$. Here, the disease dies out when the time increases. The stability of the endemic equilibrium E_1^1 is shown in Fig. 17.3. Indeed, in this second figure, we have changed one control parameter $\beta = 0.0025$, and the basic reproduction number becomes greater than the unity $R_0 = 4.8 > 1$, which means that the endemic equilibrium E_1^1 is stable. We observe that all the curves converge toward the equilibrium $E_1^1 = (8.33, 6.33, 76)$. Here, the disease persists when the time increases. One can conclude that the basic reproduction number R_0 can be considered a useful tool to predict the evolution of the disease during the time. In the next two sections, numerical simulations will be concerned to only the disease equilibria with nonzero components.

EFFECT OF CYTOTOXIC T-LYMPHOCYTES ON THE INFECTION DYNAMICS

Disease-Free and Endemic Equilibria

In order to study the interaction between the HIV infection and CTLs, we will choose model (2). In this model, another compartment

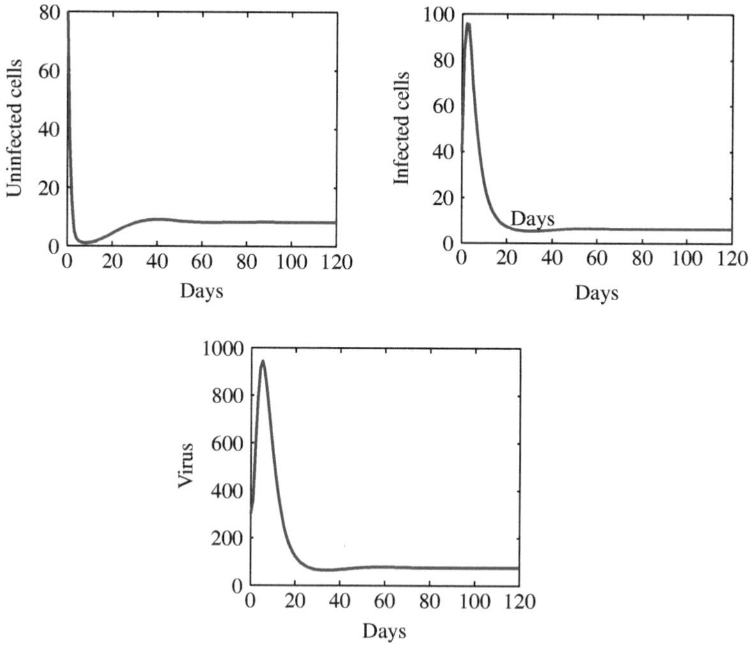

FIGURE 17.3 Behavior of the infection during the time for $\lambda = 2$, $d = 0.05$, $\beta = 0.0025$, $N = 120$, $a = 0.25$, and $u = 2.5$.

representing the cellular immunity will be added. This model has the three following steady states:

$$E_0^2 = \left(\frac{\lambda}{d}; 0; 0; 0\right),$$

$$E_1^2 = \left(\frac{\lambda}{dR_0}; \frac{\lambda}{a}\left(\frac{R_0 - 1}{R_0}\right); \frac{\lambda N}{u}\left(\frac{R_0 - 1}{R_0}\right); 0\right),$$

and

$$E_2^2 = \left(\frac{\lambda uc}{cud + \beta aNb}; \frac{b}{c}; \frac{aNb}{cu}; \frac{-acud + acN\beta\lambda - \beta a^2 Nb}{pcud + p\beta aNb}\right).$$

The difference between models (1) and (2) is that in the second one, a new endemic equilibrium appears, which means that the model with CTL immune response gives more information about the disease. It is worth noting that this endemic equilibrium exists when the basic reproduction number is greater than unity.

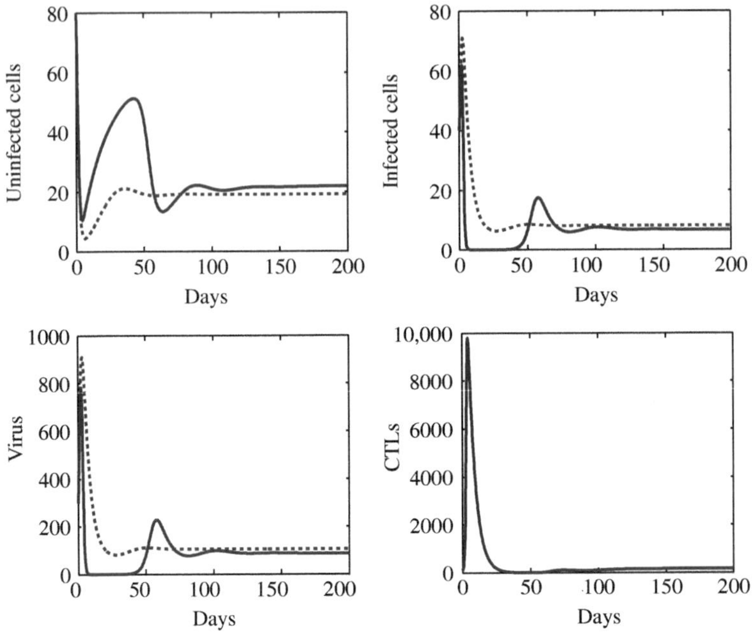

FIGURE 17.4 Behavior of the infection during the time for $\lambda = 3$, $d = 0.05$, $\beta = 0.001$, $N = 130$, $a = 0.25$, and $u = 2.5$; with CTLs $P = .0002$, $c = 0.03$, $b = 0.2$ (*solid line*); without CTLs (*dashed line*). *CTL*, Cytotoxic T-lymphocyte.

Numerical Computations

This subsection is devoted to check numerically the effect of the cellular immunity on the HIV viral dynamics. Indeed, Fig. 17.4 depicts HIV viral dynamics with and without the CTL immune response for $\lambda = 3$, $d = 0.05$, $\beta = 0.001$, $N = 130$, $a = 0.25$, and $u = 2.5$. The solid lines represent the dynamics with CTLs for $p = 0.0002$, $c = 0.03$, and $b = 0.2$, while the dynamics without CTLs is represented by the dashed lines. For these parameters the basic reproduction number is greater than unity ($R_0 = 3.12 > 1$), and all the curves with CTLs converge toward the endemic equilibrium $E_2^2 = (21.95, 6.66, 86.66, 176.82)$ which demonstrates the local stability of this endemic steady state. We clearly observe that cellular immunity reduces considerably the viral load, maximizes the healthy cells, and fights the infected cells.

ANALYSIS OF THE ADAPTIVE IMMUNE RESPONSE EFFECT

Disease-Free and Endemic Equilibria

The adaptive immunity was found in vertebrates, which is considered an important weapon against different infections.

The main cellular and humoral components of this immunity are represented by CTLs and antibodies. They are originally produced in the bone marrow, differentiated in thymus for the T cells and in the spleen for the B cells; they are distributed to the whole body organs. In order to study mathematically, the adaptive immune response along with HIV viral infection, another differential equation will be added to the previous one (2). This new equation describes the dynamics of the antibodies and their interactions with the free virions. By a simple calculation problem (3) has the following five steady states:

$$E_0^3 = \left(\frac{\lambda}{d}; 0; 0; 0\right),$$

$$E_1^3 = \left(\frac{\lambda}{dR_0}; \frac{\lambda}{a}\left(\frac{R_0 - 1}{R_0}\right); \frac{\lambda N}{u}\left(\frac{R_0 - 1}{R_0}\right); 0; 0\right),$$

$$E_2^3 = \left(\frac{\lambda uc}{cud + \beta aNb}; \frac{b}{c}; \frac{aNb}{cu}; \frac{acud + acN\beta\lambda - \beta a^2 N}{pcud + p\beta aNb}; 0\right),$$

$$E_3^3 = \left(\frac{ug + hq}{\beta Ng}, \frac{\lambda\beta Ng - dug - hqd}{\beta aNg}, \frac{\lambda\beta Ng - dug - hqd}{\beta(ug + hq)}, 0, \frac{h}{g}\right),$$

and

$$E_4^3 = \left(\frac{\lambda c(ug + hq)}{\beta baNg + dugc + dqhc}, \frac{b}{c}, \frac{baNg}{c(ug + hq)}, \frac{-a(\beta baNg - \lambda\beta Ngc + dugc + dqhc)}{p(\beta baNg + dugc + dqhc)}, \frac{h}{g}\right).$$

We observe that for this third problem, one more steady state appears, which predicts more rich dynamics than the previous ones.

Numerical Simulations

This subsection highlights the effect of the humoral response in reducing the viral load. Indeed, Fig. 17.5 depicts HIV viral dynamics with and without the humoral immune response for $\lambda = 4$, $d = 0.05$, $\beta = 0.0008$, $N = 150$, $a = 0.25$, $u = 2.5$, $p = 0.0002$, $c = 0.03$, and $b = 0.2$. The solid lines represent the dynamics with antibodies effect for $q = 0.001$, $g = 0.001$, $h = 0.01$, while the dynamics without antibodies is represented with the dashed lines. It is clearly seen that the antibodies have a significant effect in reducing the viral load. It has a logic consequence in reducing the infected cells and increasing the amount of the healthy cells.

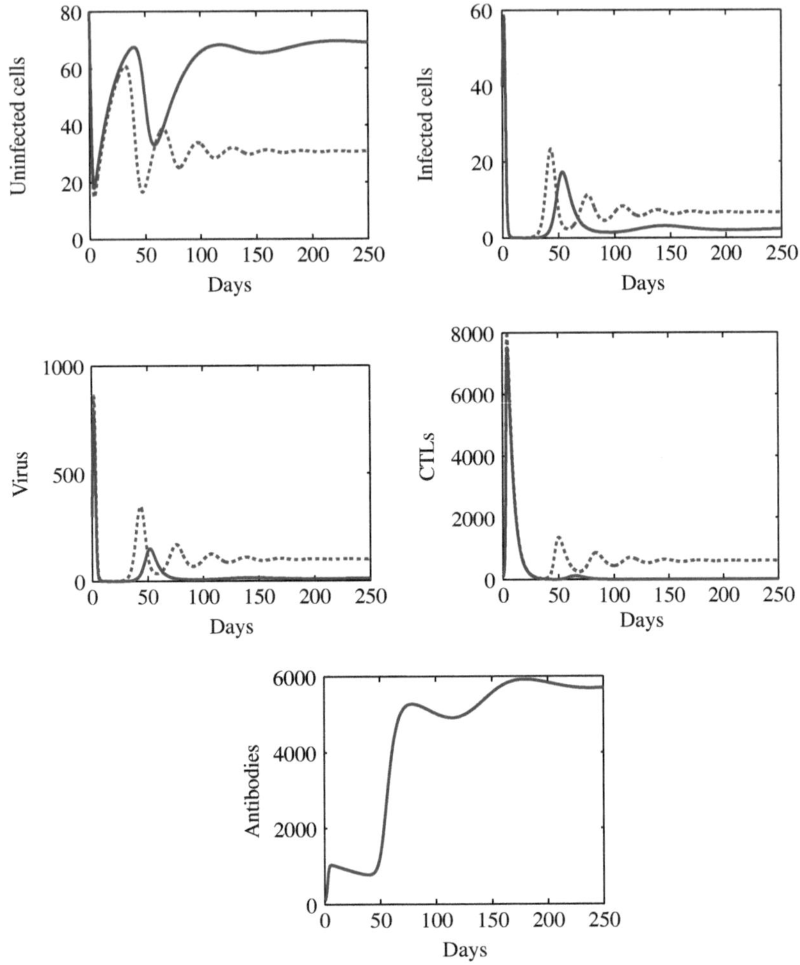

FIGURE 17.5 Behavior of the infection during the time for $\lambda = 4$, $d = 0.05$, $\beta = 0.0008$, $N = 150$, $a = 0.25$, $u = 2.5$, $P = .0002$, $c = 0.03$, and $b = 0.2$; with antibodies $q = 0.001$, $g = 0.001$, $h = 0.01$ (*solid line*); without antibodies (*dashed line*).

CONCLUSION

In this chapter, we have studied some mathematical models describing the HIV in the presence of the adaptive immune responses. This response is mainly represented by the CTL immune and the antibody immune responses. The first one is responsible to attack and kill the infected cells; however, the second one is programmed to attack and neutralize the viruses. These two kinds of immune responses were found in vertebrates, which protect them from the undesirable infections. The T- and B cells are produced in bone morrow, differentiated in thymus and in spleen, respectively. Both are distributed to the whole body in order to play their essential role of protection. The considered basic viral model is in the form of a system of three nonlinear differential equations representing the interaction between the uninfected $CD4^+$ T cells, the infected ones, and the free HIV viruses. We have added in each of the two previous sections the effect of the two considered immune responses, one by one. For each of them the basic model will be changed by adding another equation describing the interaction of added immune response with HIV viral infection. Numerical simulations were performed in order to show that the disease dies out in the presence of the adaptive response. One can conclude that the adaptive immune responses can reduce viral replication and maximize the healthy $CD4^+$ T cells that can enhance the quality life of the patient.

Acknowledgments

The authors would like to thank the Moroccan "Centre National de Recherche Scientifique et Technique" and the French "Centre National de Recherche Scientifique" for the support of the project in the form of the PICS project. The authors would also like to thank the Moroccan Ministry of Higher Education, the University Hassan II of Casablanca and Faculty of Sciences and Technologies of Mohammedia for the financial support. They would like to thank as well for the technical supports all researches, technical staff, and PhD students of the Laboratory of Virology, Microbiology, Quality and Biotechnologies/ Ecotoxicology and Biodiversity; team of Virology, Oncology & Medical Biotechnologies; and Laboratory of Mathematics and Applications.

References

Allali, K., Tabit, Y., Harroudi, S., 2017. On HIV model with adaptive immune response, two saturated rates and therapy. Math. Model. Nat. Phenom. 12 (5), 1–14.

Betts, M.R., Nason, M.C., West, S.M., De Rosa, S.C., Migueles, S.A., Abraham, J., et al., 2006. HIV nonprogressors preferentially maintain highly functional HIV-specific CD8 + T cells. Blood 107 (12), 4781–4789.

Blattner, W., Gallo, R.C., Temin, H.M., 1988. HIV causes AIDS. Science 241 (4865), 515–516.

Burnet, S.F.M., 1959. The Clonal Selection Theory of Acquired Immunity. Cambridge University Press.

Chen, L., Narasimhan, V., 2003. Human security and global health. J. Hum. Dev. 4 (2), 181–190.

Ciupe, M.S., Bivort, B.L., Bortz, D.M., Nelson, P.W., 2006. Estimating kinetic parameters from HIV primary infection data through the eyes of three different mathematical models. Math. Biosci. 200 (1), 1–27.

Culshaw, R.V., Ruan, S., Spiteri, R.J., 2004. Optimal HIV treatment by maximising immune response. J. Math. Biol. 48 (5), 545–562.

De Boer, R.J., Perelson, A.S., 1998. Target cell limited and immune control models of HIV infection: a comparison. J. Theor. Biol. 190 (3), 201–214.

Hattaf, K., Khabouze, M., Yousfi, N., 2015. Dynamics of a generalized viral infection model with adaptive immune response. Int. J. Dynam. Control 3 (3), 253–261.

Korobeinikov, A., 2004. Global properties of basic virus dynamics models. Bull. Math. Biol. 66 (4), 879–883.

Morse, S.S. (Ed.), 1996. Emerging Viruses. Oxford University Press on Demand.

Morse, S.S., 2001. Factors in the emergence of infectious diseases. Plagues and Politics. Palgrave Macmillan, London, pp. 8–26.

Nowak, M.A., Bangham, C.R., 1996. Population dynamics of immune responses to persistent viruses. Science 272 (5258), 74–79.

Pawelek, K.A., Liu, S., Pahlevani, F., Rong, L., 2012. A model of HIV-1 infection with two time delays: mathematical analysis and comparison with patient data. Math. Biosci. 235 (1), 98–109.

Perelson, A.S., Neumann, A.U., Markowitz, M., Leonard, J.M., Ho, D.D., 1996. HIV-1 dynamics in vivo: virion clearance rate, infected cell life-span, and viral generation time. Science 271 (5255), 1582–1586.

Saez-Cirion, A., Lacabaratz, C., Lambotte, O., Versmisse, P., Urrutia, A., Boufassa, F., et al., 2007. HIV controllers exhibit potent CD8 T cell capacity to suppress HIV infection ex vivo and peculiar cytotoxic T lymphocyte activation phenotype. Proc. Natl. Acad. Sci. U.S.A. 104 (16), 6776–6781.

Satcher, D., 1995. Emerging infections: getting ahead of the curve. Emerg. Infect. Dis. 1 (1), 1–6.

Wang, Y., Zhou, Y., Brauer, F., Heffernan, J.M., 2013. Viral dynamics model with CTL immune response incorporating antiretroviral therapy. J. Math. Biol. 67 (4), 901–934.

Zhu, H., Zou, X., 2009. Dynamics of a HIV-1 infection model with cell-mediated immune response and intracellular delay. Discrete Contin. Dyn. Syst. Ser. B 12 (2), 511–524.

CHAPTER

18

Mathematical Modeling in Virology

Khalid Hattaf[1,2] *and Noura Yousfi*[2]

[1]Centre Régional des Métiers de l'Education et de la Formation (CRMEF), Casablanca, Morocco [2]Laboratory of Analysis, Modeling and Simulation (LAMS), Faculty of Sciences Ben M'sik, University Hassan II of Casablanca, Casablanca, Morocco

INTRODUCTION

Viruses are microscopic organisms that need to penetrate inside a cell of their host to multiply and replicate. Many viruses infect the human body such as hepatitis B virus (HBV) and human immunodeficiency virus (HIV). For example, HBV infects liver cells called hepatocytes. This HBV infection is characterized by inflammation of the liver that is often asymptomatic, but it can progress to chronic infection and later cirrhosis or liver cancer. From the World Health Organization an estimated 257 million people are living with HBV infection (defined as hepatitis B surface antigen positive), and about 887,000 people die in 2015 due to hepatitis B infection, including cirrhosis and liver cancer (WHO, 2017a). HIV attacks the $CD4^+$ T cells and reduces their number in the body. It is known that when the number of these cells is less than 200 cells/μL, the patient enters the phase of acquired immunodeficiency syndrome (AIDS). This phase is characterized by the appearance of opportunistic infections caused by bacteria, viruses, or fungi or by the appearance of certain types of cancer. There were approximately 36.7 million people living with HIV at the end of 2016, and about 1 million

Emerging and Reemerging Viral Pathogens
DOI: https://doi.org/10.1016/B978-0-12-814966-9.00018-4

people died from HIV-related causes in the same year (WHO, 2017b). In Morocco the number of people living with HIV is estimated at 28,740, and 1097 people died of AIDS in 2014, while the cumulative number of HIV/AIDS cases reported since the beginning of the epidemic is 10,017 (Royaume du Maroc, 2015). According to the above statistical data and also despite the scientific development, we note that viral infections continue to cause a large number of deaths worldwide. To determine effective treatments against these infections requires a deep understanding of the virus dynamics.

The main purpose of this study is to model the dynamics of viral infections using different approaches in order to better understand them. To do this the next section deals with mathematical modeling using ordinary differential equations (ODEs). The sections "Modeling With Delay Differential Equations" (DDEs) and "Modeling With Partial Differential Equations" (PDEs) come after that. The final section is meant for conclusions of our analytical results.

MODELING WITH ORDINARY DIFFERENTIAL EQUATIONS

Many mathematical models using ODEs have been proposed and developed to better describe the dynamics of viral infections, including HIV, HBV, and hepatitis C virus (HCV). One of the earliest of these models was proposed by Perelson et al. (1996) in 1996 to investigate HIV infection, and later adapted by Nowak et al. (1996) to HBV infection and by Neumann et al. (1998) to HCV infection. This model is given by the following system of three ODEs:

$$\begin{cases} \dfrac{dT}{dt} = \lambda - dT(t) - \beta T(t)V(t), \\[2ex] \dfrac{dI}{dt} = \beta T(t)V(t) - aI(t), \\[2ex] \dfrac{dV}{dt} = kI(t) - \mu V(t), \end{cases} \tag{18.1}$$

where $T(t)$, $I(t)$, and $V(t)$ denote the concentrations of susceptible host cells, infected cells, and free virus at time t, respectively. Uninfected cells are produced at rate λ, die at rate d, and become infected by free virus at rate β. The parameter k represents the production rate of free virus by an infected cell; a is the death rate of infected cells, and μ is the clearance rate of free virus.

System (1) considered uniquely the interaction between host cells and virus and neglected the important role of the two main arms of the adaptive immunity in the control of viral infections. These arms are the humoral and cellular immune responses. The first one is based on the antibodies that are produced by B cells and are capable of identifying and neutralizing viruses, and the second arm is mediated by cytotoxic T lymphocyte (CTL) cells that attack and kill the infected cells. Wodarz (2003) investigated the role of both arms of adaptive immunity in HCV dynamics and pathology by proposing the following model:

$$\begin{cases} \frac{dT}{dt} = \lambda - dT(t) - \beta T(t)V(t), \\ \frac{dI}{dt} = \beta T(t)V(t) - aI(t) - pI(t)Z(t), \\ \frac{dV}{dt} = kI(t) - \mu V(t) - qV(t)W(t), \\ \frac{dW}{dt} = gV(t)W(t) - hW(t), \\ \frac{dZ}{dt} = cI(t)Z(t) - bZ(t), \end{cases} \tag{18.2}$$

where $W(t)$ and $Z(t)$ are the concentrations of antibodies and CTL cells at time t, respectively. Free virus particles are neutralized by the antibodies at rate q, while the infected cells are killed by CTL cells at rate p. The parameters g and c represent the proliferation rate of CTL cells and antibodies, respectively. The parameters h and b are, respectively, the death rates of antibodies and CTL cells. The other variables and parameters have the same biological meaning as in system (1).

For the mathematical analysis of model (2), we refer the reader to the work Yousfi et al. (2009). In this work, it was observed that the basic reproduction number of (2) is proportional to the total number of liver cells. This suggests that this model cannot be a reasonable model for describing HCV infection since it implies that an individual with a smaller liver may be more resistant to the HCV infection than an individual with a larger one. Therefore Yousfi et al. (2011) corrected this problem by replacing the mass action process βTV with a standard incidence function of the form $\beta TV/(T+I)$ in order to model the adaptive immune response in HBV infection. Hattaf et al. (2015) generalized the above models to describe other types of infectious diseases caused by many viruses such as HIV, Ebola, and Zika virus. So, they proposed the following model:

$$\begin{cases} \dfrac{dT}{dt} = \lambda - dT(t) - f(T(t), I(t), V(t))V(t), \\ \dfrac{dI}{dt} = f(T(t)), I(t), V(t)) - aI(t) - pI(t)Z(t), \\ \dfrac{dV}{dt} = kI(t) - \mu V(t) - qV(t)W(t), \\ \dfrac{dW}{dt} = gV(t)W(t) - hW(t), \\ \dfrac{dZ}{dt} = cI(t)Z(t) - bZ(t), \end{cases} \tag{18.3}$$

where the incidence function $f(T, I, V)$ is assumed to be continuously differentiable in the interior of $\mathbb{R}^3_+$ and satisfies the following hypotheses:

- (H_1) $f(0, I, V) = 0$, for all $I \geq 0$ and $V \geq 0$, (H_2) $f(T, I, V)$ is a strictly monotone increasing function with respect to T, for any fixed $I \geq 0$ and $V \geq 0$.
- (H_3) $f(T, I, V)$ is a monotone decreasing function with respect to I and V, that is, $(\partial f/\partial I)(T, I, V) \leq 0$ and $(\partial f/\partial V)(T, I, V) \leq 0$ for all $T \geq 0$, $I \geq 0$, and $V \geq 0$.

Biologically, the above three hypotheses are reasonable and consistent with the reality. For more details on the biological meanings of these three hypotheses, we refer the reader to the works (Hattaf and Yousfi, 2016a; Wang et al., 2016). The schematic representation of model (3) is illustrated in Fig. 18.1.

MODELING WITH DELAY DIFFERENTIAL EQUATIONS

Generally, the time delays appear in differential equations due to the time lag between the action on the system and the response of the system to this action, or because a certain threshold must be reached before the system is activated. So, a real system should be modeled by differential equations with time delays. These types of equations are called DDEs and have several applications in various fields such as physics, biology, ecology, physiology, and economics.

In system (3), infection process and virus production are instantaneous. In reality, there are two kinds of delays: one in cell infection, and the other in virus production. For this reason, we improve system (3) by proposing the following model:

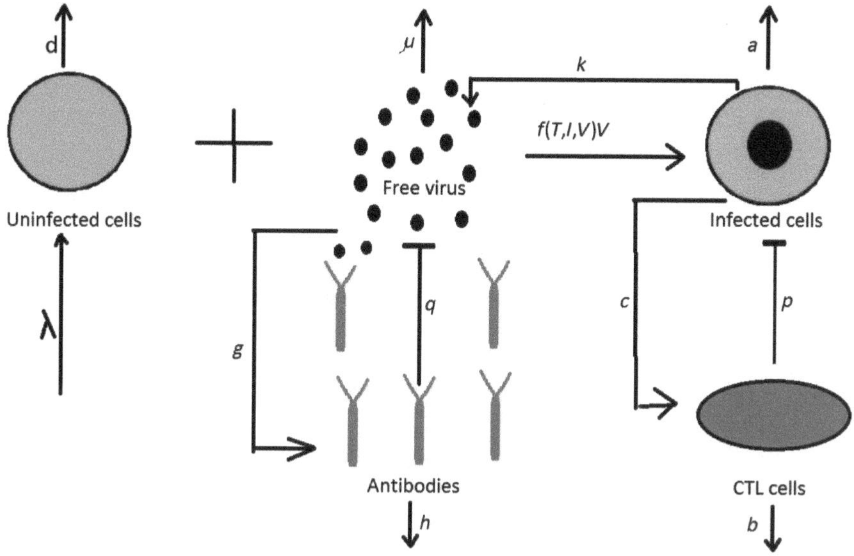

FIGURE 18.1 Schematic representation of model (3).

$$
\begin{cases}
\dfrac{dT}{dt} = \lambda - dT(t) - f(T(t), I(t), V(t))V(t), \\
\dfrac{dI}{dt} = e^{-\alpha_1\tau_1} f(T(t-\tau_1), I(t-\tau_1), V(t-\tau_1))V(t-\tau_1) \\
\qquad - aI(t) - pI(t)Z(t), \\
\dfrac{dV}{dt} = ke^{-\alpha_2\tau_2} I(t-\tau_2) - \mu V(t) - qV(t)W(t), \\
\dfrac{dW}{dt} = gV(t)W(t) - hW(t), \\
\dfrac{dZ}{dt} = cI(t)Z(t) - bZ(t),
\end{cases}
\tag{18.4}
$$

where the first delay τ_1 denotes the time needed for infected cells to produce virions after viral entry and the factor $e^{-\alpha_1\tau_1}$ accounts for the probability of surviving from time $t-\tau_1$ to time t, where α_1 is the death rate for infected but not yet virus-producing cells. The second delay τ_2 represents the time necessary for the newly produced virions to become mature and then infectious particles. In the same the factor $e^{-\alpha_2\tau_2}$ is the probability of surviving the immature virions during the delay period, where $1/\alpha_2$ is the average lifetime of an immature virus.

Therefore the term $e^{-\alpha_2\tau_2}I(t-\tau_2)$ describes the mature viral particles produced at time t. The other variables and parameters are defined as those in systems (1)–(3).

It is very important to note that the model described by system (4) generalizes many virus dynamics models. For example, we get the model of Wodarz presented by system (2) when $\tau_1=\tau_2=0$ and $f(T, I, V)=\beta T$. When $\tau_2=0$ and $f(T, I, V)=\beta T$, we obtain the delayed virus dynamics model with bilinear incidence rate (Yan and Wang, 2012). When $\tau_1=\tau_2=0$ and $f(T,I,V)=\beta T/(1+\epsilon_1 T+\epsilon_2 V)$, where ϵ_1, $\epsilon_2\geq 0$ are constants, we get the humoral and cellular immunity virus dynamics model with the Beddington–DeAngelis incidence rate (Su et al., 2015). Further, when $\tau_2=0$ and $f(T,I,V)=\beta T/(1+\epsilon_2 V)$, we get the delayed model with saturated incidence rate (Wang and Liu, 2013). In addition, when $\tau_2=0$ and $f(T,I,V)=\beta T/(1+\epsilon_1 T+\epsilon_2 V)$, we obtain the delayed HCV infection model with the Beddington–DeAngelis incidence rate (Zhao and Xu, 2014).

Moreover, the general incidence function $f(T, I, V)$ includes other types of incidence rate existing in the literature such as the Crowley–Martin functional response (Crowley and Martin, 1989), the incidence function was used by Zhuo (2012) in order to investigate the HBV infection with noncytolytic loss of infected cells, and the Hattaf–Yousfi functional response introduced in Hattaf and Yousfi (2016b) and used in Raid et al. (2016) and Mahrouf et al. (2017). Furthermore, system (4) is a special case of a class of delayed viral infection models that are proposed and analyzed in Hattaf and Yousfi (2016b).

MODELING WITH PARTIAL DIFFERENTIAL EQUATIONS

Modeling with PDEs allows one to describe the evolution in time and space of infectious diseases. Recently, many mathematical models using PDEs have been developed to better describe the dynamics of these infectious diseases.

System (4) assumes that healthy cells, infected cells, free virus, antibodies, and CTL cells are well mixed and ignores their mobility. For this reason, we assume that the motion of the above five populations follows the Fickian diffusion, meaning that the fluxes of these five populations are proportional to their concentration gradient and go from the regions of high concentration to the regions of low concentration, with the diffusion coefficients d_T, d_I, d_V, d_W, and d_Z, respectively. Hence, system (4) becomes

$$\begin{cases} \dfrac{\partial T}{\partial t} = d_T \Delta T(x,t) + \lambda - dT(x,t) - f(T(x,t), I(x,t), V(x,t))V(x,t), \\ \dfrac{\partial I}{\partial t} = d_I \Delta I(x,t) + e^{-\alpha_1 \tau_1} f(T(x,t-\tau_1), I(x,t-\tau_1), V(x,t-\tau_1))V(x,t-\tau_1) \\ \qquad - aI(x,t) - pI(x,t)Z(x,t), \\ \dfrac{\partial V}{\partial t} = d_V \Delta V(x,t) + ke^{-\alpha_2 \tau_2} I(x,t-\tau_2) - \mu V(x,t) - qV(x,t)W(x,t), \\ \dfrac{\partial W}{\partial t} = d_W \Delta W(x,t) + gV(x,t)W(x,t) - hW(x,t), \\ \dfrac{\partial Z}{\partial t} = d_Z \Delta Z(x,t) + cI(x,t)Z(x,t) - bZ(x,t), \end{cases} \tag{18.5}$$

where $T(x, t)$, $I(x, t)$, $V(x, t)$, $W(x, t)$, and $Z(x, t)$ are the concentrations of uninfected cells, infected cells, free virus, antibodies, and CTL cells at location $x \in \Omega$ and time t, respectively. The region Ω is a bounded domain in $R^n (n \leq 3)$ with smooth boundary $\partial\Omega$. $\Delta = \sum_{i=1}^{n} \partial^2/\partial x_i^2$ is the Laplacian operator with zero-flux boundary conditions:

$$\frac{\partial T}{\partial v} = \frac{\partial I}{\partial v} = \frac{\partial V}{\partial v} = \frac{\partial W}{\partial v} = \frac{\partial Z}{\partial v} = 0 \text{ on } \partial\Omega, \tag{18.6}$$

where $\partial/\partial v$ indicates the outward normal derivative on $\partial\Omega$. From the biological point of view, these conditions mean that the uninfected cells, infected cells, free virus, antibodies, and CTL cells do not move across the boundary $\partial\Omega$.

Now, we focus on the existence of equilibria of system (5) that are stationary solutions in time and space. Such solutions are in fact constant solutions of (5). Let $E(T, I, V, W, Z)$ be an equilibrium of (5). Then

$$\begin{cases} \lambda - dT - f(T,I,V)V = 0, \\ e^{-\alpha_1 \tau_1} f(T,I,V)V - aI - pIZ = 0, \\ ke^{-\alpha_2 \tau_2} I - \mu V - qVW = 0, \\ gVW - hw = 0, \\ cIZ - bZ = 0. \end{cases} \tag{18.7}$$

Obviously, system (5) has always an infection-free equilibrium $E_0(T_0, 0, 0, 0, 0)$, where $T_0 = \lambda/d$. The basic reproduction number of (5) is given by

$$R_0 = \frac{k}{a\mu} f\left(\frac{\lambda}{d}, 0, 0\right) e^{-\alpha_1 \tau_1 - \alpha_2 \tau_2}. \tag{18.8}$$

For the biological meaning, R_0 denotes the average number of secondary infections produced by one infected cell during the period of infection when all cells are uninfected, and the infection-free equilibrium, E_0, represents the extinction of the viruses.

From (7) and similarly to Hattaf and Yousfi (2016b), if $R_0 > 1$, system (5) has another infection equilibrium without immunity $E_1(T_1, I_1, V_1, 0, 0)$, where $I_1 = (\lambda - dT_1)e^{-\alpha_1\tau_1}/a$, $V_1 = \kappa(\lambda - dT_1)e^{-\alpha_1\tau_1-\alpha_2\tau_2}/a\mu$, and $T_1 \in \left(0, \lambda/d\right)$ is the unique root of the following equation:

$$e^{-\alpha_1\tau_1-\alpha_2\tau_2} f\left(T, \frac{(\lambda - dT)e^{-\alpha_1\tau_1}}{a}, \frac{\kappa(\lambda - dT)e^{-\alpha_1\tau_1-\alpha_2\tau_2}}{a\mu}\right) - \frac{a\mu}{k} = 0.$$

If both humoral and cellular immune responses have not been established, we have $gV_1 - h \leq 0$ and $cI_1 - b \leq 0$. So, we define the reproduction number for humoral immunity as follows:

$$R_1^W = \frac{gV_1}{h} \tag{18.9}$$

and the reproduction number for cellular immunity is as follows:

$$R_1^Z = \frac{cI_1}{b}. \tag{18.10}$$

Then $gV_1 - h \leq 0$ and $cI_1 - b \leq 0$ are equivalent to $R_1^W \leq 1$ and $R_1^Z \leq 1$, respectively.

For the biological significance, $1/h$ denotes the average life expectancy of antibodies, and V_1 is the number of free viruses at E_1. Hence, R_1^W represents the average number of the antibodies activated by virus when viral infection is successful and cellular immunity has not been established. Similarly, $1/b$ denotes the average life expectancy of CTL cells, and I_1 is the number of infected cells at E_1. Thus R_1^Z represents the average number of the CTL immune cells activated by infected cells when viral infection is successful, and humoral immunity has not been established.

When $R_1^W > 1$, system (5) has an infection equilibrium with only humoral immunity $E_2(T_2, I_2, V_2, W_2, 0)$, where $I_2 = (\lambda - dT_2)e^{-\alpha_1\tau_1}/a$, $V_2 = h/g$, $W_2 = \left(kge^{-\alpha_1\tau_1-\alpha_2\tau_2}(\lambda - dt_2) - a\mu h\right)/aqh$ and $T_2 \in \left(0, (\lambda/d) - (a\mu h/dkge^{-\alpha_1\tau_1-\alpha_2\tau_2})\right)$ are the unique roots of the following equation:

$$f\left(T, \frac{(\lambda - dt)e^{-\alpha_1\tau_1}}{a}, \frac{h}{g}\right) - \frac{g}{h}(\lambda - dT) = 0.$$

If cellular immunity has not been established, we have $cI_2 - b \leq 0$. For this, we define the reproduction number for cellular immunity in competition as

$$R_2^Z = \frac{cI_2}{b}, \tag{18.11}$$

where $1/b$ denotes the average life expectancy of CTL cells, and I_2 is the number of infected cells at E_2. For the biological significance, R_2^Z represents the average number of the CTL immune cells activated by infected cells under the condition that humoral immunity has been established.

When $R_1^Z > 1$, system (5) has an infection equilibrium with only cellular immunity $E_3(T_3, I_3, V_3, 0, Z_3)$, where $I_3 = b/c$, $V_3 = \kappa b e^{-\alpha_2 \tau_2}/\mu c$, $Z_3 = (ce^{-\alpha_1 \tau_1}(\lambda - dT_3) - ab)/pb$ and $T_3 \in \left(0, (\lambda/d) - (ab/dce^{-\alpha_1 \tau_1})\right)$ are the unique roots of the following equation:

$$f\left(T, \frac{b}{c} \frac{kbe^{-\alpha_2 \tau_2}}{\mu c}\right) - \frac{\mu c}{kbe^{-\alpha_2 \tau_2}} (\lambda - dT) = 0.$$

If humoral immunity has not been established, we have $gV_3 - h \le 0$. In this case, we define the reproduction number for humoral immunity in competition as follows:

$$R_3^W = \frac{gV_3}{h}, \tag{18.12}$$

which $1/h$ denotes the average life expectancy of antibodies, and V_3 is the number of the viruses at E_3. Then R_3^W represents the average number of the antibodies activated by viruses under the condition that cellular immunity has been established.

When $R_2^Z > 1$ and $R_3^W > 1$, system (5) has an infection equilibrium with both cellular and humoral immune responses $E_4(T_4, I_4, V_4, W_4, Z_4)$, where $I_4 = b/c$, $V_4 = h/g$, $W_4 = (u/q)(R_3^W - 1)$, $Z_4 = (ce^{-\alpha_1 \tau_1}(\lambda - dT_4) - ab)/pb$, and $T_4 \in \left(0, (\lambda/d) - (ab/dce^{-\alpha_1 \tau_1})\right)$ are the unique roots of the following equation:

$$f\left(T, \frac{b}{c}, \frac{h}{g}\right) - \frac{g}{h}(\lambda - dT) = 0.$$

Summarizing the above discussions, we obtain the following theorem.

Theorem 18.1.

1. *If $R_0 \le 1$, then system (5) has one infection-free equilibrium $E_0(T_0, 0, 0, 0, 0)$, where $T_0 = \lambda/d$.*
2. *If $R_0 > 1$, then system (5) has an infection equilibrium without immunity $E_1(T_1, I_1, V_1, 0, 0)$.*
3. *If $R_1^W > 1$, then system (5) has an infection equilibrium with only humoral immunity $E_2(T_2, I_2, V_2, W_2, 0)$.*
4. *If $R_1^Z > 1$, then system (5) has an infection equilibrium with only cellular immunity $E_3(T_3, I_3, V_3, 0, Z_3)$.*
5. *If $R_2^Z > 1$ and $R_3^W > 1$, then system (5) has an infection equilibrium with both cellular and humoral immune responses $E_4(T_4, I_4, V_4, W_4, Z_4)$.*

By a simple calculation, we easily get

$$R_3^W = \frac{R_1^W}{R_1^Z} = \frac{gkbe^{-\alpha_2\tau_2}}{h\mu c}, \quad W_2 = \frac{u}{q}(R_2^Z R_s^W - 1) \text{ and } R_3^W > \frac{1}{R_2^Z}. \tag{18.13}$$

Hence, the condition $R_3^W > 1$ is equivalent to $R_1^W > R_1^Z$. In this case the humoral immune response is dominant.

Next, we focus on the global stability of the five equilibria of system (5). First, we have the following result.

Theorem 18.2. *The infection-free equilibrium E_0 of system (5) is globally asymptotically stable if $R_0 \leq 1$.*

Proof. Based on our method proposed in Hattaf and Yousfi (2013), we construct the Lyapunov functional for system (5) at E_0 as follows:

$$\begin{aligned} L_0 = \int_\Omega \Bigg\{ & T(x,t) - T_0 - \int_{T_0}^{T(x,t)} \frac{f(T_0,0,0)}{f(s,0,0)} ds + e^{\alpha_1\tau_1} I(x,t) \\ & + \frac{a}{k} e^{\alpha_1\tau_1+\alpha_2\tau_2} V(x,t) + \frac{aq}{gk} e^{\alpha_1\tau_1+\alpha_2\tau_2} W(x,t) + \frac{p}{c} e^{\alpha_1\tau_1} Z(x,t) \\ & + \int_{t-\tau_1}^{t} f(T(x,s), I(x,s)V(x,s))V(x,s)ds + ae^{\alpha_1\tau_1} \int_{t-\tau_2}^{t} I(x,s)ds \Bigg\} dx. \end{aligned}$$

For convenience, we let $\varphi = \varphi(x, t)$ and $\varphi_\tau = \varphi(x, t - \tau)$ for any $\varphi \in \{T, I, V, W, Z\}$. The time derivative of L_0 along the solution of system (5) satisfies

$$\begin{aligned} \frac{dL_0}{dt} = & \int_\Omega \Bigg\{ \left(1 - \frac{f(T_0,0,0)}{f(T,0,0)}\right) \frac{\partial T}{\partial t} + e^{\alpha_1\tau_1} \frac{\partial I}{\partial t} + \frac{a}{k} e^{\alpha_1\tau_1+\alpha_2\tau_2} \frac{\partial V}{\partial t} \\ & + \frac{aq}{gk} e^{\alpha_1\tau_1+\alpha_2\tau_2} \frac{\partial W}{\partial t} + \frac{p}{c} e^{\alpha_1\tau_1} \frac{\partial Z}{\partial t} \\ & + f(T,I,V)V - f(T_{\tau_1}, I_{\tau_1}, V_{\tau_1})V_{\tau_1} + ae^{\alpha_1\tau_1}(I - I_{\tau_2)}) \Bigg\} dx \\ = & \int_\Omega \Bigg\{ dT_0 \left(1 - \frac{T}{T_0}\right) \left(1 - \frac{f(T_0,0,0)}{f(T,0,0)}\right) + \frac{a\mu}{k} e^{\alpha_1\tau_1+\alpha_2\tau_2} V \left(\frac{f(T,I,V)}{f(T,0,0)} R_0 - 1\right) \\ & - \frac{aqh}{gk} e^{\alpha_1\tau_1+\alpha_2\tau_2} W - \frac{pb}{c} e^{\alpha_1\tau_1} Z \Bigg\} dx - d_T f(T_0,0,0) \int_\Omega \frac{\partial f}{\partial T}(T,0,0) \frac{|\nabla T|^2}{[f(T,0,0)]^2} dx. \end{aligned}$$

$$\leq \int_{\Omega} \left\{ dT_0\left(1-\frac{T}{T_0}\right)\left(1-\frac{f(T_0,0,0)}{f(T,0,0)}\right)+\frac{a\mu}{k}e^{\alpha_1\tau_1+\alpha_2\tau_2}(R_0-1)V \right.$$
$$\left. -\frac{aqh}{gk}e^{\alpha_1\tau_1+\alpha_2\tau_2}W-\frac{pb}{c}e^{\alpha_1\tau_1}Z\right\}dx-d_T f(T_0,0,0)\int_{\Omega}\frac{\partial f}{\partial T}(T,0,0)\frac{|\nabla T|^2}{[f(T,0,0)]^2}dx.$$

Since the function $f(T, I, V)$ is strictly monotonically increasing with respect to T, we have $\partial f/\partial T(T,0,0)>0$ and $(1-(T/T_0))(1-(f(T_0,0,0)/f(T,0,0)))\leq 0$.

If follows from $R_0\leq 1$ that $dL_0/dt\leq 0$. Further, it is not hard to see that the largest invariant set in $\{(T,I,V)|dL_0/dt=0\}$ is $\{E_0\}$. By LaSalle invariance principle (Hale and Verduyn Lunel, 1993), we deduce that the infection-free equilibrium, E_0, is globally asymptotically stable when $R_0\leq 1$. This completes the proof.

Finally, we establish the global stability of the four infection steady states E_i $(1\leq i\leq 4)$ of system (5). To do this, we assume that $R_0>1$, and the incidence function f satisfies for each infection equilibrium E_i $(1\leq i\leq 4)$ the following further hypothesis

$$\left(1-\frac{f(T,I,V)}{f(T,I_i,V)_i}\right)\left(\frac{f(T,I_i,V_i)}{f(T,I,V)}-\frac{V}{V_i}\right)\leq 0 \quad \text{forall} \quad T,I,V>0. \quad (H_4)$$

Similarly to above and based on the Lyapunov functionals constructed in Hattaf and Yousfi (2016b) for DDE models, we easily get the following result:

Theorem 18.3. *Assume* $R_0>1$ and (H_4) holds for each E_i $(1\leq i\leq 4)$.

1. *The infection equilibrium without immunity* E_1 *of system (5) is globally asymptotically stable if* $R_1^W\leq 1$ *and* $R_1^Z\leq 1$.
2. *The infection equilibrium with only humoral immunity* E_2 *of system (5) is globally asymptotically stable if* $R_1^W>1$ *and* $R_2^W\leq 1$.
3. *The infection equilibrium with only cellular immunity* E_3 *of system (5) is globally asymptotically stable if* $R_1^Z>1$ *and* $R_3^W\leq 1$.
4. *The infection equilibrium with both cellular and humoral immune responses* E_4 *of system (5) is globally asymptotically stable if* $R_2^Z>1$ *and* $R_3^W>1$.

The conditions of the global stability of E_2 and those of E_3 given in (2) and (3) of Theorem 18.3 do not hold simultaneously. In fact, supposing the contrary, then $R_1^W>1\geq R_2^Z$ and $R_1^Z>1\geq R_3^W$. Since $R_3^W\leq 1$ and $R_2^Z>1/R_3^W$, we have $R_2^Z>1$. This is a contradiction with $R_2^Z\leq 1$.

Biologically, Theorem 18.2 means that the virus is cleared and the infection dies out when $R_0\leq 1$. However, (1) of Theorem 18.3 implies that if both humoral and cellular immune responses have not been established ($R_1^W \leq 1$ and $R_1^Z \leq 1$), then the virus persists in the host. For the other cases of Theorem 18.3 the virus persists, and the infection becomes chronic due to the domination of the humoral or cellular immune response. In the case of the humoral immune response is dominant and the number of infected cells at the state E_2 is sufficient to activate the cellular immunity, all healthy cells, infected cells, free virus, antibodies and CTL cells can coexist in vivo.

From (13) and Theorem 18.3, we have the following important remark.

Remark 18.1. *Assume* $R_0 > 1$ and (H_4) *holds for each* E_i $(1 \leq i \leq 4)$.

1. *If* $\max(R_1^W, R_1^Z) \leq 1$, *then system (5) converges to* E_1 *without immunity.*
2. *If* $\max(R_1^W, R_1^Z) > 1$, *two cases arise:*
   a. *When* $\max(R_1^W, R_1^Z) = R_1^W$, *the humoral immunity is dominant, and model (5) converges to* E_2 *if* $R_2^Z \leq 1$, *or to* E_4 *if* $R_2^Z > 1$.
   b. *When* $\max(R_1^W, R_1^Z) = R_1^Z$, *the cellular immunity is dominant, and model (5) converges to* E_3 *without humoral immunity.*

From this important result, we can define the over domination of humoral immunity when $R_2^Z > 1$ and $R_3^W > 1$, and the over domination of cellular immunity when $R_2^Z > 1$ and $R_3^W < 1$.

CONCLUSIONS

In this work, we have proposed a new spatiotemporal model that describes the dynamical behavior of viral infection and takes into account the two main arms of adaptive immunity and the two kinds of delays during infection process and virus producing. This model improves and generalizes many viral infection models formulated by ODEs (Wodarz, 2003; Yousfi et al., 2011; Hattaf et al., 2015; Su et al., 2015), by DDEs (Yan and Wang, 2012; Wang and Liu, 2013; Zhao and Xu, 2014), and by PDEs (Hattaf and Yousfi, 2015a,b; Wang et al., 2011; Yang and Xu, 2016; Kang et al., 2017). Under some assumptions on the general incidence function, we have proved that the global dynamics of the proposed spatiotemporal model is completely determined by five threshold parameters that are the reproduction numbers for viral infection R_0, for humoral immunity R_1^W, for cellular immunity R_1^Z, for cellular immunity in competition R_2^Z, and for humoral immunity in competition R_3^W. The first threshold parameter for viral infection is also called the basic reproduction number, and it can be the Richter scale for viral diseases. More precisely, the infection-free equilibrium is globally asymptotically stable if $R_0 \leq 1$, which biologically means that the virus is cleared, and the infection dies out. When $R_0 > 1$, our PDE model has four infection steady states that are the infection equilibrium without immunity which is globally asymptotically stable if $R_1^W \leq 1$ and $R_1^Z \leq 1$; the infection equilibrium with only humoral immunity which is globally asymptotically stable if $R_1^W > 1$ and $R_2^Z \leq 1$; the infection equilibrium with only cellular immunity which is globally asymptotically stable if $R_1^Z > 1$ and $R_3^W \leq 1$; and the infection equilibrium with both cellular and humoral immune responses which is globally asymptotically stable if $R_2^Z > 1$ and $R_3^w > 1$. Biologically speaking, the infection becomes chronic and the virus persists in the host when the basic reproduction number is greater than one. In this case the activation of one or both arms of adaptive immunity is unable to eliminate the virus in the host but it plays a crucial role in the reduction of virus particles and infected cells. This last result is easily deduced by comparing the components of virus particles and infected cells before and after the activation of cellular and humoral immunity.

According to the important Remark 18.1, we can conclude that the over domination of cellular immunity leads to the absence of the humoral immunity, and the over domination of the humoral immunity leads to the persistence of disease. This biological finding can be an explanation of the dysfunction of the adaptive immunity in viral infections such HBV, which is still largely incomplete (Boni et al., 2007).

Based on the above mathematical and biological conclusions, we derive a practical result that can be interesting for the future medicine. This result is based on the calculation of the basic reproduction number for each patient. If this number is less than or equal to one, then the infection will disappear, and the patient does not need treatment. However, if the basic reproduction number is greater than one, then the virus persists in the host, and the patient must follow a treatment to improve his quality of life or completely cure the disease. In addition, the results obtained from this study can help biologists and pharmaceutical laboratories to develop new medication for viral infections in order to make the number the basic reproduction number less than one.

References

Boni, C., Fisicaro, P., Valdatta, C., Amadei, B., Di Vincenzo, P., Giuberti, T., et al., 2007. Characterization of hepatitis B virus (HBV)-specific T-cell dysfunction in chronic HBV infection. J. Virol. 81 (8), 4215–4225.

Crowley, P.H., Martin, E.K., 1989. Functional responses and interference within and between year classes of a dragonfly population. J. North. Am. Benth. Soc. 8, 211–221.

Hale, J.K., Verduyn Lunel, S.M., 1993. Introduction to Functional Differential Equations. Springer-Verlag, New York.

Hattaf, K., Yousfi, N., 2013. Global stability for reaction-diffusion equations in biology. Comput. Math. Appl. 66, 1488–1497.

Hattaf, K., Yousfi, N., 2015a. Global dynamics of a delay reaction-diffusion model for viral infection with specific functional response. J. Comput. Appl. Math. 34 (3), 807–818.

Hattaf, K., Yousfi, N., 2015b. A generalized HBV model with diffusion and two delays. Comput. Math. Appl. 69 (1), 31–40.

Hattaf, K., Yousfi, N., 2016a. A numerical method for a delayed viral infection model with general incidence rate. J. King Saud Univ. Sci. 28 (4), 368–374.

Hattaf, K., Yousfi, N., 2016b. A class of delayed viral infection models with general incidence rate and adaptive immune response. Int. J. Dynam. Control 4 (3), 254–265 (2015).

Hattaf, K., Khabouze, M., Yousfi, N., 2015. Dynamics of a generalized viral infection model with adaptive immune response,. Int. J. Dynam. Control 3, 253–261.

Kang, C., Miao, H., Chen, X., Xu, J., Huang, D., 2017. Global stability of a diffusive and delayed virus dynamics model with Crowley-Martin incidence function and CTL immune response. Adv. Differ. Equ. Available from: http://dx.doi.org/10.1186/s13662-017-1332-x.

Mahrouf, M., Hattaf, K., Yousfi, N., 2017. Dynamics of a stochastic viral infection model with immune response. Math. Model. Nat. Phenom. 12 (5), 15–32.

Royaume du Maroc, 2015. Mise en oeuvre de la déclaration politique sur le VIH/sida. In: Rapport National. Available at: <http://www.unaids.org/sites/default/files/country/documents/MAR_narrative_report_2015.pdf>.

Neumann, A.U., Lam, N.P., Dahari, H., Gretch, D.R., Wiley, T.E., Layden, T.J., et al., 1998. Hepatitis C viral dynamics in vivo and the antiviral efficacy of interferon-α therapy. Science 282, 103–107.

Nowak, M.A., Bonhoeffer, S., Hill, A.M., Boehme, R., Thomas, H.C., McDade, H., 1996. Viral dynamics in hepatitis B virus infection. Proc. Natl. Acad. Sci. U.S.A. 93 (1996), 4398–4402.

Perelson, A.S., Neumann, A.U., Markowitz, M., Leonard, J.M., Ho, D.D., 1996. HIV-1 dynamics in vivo: virion clearance rate, infected cell life-span, and viral generation time. Science 271, 1582–1586.

Riad, D., Hattaf, K., Yousfi, N., 2016. Dynamics of capital-labour model with Hattaf-Yousfi functional response. Br. J. Math. Comput. Sci. 18 (5), 1–7.

Su, Y., Sun, D., Zhao, L., 2015. Global analysis of a humoral and cellular immunity virus dynamics model with the Beddington-DeAngelis incidence rate. Math. Methods Appl. Sci. 38 (14), 2984–2993.

WHO, 2017a. Hepatitis B, fact sheet. Available at: <http://www.who.int/mediacentre/factsheets/fs204/en/>.

WHO, 2017b. HIV/AIDS, fact sheet. Available at: <http://www.who.int/mediacentre/factsheets/fs360/en/>.

Wang, S., Feng, X., He, Y., 2011. Global asymptotical properties for a diffused HBV infection model with CTL immune response and nonlinear incidence. Acta Math. Sci. 31, 1959–1967.

Wang, X., Liu, S., 2013. A class of delayed viral models with saturation infection rate and immune response. Math. Methods Appl. Sci. 36 (2), 125–142.

Wang, X.-Y., Hattaf, K., Huo, H.-F., Xiang, H., 2016. Stability analysis of a delayed social epidemics model with general contact rate and its optimal control. J. Ind. Manage. Optim. 12 (4), 1267–1285.

Wodarz, D., 2003. Hepatitis C virus dynamics and pathology: the role of CTL and antibody responses. J. Gen. Virol. 84, 1743–1750.

Yan, Y., Wang, W., 2012. Global stability of a five-dimensional model with immune responses and delay. Discret. Contin. Dyn. Syst. Ser. B 17 (1), 401–416.

Yang, Y., Xu, Y., 2016. Global stability of a diffusive and delayed virus dynamics model with Beddington-DeAngelis incidence function and CTL immune response. Comput. Math. Appl. 71, 922–930.
Yousfi, N., Hattaf, K., Rachik, M., 2009. Analysis of a HCV model with CTL and antibody responses. Appl. Math. Sci. 3 (57), 2835–2845.
Yousfi, N., Hattaf, K., Tridane, A., 2011. Modeling the adaptive immune response in HBV infection. J. Math. Biol. 63, 933–957.
Zhao, Y., Xu, Z., 2014. Global dynamics for a delayed hepatitis C virus infection model. Electron. J. Differ. Equ. 2014 (132), 1–18.
Zhuo, X., 2012. Analysis of a HBV infection model with noncytolytic cure process. In: IEEE Sixth International Conference on Systems Biology (ISB). pp. 148–151.

Index

Note: Page numbers followed by "*f*" and "*t*" refer to figures and tables, respectively.

D

F

G

H

N

FSC
www.fsc.org
MIX
Papier | Fördert gute Waldnutzung
FSC® C083411

Kolibri 360 GmbH
Ferdinand-Jühlke-Straße 7
99095 Erfurt, Deutschland
produktsicherheit@kolibri360.de